Switching to VoIP

Other resources from O'Reilly

oreilly.com   *oreilly.com* is more than a complete catalog of O'Reilly books. You'll also find links to news, events, articles, weblogs, sample chapters, and code examples.

oreillynet.com is the essential portal for developers interested in open and emerging technologies, including new platforms, programming languages, and operating systems.

Conferences   O'Reilly brings diverse innovators together to nurture the ideas that spark revolutionary industries. We specialize in documenting the latest tools and systems, translating the innovator's knowledge into useful skills for those in the trenches. Visit *conferences.oreilly.com* for our upcoming events.

Safari Bookshelf (*safari.oreilly.com*) is the premier online reference library for programmers and IT professionals. Conduct searches across more than 1,000 books. Subscribers can zero in on answers to time-critical questions in a matter of seconds. Read the books on your Bookshelf from cover to cover or simply flip to the page you need. Try it today for free.

Switching to VoIP

Ted Wallingford

O'REILLY®

Beijing · Cambridge · Farnham · Köln · Paris · Sebastopol · Taipei · Tokyo

Switching to VoIP
by Ted Wallingford

Published by O'Reilly Media, Inc., 1005 Gravenstein Highway North, Sebastopol, CA 95472.

O'Reilly books may be purchased for educational, business, or sales promotional use. Online editions are also available for most titles (*safari.oreilly.com*). For more information, contact our corporate/institutional sales department: (800) 998-9938 or *corporate@oreilly.com*.

Editor:	Michael K. Loukides
Production Editor:	Adam Witwer
Cover Designer:	Ellie Volckhausen
Interior Designer:	David Futato

Printing History:

June 2005:	First Edition.

 This book uses RepKover,™ a durable and flexible lay-flat binding.

ISBN: 0-596-00868-6

[M]

To Kelly, my charming wife and the indispensable recipient of so many of my VoIP calls.

Table of Contents

Foreword

This is the year of VoIP! So was last year, and the year before that, and the year before that.... Okay, so what's the real story? Is VoIP really going to happen? The answer is that VoIP has been slowly happening all this time. Bit by bit, industry by industry, this technology has been creeping into homes and businesses everywhere.

As a technology, VoIP is a pretty simple idea—use packet-switched data encapsulation instead of the tried-and-true time-division multiplexed (TDM), circuit-switched methods that telephony has used since its creation. Since its creation? Is that an exaggeration? TDM telephony today is essentially just an electronic version of the "cord board" that the old-time operators used to connect one caller with another. Advances in packet communication technologies are now making that model obsolete, permitting more efficient use of bandwidth resources while providing mobility and the integration of voice, presence, and other information.

If VoIP is so great, why has it been so slow to catch on? First, as currently standardized, VoIP is more complicated to administer and set up than a traditional telephone line. Second, while VoIP's technical merits are impressive, TDM technology is globally deployed. "Copper may be buried, but it's not dead," the saying goes. In order to be successful, VoIP must be interoperable with this established technology. Finally, until recently, VoIP was more expensive to deploy than traditional telephony. Today, thanks to competition enabled by open standards and more recently open source technologies such as Asterisk, the cost of a VoIP system is not only competitive with TDM, but in many cases less expensive (as well it should be).

The key to evaluating when VoIP may or may not help you, and to having successful VoIP deployments where it does help, is understanding the technologies, protocols, and tools (including open source platforms such as Asterisk, for which I can't resist including a second, shameless parenthetical plug). It is the goal of this book to educate and inform you, to put control in your hands, and to save you from as much frustration as possible!

—Mark Spencer
Original author and maintainer of Asterisk
Founder and president of Digium

Preface

Voice over IP is a family of technologies that has sweeping implications for everybody who uses telephones, the Internet, fax machines, email, and the Web. VoIP borrows from, and enhances, many disciplines of communications technology; it promises to revolutionize the most familiar of these technologies, the telephone. The Internet Protocol, analog telephony, digital telephony and T1 circuits, digital audio signal processing, high-availability networking, and a host of other concerns are all touched by growing borders of the vast, ambitious realm of VoIP.

VoIP has found its way into business phone systems, desktop messaging software, and residential telephony service. Your mortgage or insurance company's web site may offer you the ability to communicate by VoIP with a customer service rep using your computer. You may subscribe to VoIP-based telephone services like Packet8, AT&T CallVantage, Vonage, or Broadvox Direct as a replacement for your old, traditional phone service.

In the late 1990s, VoIP was lauded as a way to save on long-distance charges by calling Grandma and Grandpa using a PC with a headset and a microphone. But today's definition of VoIP is far broader. Hundreds of thousands of VoIP-based devices are in use in the United States, and fast-growing shipments of VoIP-compatible telephone systems have revitalized the data networking industry. The next evolutionary step for the Internet is to become reliable enough to replace the global telephone network as we know it.

So, VoIP is a disruptive technology family that promises to revolutionize the way we communicate, while driving decreases in the cost of that communication and increases in the speed, reliability, and availability of the Internet itself. That's what VoIP *does*. But you're reading this book because you want to know *how* it does it. Here's what you'll encounter while reading *Switching to VoIP*:

- A definition of public and private legacy voice systems and a quick review of their evolution

- A description of the core technologies of digital telephony—sampling, pulse code modulation, time division multiplexing, trunking, and call signaling

- Examples of how VoIP networks overlay IP networks and how voice applications reside within the OSI network model

- An explanation of why many early VoIP adopters' implementations didn't live up to expectations

- An introduction to the two subsets of VoIP standards: audio transmission and signaling

- Introductions to SIP, H.323, and other signaling specifications that are commonly used in IP telephony

- Some practical strategies for justifying VoIP adoption in a business scenario

- An introduction to IP hardphones, analog telephone adapters, softPBX servers, and the other devices that make VoIP systems tick

- Strategies for dealing with quality-of-service issues, including policy-based, protocol-based, and practice-based approaches to ensuring tip-top operational quality of a VoIP network

- Information about interfacing VoIP systems with traditional phone equipment and the public telephone system by way of analog trunks and T1 lines

- The basics of enforcing low-layer security in a VoIP environment, with examples for IP Tables firewalls

- Tips for gauging VoIP readiness on an enterprise network or over an Internet pathway

- Scenarios that describe what network technologies work best for voice systems, and how to compensate with quality-of-service measures across those that don't work so well for voice

- Dozens of practical, hands-on projects that you can use to build the pieces of your VoIP network while learning which standards and practices will work best for your particular environment

- A number of troubleshooting scenarios to help you deal with, and avoid, the most common mistakes that are made when implementing this technology

- A handy vendor reference that will put you in touch with the key software, hardware, and service players in the VoIP business

- A reference for the basic configuration, SIP interfacing, and legacy interfacing of the world's most popular open source telephone server, Asterisk

Audience

Though this book presents a fair amount of theory, we have gone to great lengths to keep the material as practical as possible. It has been written so that you can read a chapter, apply that chapter, and come back to learn more. IT professionals, a large piece of this book's target audience, are often the type of people who learn best by *doing*, so that's how the material is presented. In fact, you'll notice that the book looks at telephony more from a packet networking engineer's perspective than from a telephone network engineer's perspective, even going so far as to apply the OSI model to telephone systems, which were invented decades before the OSI model was.

But anybody who's new to VoIP should get a handle on its terminologies, implications, and scope by reading this book. IT managers, telecommunications directors, and hands-on CIOs will all benefit from the proper (and one might say refreshing) perspective this book has toward Voice over IP—not the textbook perspective of a legacy telephone system engineer, but the hands-on viewpoint of an IP networking pro who recently completed a number of legacy-to-VoIP system conversions. In this case, it can truly be said that I am cast from the same mold as my readers.

Prospective VoIP adopters, Cisco-certified network specialists, and those with a persuasion toward open standards may want to read this book on account of its vendor neutrality. Linux, Windows, and Mac solutions are discussed, while Cisco, Grandstream, Digium, and other hardware vendors play prominent roles in the hands-on projects. This is no accident. The world of telephony is a multivendor domain in which interoperability is critical and failure of competing vendors' systems to work together, unlike desktop computing, is unacceptable. This book therefore advocates strongly for international standards ratified by the International Telecommunications Union, the Internet Engineering Task Force, and other bodies.

Regarding Asterisk

Asterisk is an open source telephony server that runs on Linux, FreeBSD, and Mac OS X. It was chosen as a platform for illustrating many of the examples in this book for several reasons. It supports many of the standards needed to teach VoIP signaling concepts; it runs on regular, easy-to-get PC equipment; and perhaps most importantly, it's free. If you want to experiment with VoIP without spending a lot on hardware, Asterisk is a great starting point. And while commercial solutions have their place, you might also find that Asterisk is a worthwhile end solution. You'll find a partial Asterisk reference and appendixes at the end of the book.

Assumptions Made in This Book

This book assumes some familiarity with data networking. It reviews TCP/IP addressing and the seven-layer OSI network model, but proceed with caution if these topics aren't already familiar to you. Each chapter builds upon the concepts described in the previous chapters, so the book is intended to be read in order.

Conventions

Most definitions are given upon first encountering a new term. There's a glossary included in case you miss a definition or just need a refresher.

Carrier is a word that is referred to in many different ways throughout the book. Sometimes it means a phone company, other times it refers to a data link system. Wherever a word has two potential meanings, the context should tell you which meaning is appropriate.

In most cases, examples are provided in terms of North American standards and practices. Where appropriate, European or Asian equivalents are mentioned.

Kilobit is abbreviated as *k*, while kilobits per second shows up as *kbps*. The same convention is used for *mega-* and *giga-* prefixes where applicable. For the sake of clarity, many authors will define whether *kilo* means 1,000 or 1,024. This is an important distinction, but you don't need to worry much about it here. It's safe to assume a multiple of 1,000 when *kilo* is used. That is, 64 *kilo*bits is exactly 64,000 bits and so on.

The following typographical conventions are used in this book:

Plain text
> Indicates menu titles, menu options, menu buttons, and keyboard accelerators (such as Alt and Ctrl).

Italic
> Indicates new or technical terms, system calls, URLs, hostnames, email addresses, filenames, file extensions, pathnames, and directories.

`Constant width`
> Indicates commands, options, switches, variables, attributes, keys, functions, types, objects, HTML tags, macros, the contents of files, or the output from commands.

`Constant width bold`
> Shows commands or other text that should be typed literally by the user.

`Constant width italic`
> Shows text that should be replaced with user-supplied values.

 This icon signifies a tip, suggestion, or general note.

 This icon indicates a warning or caution.

Where to Get More

There are 1,000-page books about telephone systems, and there are 50-page reference pamphlets about a single protocol, like SIP. When writing a book like this, it's a challenge to strike a balance between the unwieldy tome that's filled with countless details of trivial import and a crash-course FAQ that tells you just enough to get started. On one hand, there's nebulous reference material, and on the other, there's not enough detail but lots of conciseness.

To address this, the book is rather detailed on certain subjects and rather brief and concise on others. If you're looking for more information in a special area, O'Reilly offers other books that may help you, including narrow but detailed books about T1 circuits, data communications theory, network security, and Ethernet. Safari Online is O'Reilly's web-based library of books, an invaluable tool that can also provide more depth. Information on O'Reilly's printed books and Safari are available at *http://www.oreilly.com*.

Safari Enabled

 When you see a Safari® Enabled icon on the cover of your favorite technology book, it means the book is available online through the O'Reilly Network Safari Bookshelf.

Safari offers a solution that's better than e-books. It's a virtual library that lets you easily search thousands of top technology books, cut and paste code samples, download chapters, and find quick answers when you need the most accurate, current information. Try it for free at *http://safari.oreilly.com*.

How to Contact Us

Please address comments and questions concerning this book to the publisher:

O'Reilly Media, Inc.
1005 Gravenstein Highway North
Sebastopol, CA 95472
(800) 998-9938 (in the United States or Canada)
(707) 829-0515 (international or local)
(707) 829-0104 (fax)

We have a web page for this book, where we list errata, examples, and any additional information. You can access this page at:

http://www.oreilly.com/catalog/switchingvoip

To comment or ask technical questions about this book, send email to:

bookquestions@oreilly.com

For more information about our books, conferences, Resource Centers, and the O'Reilly Network, see our web site: *http://www.oreilly.com*.

Acknowledgments

My first noncurricular writing was a short science fiction story I wrote after having been inspired by my third-grade teacher's reading aloud to my class. The book he read to us was *The Lion, the Witch, and the Wardrobe* by C. S. Lewis. So, I owe a thank you to Mr. Dennis Streich, my third-grade teacher, for giving me a respect for the social and educational importance of the medium of books. While I can't blame any of my bad writing habits on these two gentlemen, I can certainly credit them with helping to cultivate my love of the written word—even when it's about a geeky subject like VoIP.

More recently, my reading and writing have been of an entirely different sort--the vocational sort. When I worked for J. Walter Thompson in Detroit, I was flipped a copy of O'Reilly Media's epic masterpiece *sendmail* by one of the guys I worked with. This book was a godsend. If anybody could translate a deadly topic like sendmail configuration into terms I could understand, it was Bryan Costales and Eric Allman, the authors of that highly important book. The accessibility and organic approach of *sendmail* were common to most O'Reilly books, I later noticed. Whether it was *Practical PostgreSQL* or *RealBASIC: The Definitive Guide*, I had an easier time learning from these "animal books" than I had from others.

When given the opportunity to write about Voice over IP for O'Reilly, it was an easy decision. I had just completed the first phase of an ambitious telephony conversion on a large construction contractor's network, and I was looking for standards-advocating documentation to help me architect the next phase. The problem was, all the decision-making intelligence for my project was provided by the VoIP equipment vendors and their salespeople—Cisco, Avaya, Nortel, Mitel, NEC, and so on. Cisco was bashing Avaya, and Avaya was bashing Nortel, ad nauseum. I was looking for the neutral, standards-respecting VoIP authority in book form, and I couldn't find it. So I decided to write it myself, more or less as an exercise to aid my project. Fortunately, Tim O'Reilly and Mike Loukides, the editor of this book, thought it would be a good book to publish.

I learned a lot about VoIP while writing this book, and I hope it engages you in the subject as much as the writing process engaged me in it. VoIP is a technology family that I feel very passionately about. It has the legitimate potential to truly revolutionize distance communication. VoIP is an expression of the Internet's promise of allowing better, faster, more accurate communication between people, and it's a culturally impactful next step for the international telephone system. In other words, Voice over IP is *important*. It is becoming a sustaining technology due to the growing adoption of enthusiastic implementers such as you and I. My sincerest hope is that this book provides you with an arsenal of fair, even-handed technical information and advice that will help you succeed in building your VoIP systems.

Switching to VoIP was made possible only with the efforts of quite a few contributors, from editorial supervision to illustrators to technical reviewers. In particular, Mike Loukides, this book's editor, kept me focused like a laser on the things that mattered and steered me away from the things that didn't (such as my original proposal for a detailed description of the telegraph). Interesting—sure. But not at all useful. The technical reviewers who participated were outstanding as well. Every one of them a master of gracious criticism, the review team improved this book immensely, fixing my technical faux pas, and offering ideas I hadn't even thought of. The review team included these immensely talented networking pros: Bernard Hayes, Ryan Courtnage, Jason Becker, Rich Adamson, Jim Van Meggelen, Jason Gintert, Jared Smith, and Todd Nathan.

A number of corporate associates afforded me excellent feedback or material support while I wrote the book, including Greg Boehnlein of N2Net Inc., Paul Mitnick of TEC Communications, Inc., Jeff Fanelli of Network General, David Zupan of Visium, Inc., Mark Spencer and Malcolm Davenport of Digium, Inc., Benjamin Kowarsch of Sunrise Telephone Systems of Tokyo, Duane Leinninger of Quicken-Loans.com, Dr. Gina Leinninger, John Huang of Grandstream Networks, Brian and Phaedrah Downey of the Linux Fix, Robyn Roberts and Jonathan Varman of Avaya Corp., Kevin Young of Polycom Corp., Kelly Larabee representing Skype Technologies S.A., Len Fernandes respresenting the SOYO Group, Inc., Chris Liu of VoicePulse, Inc., Dave Karpaty of Tiger Direct B2B, Ryan Courtnage and Jason Becker of Coalescent Systems, and Doug Fraim representing Zoom Electronics. Of particular importance was the input of Jason Gintert at Fidelity Access Networks, whose company is on the front line of hosted PBX and data connectivity services. Jason helped me stamp out some "prose bugs."

My son Jacob, daughter Madelyn, and wife Kelly were all instrumental participants in the process, too. Our family's switch to VoIP occurred at the beginning of the writing process, and observing our domestic use of the technology on a day-to-day basis was interesting in the least.

## CHAPTER 1

Voice and Data: Two Separate Worlds?

Telephony is the communication of spoken information between two or more participants, by means of signals carried over electric wires or radio waves. Ever since Alexander Graham Bell invented the telephone circuit and first envisioned the public
telephone system, consumers and businesses have relied on telephony as a staple of
human interaction.

With the advent of Internet technologies and high-speed data connectivity in the
enterprise, a new family of telephony technologies began taking hold. Voice over IP,
or VoIP, has significant appeal for the enterprise, for service providers, and for end
users, because it allows the Internet and commonplace data networks, like those at
offices, factories, and campuses, to become carriers for voice calls, video conferencing, and other real-time media applications. VoIP-savvy organizations are discovering that they can apply the paradigm of distributed, software-based networking to
voice applications and enable a new generation of telecommunications features, cost-
savings, and productivity enhancements.

VoIP can replace business telephone systems, or it can add value to existing traditional telephony devices. For instance, long-distance connectivity between two
offices with traditional telephone systems can often be accomplished with a lower
cost per call when VoIP is employed.

VoIP network protocols can serve as a platform for other communication media like
text messaging and video conferencing. In fact, you've probably used a flavor of VoIP
for such an application by now; they've been popular as an Internet pastime for several years. Yahoo! offers a "party line"–style service that features Voice over IP chat
rooms (*http://chat.yahoo.com*). Apple's iChat and Microsoft's NetMeeting applications also offer text, voice, and video calling delivered through VoIP protocols.

Dozens of standards define how Voice over IP works, but little documentation exists
on best practices for implementing and maintaining the technology in the enterprise.
There's not much introductory instruction for VoIP, so beginners may have a hard
time taking their first steps with it. There have been several high-profile

implementation failures among large enterprise adopters, and this may be why IP telephony has such an intimidating reputation.

Nonetheless, if it's done right, Voice over IP can transform the cost model of tele-communications by combining the overhead of voice and data expertise and infra-structure. It can also enhance productivity for end users by introducing new features and for telecom administrators by centralizing management functions. Voice over IP can decrease the expense of future computer-telephony integration projects, while making it easier to link voice systems with web servers and database applications.

This book will give you practical guidance on switching to Voice over IP from tradi-tional telephony systems. It gives a brief introduction to traditional telecom systems, and correlates their features and fundamentals to those of IP telephony systems, while showing ways of integrating traditional telephony assets into an IP-based voice network. It will describe the standards involved, so you can make educated choices among the large selection of components and vendors. It will also help you conquer some of the most problematic issues that people face when building telephony sys-tems with Voice over IP.

The PSTN

In 2004, there were close to 200 million traditional, landline telephones in use in the United States.* The network that connects all these phones together wraps around the globe. The job of the PSTN (Public Switched Telephone Network) is to reliably facilitate telephone conversations at any time of day, year-round. The PSTN com-bines analog, digital, and electromechanical data links that strive to make sure that every time you pick up your phone receiver, you hear a dial-tone, and every time you're hungry, you can reliably dial up the local pizza parlor.

Mesh Versus Switched

When designing a network that connects many phones (called *endpoints* in the world of telephony), there are two approaches: mesh networks and switched networks. In a *mesh network* design, every endpoint has a permanent connection to every other end-point, so all can communicate with one another. The first experimental networks developed by telephone pioneers worked this way.

Meshes are not very practical, because once you add more than a few endpoints, the number of permanent links between endpoints becomes absurd. For example, in a mesh with 10 endpoints, 100 separate links have to be maintained. In a mesh with 100 endpoints, 10,000 links have to be maintained. It doesn't make much sense to maintain so many network links, because there's a better way: the switched network.

---

* International Telecommunications Union (ITU) statistics provided by the World Bank.

In switched networks, links between endpoints don't need to be permanent because they aren't needed constantly. The only time a link between two endpoints is needed is when a call is in progress between them; the rest of the time, the link is idle, unused. Switching is a method whereby links are established and removed as needed, eliminating the need for a mesh. The PSTN is a switched network.

The PSTN carries each phone call by setting up and tearing down a temporary link, usually an electrical circuit, between a caller and a callee. The links that carry the calls may be comprised of copper wires, fiber optics, or radio systems, depending on the network infrastructure that exists between the caller and callee.

While the phone connections used in most homes are traditionally analog, electro-mechanical ones, connections between large telephone switches—for instance, calls between subscribers of competing phone companies—tend to be digital.

Signaling System 7

SS7 is the PSTN's signaling component: a second network that runs alongside the PSTN, whose purpose is coordinating communication between switches, telephone company databases and billing systems, and other parts of the public voice network. Toll-free calling and long distance call routing are both functions of SS7. While SS7 is secondary to the main function of the PSTN (setting up and tearing down links on the switched network), the modern-day PSTN couldn't function correctly without it.

Plain Old Telephone Service

When you pick up a traditional telephone, like the one in your house, the dial-tone you hear is a signal transmitted over the simplest kind of voice circuit: an analog phone line, also called a POTS (plain old telephone service) line. The line is a simple electrical loop. On the other end of the line is the telephone company's switch, or exchange. It is this switch that gives your phone its electrical power, sends the sound signals emitted by its audio transducer, and knows how to handle the numbers you dial on its keypad.

The switch also knows how to provide POTS calling features like call-waiting, a service that allows your call in progress to be placed on hold while you answer another caller who is trying to call your line—when you tap the receiver's hook, the switch recognizes the electromagnetic signal you created (called a *flash*), connects your phone line to the call that's waiting on another line on the switch, and places your first caller on hold. Later, when you send the flash signal again, the first caller is reconnected with your line and you can resume the original conversation. On the PSTN, calling features are always provided by the telephone company switch—not by the phone itself.

Despite historically increasing demand for features and gains in efficiency, POTS has been the most popular type of service provided by phone companies for the last several decades. Its core of a copper loop with an electromagnetic telephone receiver on one end and a central office switch on the other end hasn't changed much during that time. POTS' stalwart reputation and simplicity have been key reasons for its longevity.

So why not keep that copper telephone loop forever? POTS works great in low-density scenarios; that is, places with a small number of simultaneous callers, such as a one-family home. POTS can also excel where extensive calling features aren't important or where the telephone is rarely used. But when density and features become more critical, POTS isn't always the best solution.

Many phone companies offer an enhanced type of analog phone service known as Centrex to permit more calling features, such as call transfers and call-forwarding. Since Centrex and POTS lines are owned (and billed) by the phone company, they can create high reliance upon the phone company. This can be cost-prohibitive and inconvenient. To solve this problem, organizations on the grow usually choose to bring their telephony facilities in-house. By facilitating its own calling features, the enterprise can make telephony less costly and even leverage it as a productivity enhancer.

Key Systems and PBXs

As the telephone became more important in the business world, innovators extended its capabilities and made it more convenient. They did so using enterprise telephony devices—gateways that connect privately owned phones together into a private voice network with self-managed calling features. When the enterprise gained ownership over its own voice network, it set about building telephony applications specific to its business.

One such device is a KTS, or key telephone system. In many small businesses, telephones can share a group of telephone company POTS lines through the use of a KTS. Each phone in a key system has direct access to one or more of the telephone company's lines, just as a simple residential phone has access to a single line. Unlike a single-line phone setup, KTSs allow a group of phones to use more than one telephone line at a time. This allows a single operator to place a call on hold while answering a call on another line, among other things, without using any phone company calling features. Generally, KTSs are not referred to as switches, because they rely upon the circuit-switching abilities of the central office in order to connect calls.

In many larger offices, telephones connect to a private, on-premises switch that interfaces with the telephone company's lines. This switch is called a PBX, or private branch exchange. PBXs are smaller, enterprise-friendly versions of the heavy-duty switches used by the telephone company, and they allow businesses to run their own telephony applications in-house. Unlike with a key system, PBX phones in the office can call other phones in the office without tying up an external telephone line. So several simultaneous conversations between parties in the same office can occur without making use of the PSTN at all. One job of the PBX is to determine how to "route" calls—that is, how to ascertain whether the calling party is trying to reach another person within the same office or trying to reach somebody via the PSTN. Most PBX vendors refer to the call-routing scheme as the *dial-plan*.

Single-line phones, key systems, and PBX systems all connect to the PSTN—but for various reasons. Single-line phones and key systems connect calls to the PSTN even if they are from one phone to another in the same office, but a PBX connects calls to the PSTN only if they are bound for an outside organization.

Despite their various capabilities, POTS, KTSs, and PBXs are all based on the same circuit-switching, electromagnetic-signaling technologies. Even POTS' higher-capacity digital cousins, ISDN and T1, which are able to squeeze many simultaneous phone calls onto a single copper loop, are members of the same technology family. All the standards that govern these traditional telephony systems stem from the International Telecommunications Union (ITU), and many have been unchanged for decades because they are incredibly reliable. They'd have to be reliable in order to run the global telephone system, wouldn't they?

Lines and Trunks

Connections from the phone company switch, like the ones that connect to residential analog telephones, are called *lines*. POTS, Centrex, ISDN, and T1 connections are referred to as lines when their purpose it to deliver dial-tone to a phone or group of phones.

But when the connections deliver services to a PBX, they are referred to as *trunks*. Trunks are phone company lines that run from one switch to another, so a connection from the central office switch to a PBX switch would be called a trunk. Connections between two PBX switches are called trunks, and so are connections from one central office switch to another, coincidentally.

Limits of Traditional Telephony

The PSTN's capabilities are largely proportional to its physical connections, because every call must have a circuit, or loop, set up at the beginning of the call and torn down at the end. While the PSTN's switching equipment does a great job of this, some hard limitations are associated with its "circuit-switched" nature.

New features can take a while for the phone company to roll out. It took many years to upgrade central office switches to support features like call-waiting and three-way calling. Even now, some parts of the PSTN still don't support caller ID.

Capacity limits are another engineering challenge on traditional telephony networks. The fidelity of a call's sound reproduction is limited to the available bandwidth between the caller and the recipient, and the maximum number of calls between two offices is limited to the availability of voice circuits that exist between them. The problem posed to the enterprise is one of cost: every PSTN circuit used by the enterprise, be it a POTS line or a T1, adds to its telecommunications expenses.

The telephone companies and phone equipment vendors have made great strides to identify and resolve capacity and cost problems. High-density digital circuits like T1s and T3s have brought the cost of high-density telephony down, and PBX features such as least cost routing (LCR) allow the enterprise to minimize its long-distance calling expenditures. Long-distance calling has become cheaper, and the cost of on-premises PBX equipment and feature-rich business telephones has dropped over time, too.

At one time, telephony features were considered a competitive advantage. As businesses adopted them, they became part of the cost of doing business, and users began seeking a new telephony paradigm—one that could inspire big competitive advances again. The question the telecom industry sought to answer was, "Where do we go from here?"

Enterprise telephony innovators began looking to the Internet for the answer. Because of core differences in engineering philosophy and many years of additional discourse on the matter, the Internet is superior in many ways to traditional voice networks.

On the Internet (and IP networks in general), communications protocols are in a constant state of improvement, so more and more features can be delivered while bandwidth efficiency steadily improves and the cost of the network shrinks. On the Internet, capacity is tied closely to the efficiency of software, rather than to the physical capacity of a telephone switch, as with a PBX and the PSTN. As software improves, IP networks grow in capacity, but traditional switches need additional (often expensive) hardware to add capacity.

IP networks have always had another advantage over the traditional public voice network: their software uses standardized hardware components like low-cost PCs. This means that even when hardware upgrades are necessary, they can be procured less expensively. Unlike traditional PBXs, hardware upgrades on IP networks intrinsically improve software productivity and enable more and more capacity. Generally, capacity is easier to scale on IP networks than it is on circuit-switched networks like the PSTN.

While the PSTN is quite reliable, it is far less disaster-proof than IP networks. The Internet Protocol permits redundancy and failover capabilities that are inexpensive and relatively easy to implement. Geographic diversity, a technique used on data networks to circumvent local connectivity interruptions, is very easy for the enterprise to achieve with the Internet, but more difficult on the PSTN. For example, you can connect to two Internet service providers and use the same set of IP addresses with both, thanks to the BGP* standard, but it's nearly impossible to use the same set of phone numbers with two telephone companies.

Because most modern enterprise networks use the same protocols as the Internet, it was only a matter of time before the advantages of those protocols began appealing to designers of voice networks. The result of that appeal is an immense technology family called Voice over IP, or VoIP.

VoIP is loosely defined as using the TCP/IP Protocol suite to facilitate voice conversations, but it's really much more than that. It can be used to replace traditional telephony in the enterprise or in the home or merely to add features to a traditional telephony system. VoIP can also solve connectivity challenges, like linking traditional PBXs at remote sites together, linking private telephone extensions at a single site together like a PBX, or simply aggregating calls among a few analog phones like a key system.

VoIP can be used to facilitate voice communications on many different application substrates, too. It can provide on-demand voice-calling capabilities to users of a customer service web page and allow people to use their personal computers as fully featured business telephones. It can play a role in bridging cell phone and landline systems, too.

VoIP technologies can even run the entire enterprise voice telephone network. The VoIP technology family can even facilitate video streaming, conferencing, whiteboard applications, and instant text messaging, challenging the traditional distinction between data and voice networks.

Of course, VoIP needs a TCP/IP network in order to operate. Until recently, TCP/IP networks were less abundant than were connections to the PSTN. And even when TCP/IP began to rise in popularity, many private networks still weren't connected to the Internet, so the PSTN was always more appealing than the Internet for voice applications. But today, all of that has changed. There are an estimated 36 million permanent, private TCP/IP networks in the United States, and about 30 million of them are linked via high-speed connections to the Internet.†

 When data networks like the Internet are used to carry real-time media traffic, they are called *converged* networks. The process of implementing real-time media applications, such as IP telephony, on a data network is called *convergence*.

Many businesses and campuses consider migrating their telephony applications from traditional circuit-based voice networks (such as the PSTN and PBXs) to VoIP

---

* Border Gateway Protocol, RFC 1105.
† FCC statistics, June 2003.

Reasons to Choose Voice over IP

- VoIP devices are easier and cheaper to maintain because they leverage the corporate data network rather than their own single-purpose voice-only network.
- VoIP increases the value of the Internet by using it for voice communications.
- Integrating telephony with computer applications is much easier with VoIP than with traditional voice systems, because in VoIP settings, call management tends to be more open, standardized, and software-driven.
- VoIP can scale more cheaply than traditional voice systems because it often uses off-the-shelf, frequently updated PC hardware.
- It allows for more centralized administrative control than a traditional PBX.
- Managing a VoIP network is accomplished using the same network that carries the voice information itself, unlike the PSTN with its SS7 protocol.
- Survivability against disasters and network outages is easier to achieve with TCP/IP than with traditional voice systems, due to its basis in software and TCP/IP's remarkable resilience measures and routing protocols.
- Much of your existing phone equipment can interface with VoIP systems using analog terminal adapters, or ATAs.
- VoIP's acronym is fun to say ("voyp").

networks (such as the Internet and software-based call management). VoIP's biggest advantages over traditional telephony are scalability and infrastructural cost savings, though easier integration between telephony and computer applications is a big attraction, too.

Enterprise implementers considering VoIP are comparing scalability against that of traditional PBX equipment and discovering that VoIP's basis in software gives it a big advantage. While a small PBX built for 100 telephones is quite costly to scale up to 500 phones, for example, a VoIP call-management solution can usually scale up just as servers and network hardware do—and 500 phones is not a particularly tall order for even a moderately equipped VoIP setup.

Cost savings are attracting businesses to VoIP. Since VoIP runs over a data network, it has the same facilities requirements as a data network. Whereas traditional telephony devices such as a PBX normally require separate local area wiring for analog and digital phone connections, this wiring usually can't be used for local area data networking because it doesn't provide ample resistance to interference and attenuation. With VoIP, the same wiring is used for both data and voice, since the voice is carried within the data network.

The cost of future expansion is almost always less expensive with VoIP than with traditional telephony. Because it can be centrally administered more easily than traditional systems, VoIP allows system expansion, ongoing security enforcement, and back-office call accounting to be cheaper over time.

VoIP in the Home

A number of companies offer VoIP calling services that can be used in the home, more or less replacing conventional PSTN service. They deliver telephone calling capabilities using a broadband Internet connection. Not all of them permit placing calls to or receiving calls from the PSTN, but almost all allow you to call other users of the same service using the Internet instead of the PSTN. Some providers even have "peering" arrangements that allow you to call subscribers to other providers' services using the Internet.

Some of these services work only with proprietary telephone-calling software and don't allow you to use a hardphone. Certain providers can support the use of a special hardphone that connects to your PC's USB port and uses the PC as a gateway mechanism for accessing the network. Others provide an ATA device so that you can use one or more analog phones to place and receive calls using the service. Still others offer the ability to use IP phones.

Many of these services offer competitive calling rates, decent sound quality, and features that are close to that of the traditional phone company. There are solutions for adding more features and interesting hacks to a home-based VoIP network, too. Some of them are covered in this book.

VoIP in Business

Many vendors are producing cost-effective VoIP server devices for small Ethernet LANs. These devices connect together endpoints in a small office like a PBX, either through the use of conventional analog and digital phones (more on this later) or new-generation IP phones. In either case, connecting calls to the PSTN and between local phones is usually the responsibility of these server devices—like a PBX or KTS—and there are several ways to make that PSTN connection.

Using VoIP in small business environments is easier when there's some network savvy around the office. Some traditional phone vendors are now implementing VoIP systems, so the availability of third-party networking expertise is accelerating the adoption of VoIP in small businesses.

VoIP's Changing Reputation

There are potential pitfalls along the path to Voice over IP. Implementing VoIP is like any infrastructural investment—it has hard costs and implications for the enterprise user community. How you deal with these costs and social issues is largely defined by *how much* VoIP you implement at a time. That's why the concept of migrating is important.

Challenges also arise from using a relatively young technology for a task that has been reliably delivered "the old way" for decades. If the data network hosting the VoIP system isn't provisioned correctly, the results can be disastrous. Security, stability, and call quality in a VoIP system are all tied to their counterparts on the underlying data network. If your network is insecure, unstable, or lossy, your IP-based voice system will be, too.

A leading cause of failed VoIP implementations is poor perceived call quality, which usually stems from administrator misunderstanding of VoIP's requirements. VoIP is more than just call management and voice conversations; it is also a comprehensive set of methods to deal with quality of service. Lack of attention to these aspects of VoIP will doom even the most well-intentioned implementer.

These issues have contributed to IP telephony's reputation as difficult to manage, inferior in quality, and even damaging to corporate image. These perceptions can be avoided, and the opposite outcomes achieved, if VoIP is done right.

This book will help you implement and understand VoIP networking, call management, telephony features, and call accounting within the context of an enterprise data network. Along the way, you'll build a useful, real-world call-management system, a voice mail server, and more. You'll employ next-generation VoIP hardware and software, use open source telephony tools, and leverage traditional telephony components. You'll even be able to use that old residential-style analog phone for VoIP calls across the Internet.

You'll understand the differences between old-school and next-generation telephony and be able to implement a software-based PBX, maximize quality of service, and know many of the standards that govern the world of converged networks. You'll be able to identify situations in which traditional telephony can be integrated with IP telephony, and you'll learn how to provision emergency calling services that have always relied on the public telephone system. This way, your switch to VoIP will be a success.

Key Issues: Voice and Data: Two Separate Worlds

- The PSTN developed as a result of early telephony pioneers' efforts at building meshed and switched networks.
- SS7 is the switch-to-switch signaling network using to connect and bill calls placed across the PSTN.
- POTS, or plain old telephone service, is the basic single-line analog voice service from a phone company.

- Key systems and PBXs provide businesses a way of operating their own feature-rich voice networks at a lower cost than dealing only with PSTN-provided services.

- Lines and trunks are links between telephony devices. A line is a link from a switch to a phone. A trunk is a link from a switch to a switch.

- Voice over IP encompasses a large family of interface technologies, protocols, and standards that enable real-time media applications using IP networks.

- Traditional telephony isn't scalable like VoIP because it isn't software-driven like VoIP. Its ratio of calling capacity to network infrastructure is fixed, while VoIP's has lots of room for optimization.

- VoIP is suitable for deployment in homes and businesses, and, thanks to broadband and deregulation, is earning a reputation as a more practical way of delivering telephony services.

CHAPTER 2
Voice over Data: Many Conversations, One Network

Conversations are the basis of human communication. Conversations can be spoken, written, or gestured. Conversations can even be one directional, such as a coach bawling out his star quarterback after an uncharacteristic interception. Conversations may be "one-to-many" (such as a political candidate giving a stump speech) or "many-to-one" (such as a constituency lobbying that candidate after she's in office). Conversations are more than just an analogy for networks—they literally *are* modern networking.

The underpinnings of enterprise networks are also conversations. IP data networks run on protocols that use a conversational approach to data exchange. The most common protocols for web browsing (HTTP) and email (SMTP) use a two-way "data conversation" in order to communicate. The process is simple: a client host sends an inquiry to a server host or a peer host, and then the server or peer sends a response back to the client.

Conversations between hosts on an Internet Protocol (IP) network are similar to those between people, except that instead of using words, the messages are communicated across the networks using units called datagrams. A *datagram* is like a letter in an envelope. Once it has the proper markings, namely the recipient's address and return address, and a stamp, the entire letter can be delivered by the postal service. A datagram's markings are called *headers*, and they contain delivery information, like postal letters: instead of postal addresses, datagrams use something called *host addresses*. Different networking technologies have different names for datagrams, including *cells*, *frames*, and *packets*. Having a good understanding of IP networks is crucial to your success with Voice over IP. An excellent reference on the subject is *TCP/IP Network Administration* (O'Reilly).

When voice sounds are transmitted using datagrams on the IP network, telephony gains all the same characteristics as the data network itself. Just like applications for file sharing and printing via the network, software can be made to perform useful tasks using the datagrams of voice streams and signals—tasks like conference calling and voice mail. These tasks are the *applications* of Voice over IP. Voice applications delivered using IP datagrams is the essence of VoIP.

VoIP, like the network that carries it, is therefore not an application by itself, but a way to build applications using myriad software tools and devices. These building blocks can be specialized VoIP server hardware like an analog telephone adapter (ATA), or they can be highly programmable servers that do the job of a PBX. Regardless, all VoIP components must participate in the protocol conversations that make the audible, human phone conversations possible. That means that all VoIP components must be speaking the same language.

In human conversation, people can speak many different languages. Even among different dialects of the same language, people can have a hard time understanding each other—a Bostonian and a Texan sound about as different as a Canadian and an Australian, even though they all speak English. Unfortunately, telephony standards have had similar challenges.

Many standards govern the world of Voice over IP, and some have interoperability problems, just as people with local accents sometimes confuse each other. One such annoyance lies in the definition of the word *VoIP* itself.

VoIP or IP Telephony

Are "VoIP" and "IP telephony" two different technologies, or do they both describe the same thing? Well, it really depends on whom you ask. Some vendors prefer IP telephony when referring to their IP-based voice offerings, arguing that VoIP refers to the specific act of transmitting digitized sound data on an IP network and IP telephony refers to the overall technology family. Others give VoIP the broader definition, identifying it as inclusive of IP telephony, and referring to IP telephony only as the act of mimicking traditional telephony applications.

For the purposes of this book, we'll take the latter tack: VoIP refers to the overall technology family, while IP telephony means specific application functions such as call signaling and voice mail. So when we talk about conference calling, we might call it telephony, but when we talk about conference calling, call-waiting, and voice encoding, we will refer to them collectively as VoIP. In general conversation, though, VoIP and IP telephony can be used interchangeably.

VoIP's Pros and Cons

VoIP certainly has a few disadvantages when compared to old-school phone hardware. High-utilization service guarantees are harder to deliver with VoIP than with an old-fashioned PBX. The same scalability characteristics that attract people to VoIP can ultimately be the reasons their implementations fail: a VoIP network can be so extensible that service-level guarantees are hard to make, whereas a traditional circuit-switched voice network has hard capacity limits, around which levels of service tolerance can be guaranteed easily. Certain broadcast audio applications, like overhead paging, can be difficult with VoIP, too.

The gains VoIP brings to the table far exceed the few difficulties it imposes, though. There's nothing that old PBX can do that a VoIP telephony system can't, even if VoIP makes a few things tougher.

One thing VoIP makes *easier* is physical provisioning. While a PBX requires a network of electrical, usually copper wire, loops, VoIP requires an IP network. Since IP networks are a staple of every modern business, the logistics of building a network for voice is largely simplified because the required physical elements are already in place for other common business applications: databases, messaging, Internet access, and so on. VoIP is carried on the network the same way those are.

If you're an Internet user (and who isn't these days), then you know TCP/IP is the core protocol that defines the architecture of the Internet. In most organizations, and even in many homes, a TCP/IP local area network is an important interpersonal communications tool, used for email, web surfing, and instant messaging. When VoIP replaces the traditional telephone using TCP/IP, the local area network becomes *the key piece* of telecommunications infrastructure.

Once that key piece is standardized within the enterprise, VoIP administrators have only one network to maintain—the one that supports TCP/IP. This means supporting a single network cabling system, rather than separate ones for voice and data. If you use wireless Ethernet, you don't need local area cabling at all—VoIP will still work. Meanwhile, old-school PBX administrators still have to maintain a separate local area cabling plant that serves only the PBX system.

But that key piece of telecom can be a key failure point, too. When the voice and data networks are separated, as they are in traditional telephony, their physical paths lie separately, protecting the voice system from failures isolated on the data network, and vice versa.

But with VoIP, these paths converge. When the path is broken by an equipment failure, a power failure, or a construction crew accidentally slicing underground cables, the data network fails. When a computer virus swamps your data network, VoIP phone calls may no longer be possible. When data fails, voice fails, too.

Even in the home, where you might rely on a cable or DSL Internet connection, your VoIP calling capability will swiftly disappear when your broadband provider's service fails or your power goes out.

VoIP Network Fundamentals

Since VoIP is layered on top of TCP/IP, you must have some form of TCP/IP network in order to use it. For small VoIP experiments, any Ethernet LAN will do, even a hub-based or wireless one. But for larger, critical VoIP implementations, your choice of network infrastructure will be critical. For starters, using broadcast Ethernet devices like hubs is a poor choice, as is using early-generation Ethernet switches that lack quality-of-service features. Wide area equipment, like routers, will need to support these features, too. (Quality-of-service features are covered in detail in Chapter 9.)

Generally speaking, the faster your switches, routers, and network links are, the better your VoIP network will perform. Nothing beats good ol' speed for increasing the performance of a wide area network, but sometimes a slow network link is an economic or geographic necessity. VoIP is a speed-sensitive business, as you'll find out.

The Layers of a VoIP Network

Like other networks, VoIP networking can be described using the Open Systems Interconnect (OSI) reference model, a standardized way of describing the different parts of the data communications process. The OSI model has seven layers that represent each part: physical, data link, network, transport, session, presentation, and application. The purpose of the OSI model is to simplify connections between different types of networks and to allow engineers who design network applications to assume a standardized platform upon which to build.

The physical layer

The OSI physical layer is the most fundamental part of the datacom process. It's the layer that provides for the electrical, mechanical, radiant, or optical signaling pathways that are required in order to move data across any data network. In an IP network, the physical layer can include twisted-pair LAN cabling, plugs, cross-connects and patch panels, power sources, V.35 cables like those often used with serial interfaces on routers, and so on.

Though the physical layer is itself intended to be permanent and stable, its assortment of connective technologies (copper twisted-pair, fiber-optic cables, etc.) are prone to noise and distortion, two problems that cause data transmission errors. The physical layer has no way of dealing with these problems, and that's why certain guidelines related to distance and interference exist at the physical layer. For example, a 100BaseT Ethernet connection on twisted-pair copper cable cannot be longer than 100 meters.

The data link layer

Since the physical layer is not immune to the laws of physics and the signal degradation they incur, the data link layer provides a medium for detecting errors in data transmission. Error detection at the data link layer works on behalf of a single physical link, such as an Ethernet segment or a single T1 circuit.

The data link layer "frames" the continuous stream of signals flowing across a link. *Framing* means delimiting that signal into manageable pieces, called frames (imagine that). For error detection, each frame can be subject to a CRC, or cyclic redundancy check. With certain types of connections, error correction can be attempted.

The data link and physical layers are often viewed as one and the same, and in many network substrates, such as Ethernet, their functions tend to be inseparable. That is to say, you can't build an Ethernet physical layer without building its data link layer too—both layers are facilitated by the same device, which is usually an Ethernet interface, a hub, an Ethernet switch, or an Ethernet coax bus. The data link layer is the lowest layer that VoIP applications can reference, and usually only in an indirect manner (only quality-of-service functions interact with the data link layer—more on this in Chapter 9).

The network layer

While the data link layer provides data framing over a single physical connection, such as an Ethernet segment, the network layer provides the logistical intelligence required for a single network to exist across many physical connections—say, two Ethernet segments connected by a wide area connection. The network layer is more visible to applications than the data link and physical layers. Its job is to provide:

- A scheme for data routing across wide area links
- An addressing scheme, so that disparate physical connections can be referenced by higher-layer services and by each other
- A definition for connection-oriented and connectionless datagram structures

The network layer isn't the lowest layer that is relevant to VoIP, but it is the lowest layer that VoIP applications must reference in order to function. For example, datagrams and addresses—things implemented at the network layer—are critical to the functioning of VoIP applications.

The addressing scheme used by VoIP is inherited from IP. Each device on an IP network has an IP address, so each VoIP endpoint has one too. An IP address consists of 32 bits, commonly presented by four 8-bit figures separated by dots:

    10.1.1.204

Each figure in the address allows for 256 values, so the overall address space (32 bits) of IP's addressing scheme allows for about 4.3 billion addresses. The newer version of IP, Version 6, allows for a 128-bit address space, but adoption of IP Version 6 has been slow, and this book deals strictly with the 32-bit address scheme of IP Version 4—the protocol that today's Internet runs on. In the context of the Internet and IP networks, the network layer is sometimes referred to as the *Internet layer*.

Using IP addresses, the network layer can facilitate wide area networking over dozens, hundreds, thousands, or millions of physical links. Consider the Internet, which uses

IP to connect millions of disparate networks. Individually, each of these networks tends to share a group of related addresses. Each group is what IP calls a *subnet*.

Every datagram sent across an IP network contains a source and destination address so that the devices responsible for maintaining the network layer know where to route the datagram. However, the network layer isn't responsible for any kind of error control—that's the job of the next layer up.

The transport layer

Even though the data link layer provides error detection on an individual network link, this alone isn't enough to satisfy the needs of a large, application-intensive network. That's why the transport layer provides error control across the entire network—from sender to receiver—regardless of the number of physical links between them. Transport layer error control operates independently of the measures provided by the data link layer, which tend to be specific to the type of link they are responsible for.

On the transport layer, protocols have been designed for two kinds of service:

- Datagram delivery is highly reliable, complex, and has high overhead.
- Datagram delivery is less reliable, less complex, and has lower overhead.

The kind of service elected depends on the needs of the application. Some applications don't need a high degree of reliability (video gaming, for example), while others must have absolute reliability (bank transactions). Within the transport layer, IP provides protocols—UDP and TCP—that handle both needs.

To connect or not to connect

Within IP, datagrams can be delivered using a "best effort" approach—that is, the host transmitting the datagram will not know whether it was received by the intended recipient. Also called *connectionless networking*, this method is employed by the User Datagram Protocol (UDP).

If you've ever played the Quake series of video games over a network, you've used the UDP Protocol. UDP excels in situations in which very fast delivery of data is a requirement, and reliability features, like confirming that the data has been delivered, would be a waste. In a multiplayer network game such as Quake, you and the other players each control an armed character that is trying to kill the others in a virtual 3D world. Real-time delivery of characters' location and trajectory data within the virtual world is critical to the game play. Even a slight delay in delivery of these datagrams could mean life or death for your Quake warrior. Delivery guarantees impose too much overhead—because dozens of UDP datagrams can be used by Quake in a second.

The same is true of the traffic carried over the network during VoIP phone calls. This traffic is carried across the network at a rate of between 30 and 50 datagrams per second. To verify delivery of each one would introduce a performance bottleneck that is unacceptable in a voice application. Therefore, almost all voice data flowing across a VoIP network is considered connectionless and carried by UDP.

The more reliable protocol for data transmission in an IP network is called Transmission Control Protocol, or TCP. Like UDP, TCP is encapsulated within IP. TCP's distinguishing characteristic is that transmitters using TCP must set up a transmission channel, or connection, before they send data to their receivers. For this reason, TCP is considered a connection-oriented protocol.

Error control takes place during a TCP transmission. At the end of the transmission, the sender and receiver agree to end their conversation, and the connection is closed. TCP guarantees that packets will arrive in the correct order, too. Because TCP is so cautious compared to UDP, it isn't normally used to carry voice data, but it can be used to carry call-signaling data: the bits of information that a VoIP network uses to establish, monitor, and end calls. TCP datagrams are called packets, though you often hear people refer to UDP datagrams as packets too.

IP provides both connection-oriented (TCP) and connectionless (UDP) network protocols at the transport layer, which allows it to replace both functions of the PSTN: voice transmission and call signaling.

The session, presentation, and application layers

Operating systems, end user applications, application services (like DNS), and user interfaces are provided at the topmost layers of the OSI reference model. Your interaction with a computer system or a network is most directly affected by the systems running at the application layer. The application layer's job is to take input and drive underlying functionality down through the other six layers without you, the user, having to know the details of what's going on down there.

In a VoIP network, the user interface to the telephony functions—often just a telephone receiver with a 12-key dial-pad—is provided at the application layer. A VoIP-adapted OSI model is shown in Figure 2-1.

A VoIP network is a set of networked applications and endpoints (agents that allow humans to use the applications), just as the World Wide Web is a set of networked applications and endpoints. In the case of the World Wide Web, the applications are web sites, and the endpoints are web browsers that request and display web pages. But in the case of a VoIP network, the applications are telephone calls, conference calls, voice mail, automated attendants, and even video conferencing or text messaging, while the endpoints are traditional telephones, IP phones, and software phones (softphones) that run on PCs.

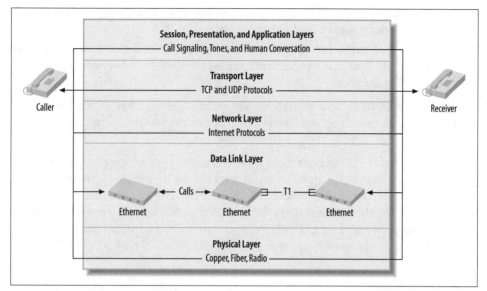

Figure 2-1. OSI reference model layers

On the WWW, web sites are hosted using software like Apache, a web server. This software communicates with the endpoints (the browsers) in order to facilitate user interaction with the application (web sites).

This model works much the same with Voice over IP. In a VoIP network, specialized servers, which we'll call VoIP servers for now, communicate with the IP or traditional phones (endpoints) in order to facilitate calling (the application).

VoIP Servers

VoIP servers, which are software-based devices that direct or participate in Voice over IP data conversations in order to facilitate calling and other VoIP applications, are usually connected to the network using Ethernet. An ATA is itself a highly specialized VoIP server and so is a VoIP-enabled voice mail system.

Carrier-class users of VoIP might have reasons to connect VoIP servers to a different type of data link, such as ATM (asynchronous transfer mode), but most enterprise implementers will use Ethernet.

VoIP servers fulfill many telephony roles:

- Call switching and connection management, such as a traditional PBX. A VoIP server in this role is usually called a *softPBX*.
- Call recording and autoattendant functions, such as a traditional voice mail system.
- Call conferencing, such as a traditional teleconferencing service.

- Access interfacing, so that traditional phones and PBXs can participate in the VoIP network by way of media conversion.
- Translation of audio encoding standards (codecs) in real time to facilitate calls between endpoints that have different audio capabilities or between analog, digital, and IP endpoints.

When VoIP endpoints and servers are connected to the same IP network, VoIP becomes the call-switching and voice-transmission mechanism, replacing the traditional PBX.

What differentiates VoIP servers from voice endpoints is whether they provide a user interface for the telephony application. Phones do, so they are endpoints. Switches, ATAs, PSTN gateway devices, and other specialized VoIP devices don't, so they are VoIP servers. Another differentiator between endpoints and servers is their abundance on the network. Like the WWW, there are more endpoints than servers in a VoIP system, sometimes even in thousands-to-one ratios.

Voice Endpoints

Endpoints that are TCP/IP aware (that is, they are valid hosts on the IP network) and connect directly to a data link that carries TCP/IP (such as Ethernet) are usually called IP phones. *IP phones* resemble feature-enriched business telephones but differ in that they usually have an RJ45 twisted-pair Ethernet connection rather than an analog or digital loop connection. IP phones can be plugged directly into an Ethernet hub or switch using an Ethernet patch cable or through a cabling distribution frame like those found in many offices. Usually, IP phones have a 10/100BaseT auto-negotiating interface, much like a desktop PC Ethernet adapter.

The voice applications that run on an IP phone facilitate calling in a manner similar to a traditional phone, but the mechanics of call signaling and voice transmission are worlds apart from the old-school telephony world, as you'll discover.

Even though they don't use RJ45 twisted-pair connections or have Ethernet smarts onboard, traditional analog telephones can be connected to Ethernet, through the use of an ATA (analog telephone adapter). The ATA is a device that converts the single-pair RJ11 analog connection into a four-pair 10/100BaseT Ethernet interface, as in Figure 2-2. ATA devices tend to be less expensive than IP phones. They provide fewer telephony features, too—after all, that old analog phone can't really run sophisticated VoIP applications, even with an ATA attached, because it doesn't contain integrated circuits or programmable components. In some cases, the limited functionality provided by a traditional analog telephone is enough.

IP telephones and ATAs are both hosts on an IP network. Like other TCP/IP hosts, they must have IP addresses and be compliant with the design of your IP network.

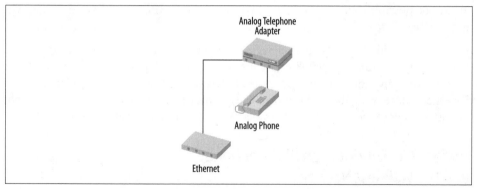

Figure 2-2. Analog phones are voice endpoints, and they can be used with VoIP networks in tandem with an ATA device

IP phones or traditional phones

The phones you choose are often determined by your project budget—traditional phones are often already there; it's just a matter of VoIP-enabling them. This can mean less capital investment during your transition to VoIP. But your choice of endpoints is also affected by your feature requirements: IP phones tend to have greater feature sets and programmability than do traditional phones.

Another big advantage of IP phones is that they can be completely software-based, so that they run as an application on a TCP/IP-aware Windows, Mac, or Linux PC. This can be appealing to mobile users who want to maintain a familiar look and feel on their telephone wherever they go. As long as the user has access to the Internet, the "softphone" can be made to function just like an IP "hardphone."

If you choose to integrate existing traditional phones into your VoIP network, then you must continue to support the wiring that is required for those phones. Often, this wiring is very simple, like the one-pair copper wiring used to power a simple analog phone, but not always. Most traditional PBX-type phone systems use telephones that require two pairs of copper and a digital signal bus—these are called, simply enough, digital phones.

Wireless IP phones can solve the challenges of wiring, but they introduce some challenges of their own. Since wireless Ethernet does not yet offer quality-of-service mechanisms or a high yield for simultaneous use, the abundance ceiling of wireless IP phones is much lower than their wired cousins. Wiring aside, using traditional cordless phones is one way to maintain mobility without the restrictions of wireless Ethernet.

If you choose to use wired IP phones, then your Ethernet wiring system must be sufficient to support them. 100BaseT Ethernet and Category 5e cabling are considered the minimum for connecting IP phones. For this reason, exclusively using IP may not always be possible. Not all sites where you want to use IP phones have Ethernet wiring in place.

If you had an old 10Base2 Ethernet segment at a certain site, you wouldn't be able to connect any IP phones at that site, because there are no IP phones on the market that support the physical layer interfaces (BNC coax connectors) required by 10Base2 Ethernet. Moreover, 100BaseT is the only Ethernet spec that allows for the right quality-of-service required to support a large rollout of IP phones.

Project 2.1. Configure an IP Hardphone and the VoIP Test Network

To complete this project, an IP hardphone is needed. We'll use the Grandstream Budgetone 100 series phones in this example.

What you need for this project:

- Grandstream IP phone
- LAN

IP phones are really just software applications that speak VoIP's protocols: SIP, SCCP, H.323, or MGCP for call signaling; RTP for audio transmission; and sometimes LDAP for directory integration. They may also include XML or Java services so that their displays and buttons can be used to further enhance the end user's telephony experience. When bundled together, this suite of protocol software that comprises an IP phone can run either on a PC, what we call a softphone, or on a specialized chassis whose enclosure looks like a traditional telephone, what we call a hardphone (see the previous section for more details).

Hard or soft, IP phones all require a TCP/IP stack to support data networking as well as a physical interface to the network. In a softphone, these are provided by the PC's operating system and networking hardware. In a hardphone, they are embedded more tightly and are less visible to the user.

Like a PC with a 10/100 Ethernet interface, an IP hardphone has an RJ45-compatible jack, so the first step in getting a hardphone online is connecting a patch cable between that jack and an Ethernet switch. This step will be largely the same regardless of the make and model of IP phone you're using.

Next, the IP phone must be given a TCP/IP host configuration that is workable on the network to which it's connected. To configure an IP phone for the network, you'll need:

- A DHCP-assigned or statically assigned IP address
- A DHCP-assigned or administrator-designated IP subnet mask

- A default gateway address (optionally assigned by DHCP)
- The address of a DNS (domain name service) server that serves this network

The IP address used by the phone can be static, or it can be dynamically assigned using DHCP (Dynamic Host Configuration Protocol) if you have a DHCP server on this Ethernet segment. DHCP is not necessary in a small environment with very few IP phones to keep track of. It becomes a necessity in larger environments where an administrator mistakenly assigning the same address to two different phones can cause a disruption, just as in PC networking. For now, we'll use static addresses.

The specific steps required to configure each make of IP phone varies depending on the administrator features and firmware of each. Most permit rudimentary network configuration using the buttons on the phone itself. The Grandstream Budgetone 101 telephone is an entry-level SIP-based IP phone, and its initial configuration is done in this manner.

 SIP is the Session Initiation Protocol, a standard for call signaling and capabilities negotiation. It is covered more extensively in Chapter 7.

Configuring a Grandstream Budgetone 101 IP phone

The Budgetone 101 phone has a Menu key, two arrow keys, and an LCD display, which are used to navigate its configuration menu options: DHCP, IP Address, Subnet Mask, Router Address, DNS Server Address, TFTP Server Address, Codec Selection Order, SIP Server Address, and Firmware Versions (called *Code Rel* on the phone's screen). When you get to the option you want, you press the Menu key to select it, and then enter the numeric data required for each option using the keypad. Use this menu only to set up the IP address, subnet mask, and router (default gateway) address.

To get the phone enabled for the next configuration step, turn DHCP off, and assign an IP address, subnet mask, and router address.

More advanced configuration is performed using the Budgetone's built-in web configuration tool. When you access the IP address you assigned to the phone using your web browser, you'll be prompted to log in to the phone, as in Figure 2-3. The default password is "admin."

Then, you'll be confronted with a big page of configuration options like the one in Figure 2-4. Many of the options are available only through this interface, not from the phone's keypad menu. For this project, the only settings we're concerned about are the codec selection ones. Configure the first (highest-priority) codec to be "PCMU" if you're in North America or "PCMA" if you're elsewhere in the world. That's all we're going to cover about codecs for now. After applying any configuration changes, the Budgetone needs to be power cycled.

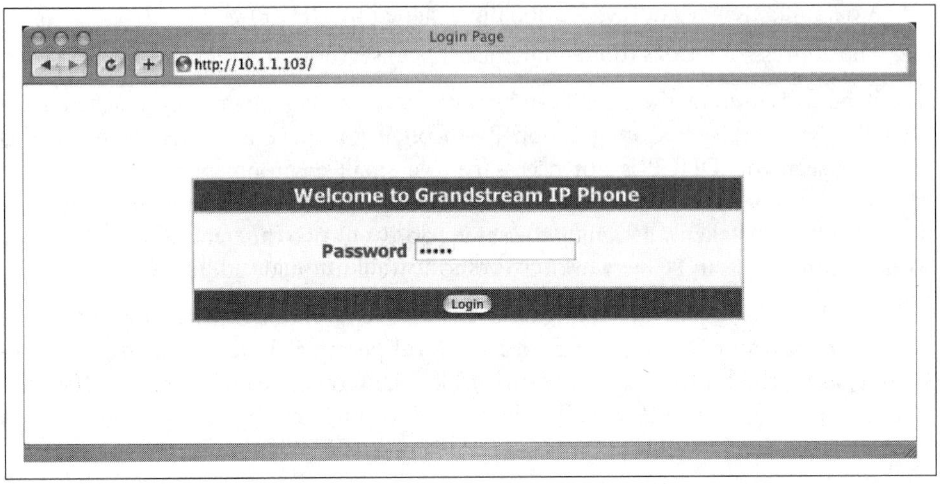

Figure 2-3. *The Budgetone's web configuration login page*

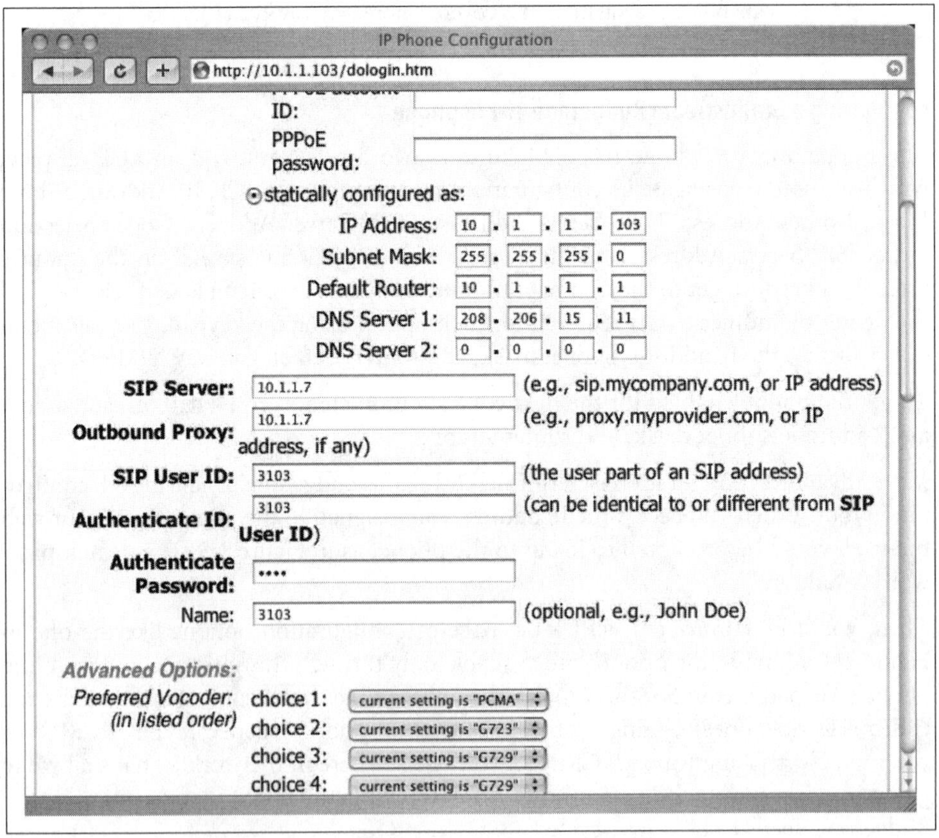

Figure 2-4. *The Budgetone's main configuration page*

Some IP phones offer a Telnet interface rather than a web-based one. To use these tools, one must connect to the phone with a Telnet client instead of a web browser. In any event, once the network configuration is set on the IP phone, ping its address from another host on the same network subnet to make sure it's speaking TCP/IP.

A simple VoIP test network

Throughout this book, a TCP/IP network is used to illustrate Voice over IP concepts through projects and hacks. This network, illustrated in Figure 2-5, is structured as follows:

- IP hardphones have an IP address of 10.1.1.100–150.
- IP softphones and ATA devices have IP addresses of 10.1.1.200–250.
- VoIP servers and nonendpoint devices, like proxies, have an IP address range of 10.1.1.10–29. The Asterisk server we use will always be 10.1.1.10.
- The default gateway router's address is 10.1.1.1.
- The subnet mask for all devices is 255.255.255.0, giving our test network a maximum size of 254 possible devices or an 8-bit subnet.
- DHCP will not be used, except as noted in specific projects.
- The test network will always use wired, switched Ethernet, unless specifically noted.
- It will consist of one segment, or one Ethernet LAN, unless specifically noted.
- This test network requires access to the Internet for many projects. To accomplish this, use a NAT firewall or Internet access appliance.

Many VoIP devices need access to a time clock. The NTP (network time protocol) server we've chosen is time.nist.gov. More NTP servers are available from the list at *http://www.nist.gov*.

Project 2.2. Make an IP-to-IP Phone Call

For this project, you'll need two IP phones. Our scenario uses two Grandstream Budgetone 100 series phones configured as directed in Project 2.1. Most IP phones permit a type of IP-to-IP calling similar to what's described here, so you can replicate an IP-to-IP call using a different make of IP phone.

What you need for this project:

- Two Grandstream IP phones
- LAN

With both IP phones connected to the same Ethernet switch or directly connected (to each other) using a crossover patch cable, make a note of the IP address you've

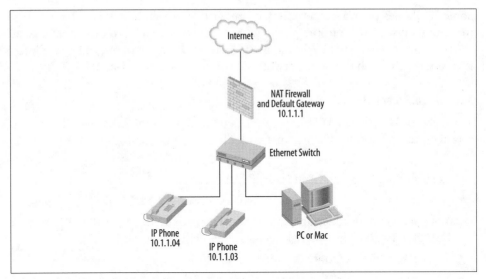

Figure 2-5. The VoIP test network at the end of Project 2.1

established for each. In this example, we'll use 10.1.1.103 for the receiver and 10.1.1.104 for the caller. If you have your phones configured for DHCP, give them this static configuration instead.

The Budgetone can place IP telephone calls from one IP endpoint directly to another without the need for a VoIP call-management server. This is called *IP-to-IP calling*. Since each IP phone has a uniquely identifying characteristic within the scope of the network—an IP address—one phone can call the other by IP address as if it was a phone number.

To do this, first make sure there is nothing set for User Name or SIP User in the Budgetone's configuration page. That is, make sure they are both blank, apply the changes if necessary, and then power cycle the phone.

Now, to dial by IP address. All IP addresses are 12 decimal digits long, even if leading zeros aren't written out. Conversely, the dots normally included in an IP address are *not* dialed. So, on the Budgetone phones, 10.1.1.103 is dialed as:

```
010 001 001 103
```

To dial, take the phone off the hook so you hear a dial-tone. Then, press the Menu key, dial the address of your second phone according to the convention shown earlier, and press the Send or Redial button. Of course, nobody would want to dial 12-digit IP addresses in order to place phone calls all the time; call-management servers, like SIP registrars, provide more elegant dialing conventions. Dialing by IP address, in this case, allows you to circumvent call management and make a direct VoIP connection between two endpoints.

When the receiving phone rings, have somebody answer the call. If you can hear the other person talk through your IP phone's handset, you've just made your first successful VoIP phone call—sort of the IP equivalent of Bell's and Watson's first phone call back in 1876.

If the receiving phone doesn't ring, then you should check the IP address you dialed, check the phone's configuration to make sure it is listening on the default port for SIP (5060), and make sure SIP registration is turned off. These options, which are accessed in the Budgetone's web configuration page, will be covered in greater detail later.

Dialing by IP address isn't user-friendly, and it isn't practical at all in a DHCP environment, let alone an enterprise or home phone system. Outside your test lab, you'll use it only for troubleshooting.

Distributed Versus Mainframe

In the world of traditional telephony, endpoints and PBXs interact in a manner similar to dumb terminals and mainframe computers. That is, the PBX (or mainframe) has all of the application functionality built in, and the user interface functions of the endpoints (or terminals) are dictated by the PBX.

With IP telephony, voice endpoints are far more programmable, lessening the requirement for centralization. VoIP endpoints don't always have their functions dictated by a particular VoIP server. In fact, VoIP endpoints may interact with many services on many different physical servers: DNS, LDAP, SIP, and RTP are all VoIP-related application protocols that may be facilitated by separate servers or by no servers at all (some operate between two endpoints and don't require a server in between). The IP-to-IP call placed in Project 2.2 is a good example of that.

Compared to a traditional telephone call, which must always be routed through a telephone switch such as a PBX, this is a significant difference. A traditional telephone call is set up, torn down, and accounted for using the same piece of hardware—the PBX. Moreover, the sounds of the conversation are routed through the PBX, because the PBX is the circuit-switching mechanism that provides the voice loop between caller and receiver. This is illustrated in Figure 2-6.

But in a VoIP network, the call-management functions are separated much more from the voice transmission functions. This allows each function to be enabled through separate network resources, as shown in Figure 2-7. Call management could occur over a wide area link, while the voice transmission could occur directly between two endpoints on the same local area link, in order to preserve capacity on the wide area link. The net result is that a single, powerful call-management server could work on behalf of many remote sites, increasing the value of the WAN and possibly saving money that might ordinarily be spent on remote PBX systems to support each site.

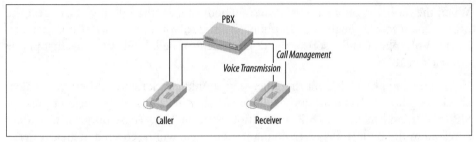

Figure 2-6. With a traditional PBX, voice transmission and call management are dependent upon a route through the voice switch

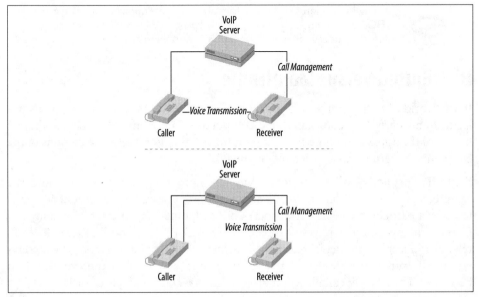

Figure 2-7. In IP telephony, call management and signaling can be separated from voice transmission

The distributed nature of VoIP applications makes them preferable to traditional telephony on a wide area network—but that's not the only advantage of VoIP on a WAN. The other great benefit of VoIP—especially in a bandwidth-conservative WAN—is compression.

The Core and the Edge

At the heart of a network resides the *core*, or network backbone. In modern IP networks, the core serves the purpose of transporting high levels of aggregate traffic between nodes that are probably not endpoints—that is, they aren't the hosts where the traffic originated or the hosts where it is headed, but rather hosts whose purpose is to forward that traffic along the core network until it needs an exit along the route to its destination.

The core is kind of like the 10-lane interstate highway: a lot of people drive on it, but nobody's driveway is an entrance ramp to it. So, while billions of hosts may send and receive data that crosses the Internet core (backbone), almost none of those hosts are directly connected to the core.

Instead, IP network endpoints connect to disparate network links that share high-capacity aggregate connections to the core. These links are collectively known as the *edge*. The edge is like the surface streets that surround the 10-lane interstate highway. Most traffic that ends up on the big highway originates from the surface streets.

A key difference between distributed and mainframe computing follows this analogy: in a mainframe environment, such as the PSTN, all the endpoints have a direct connection to a core—the central office switch. Likewise, in a PBX system, all the endpoints have a direct connection to a core—the PBX switch. So, all the driveways in a mainframe town are actually entrance ramps right onto the big highway.

VoIP facilitates the build-out of the networking smarts that normally exist at the traditional PSTN core, so that application functionality gets closer and closer to the edge of the network. This is similar to the way distributed PC applications have been displacing mainframe client/server applications over the past 20 years.

With VoIP, the core network is still there, and very much required, but it serves a different purpose than the core network of the PSTN. In a VoIP environment, the core is mainly used to move data back and forth, and the programmatic functionality of voice applications exists in a distributed model of peers: VoIP servers and endpoints. These peers can reside anywhere on the edge and offer new and changing features, without requiring changes at the core.

In traditional telephony, that isn't the case. The PSTN's core is itself responsible for all of the features available to you as a telephone company customer or enterprise PBX user, and offering new features can require the phone company or enterprise PBX administrator to alter the core network.

VoIP in Enterprise Networks

VoIP can be used to connect IP phones on an Ethernet segment to a VoIP server that is used for call management, and that VoIP server can be used to provide a connection for those phones to the PSTN, as in Figure 2-8.

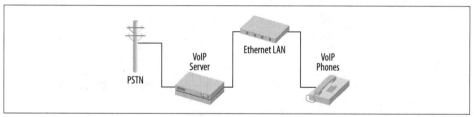

Figure 2-8. A VoIP server can be a PSTN gateway for IP phones connected via Ethernet

A single VoIP server can act as a PSTN gateway for IP phones on Ethernet segments located at remote offices, as long as WAN connectivity exists between them. This way, the IP phones at all the sites can call one another, and the VoIP server routes calls between the offices and to the PSTN, as in Figure 2-9.

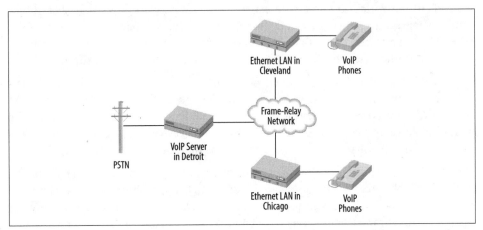

Figure 2-9. A VoIP server can be a PSTN gateway for many IP phones on a wide area network

If a large company uses a conventional PBX at every site around the country, all can be linked together using VoIP over a WAN. This way, each PBX can connect calls within its local network of traditional phones, as well as calls between them and the PSTN, but calls placed between phones on opposing PBXs can be routed over the WAN using VoIP, as in Figure 2-10.

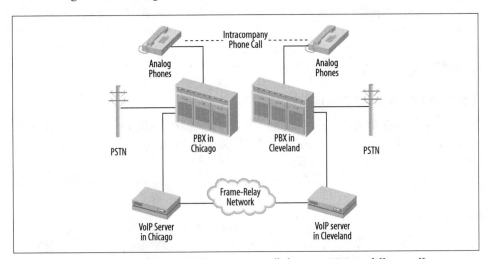

Figure 2-10. VoIP servers can use a WAN to connect calls between PBXs in different offices (switch-to-switch trunking)

At a minimum, at least two VoIP devices (such as an IP telephone and a VoIP server or two VoIP servers) and at least one form of connectivity are required by all VoIP solutions.

Like the network, VoIP is a conversation-oriented technology. Its protocols are simply rules that devices and software must follow in order to carry on the conversations required to make VoIP applications work—that is, to carry each human speech conversation. Each VoIP protocol set (H.323 and SIP are the two big sets) has its own rules that enforce proper conversation, just as Congress has a parliamentary procedure that enforces its debates. The biggest rule is the definition of VoIP's minimum requirements: two or more TCP/IP hosts using one common protocol and connected data links.

Network Convergence

When you support only one transport (in VoIP's case, TCP/IP) for all networked applications, including telecommunications, you've achieved complete convergence. The more you leverage your TCP/IP network to support voice and multimedia telecom apps, the more you converge. Theory tells us that convergence increases administrator productivity, and experience tells us that support costs drop the more voice and data networks are converged.

Convergence isn't something that has to happen overnight. There may be plenty of reasons you don't want total convergence: capital that is tied up in perfectly good legacy hardware is one; network readiness is another. As with many past paradigm shifts in networking, a migration path exists to get you from partial to total convergence. One of these paths is the "hybrid" voice switch.

Pure IP or IP Enabled

Pure IP voice switches can't make direct use of traditional circuit-switched telephones and trunks. Vendors that refer to the VoIP solutions as *pure IP* mean that the phones and trunks connected to their switch are totally packet-based. Connections to outside non-IP systems, like the PSTN, are accomplished through outboard hardware that facilitates transmission of call signals to the call-switching server using IP. In this fashion, vendors whose switching servers support only IP endpoints and not traditional endpoints tend to use the pure IP moniker. Cisco's CallManager 4.0 is a good example of what pure IP means—it's a completely software-based switch that requires outboard hardware, called a media gateway, in order to support non-IP endpoints. As you can see in Figure 2-11, any devices that communicate with a pure IP PBX do so using the TCP/IP Protocol trunked over Ethernet.

IP-enabled voice switches, unlike pure IP systems, offer support for all kinds of voice endpoints and make no bones about connecting to analog phones and trunks like those from the PSTN. Analog, digital, and IP devices can all connect, as shown in

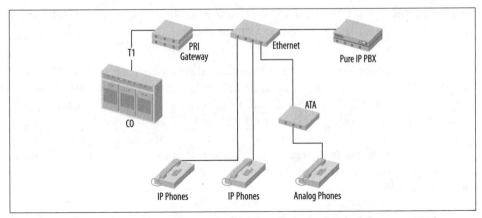

Figure 2-11. A pure IP switch has only IP-based trunks; all trunks that feed the same switch are carried by TCP/IP

Figure 2-12. The media interfacing required to use traditional telephony devices with an IP-enabled switch is all contained within the switch chassis, often using a single digital bus and microprocessor, much like a conventional PBX. Good examples of software-based IP-enabled switches are Avaya's Communication Manager 2.0 and Digium's Asterisk (an open source solution), both of which run on Linux. Sometimes VoIP implementers refer to IP-enabled switches as *hybrid* switches.

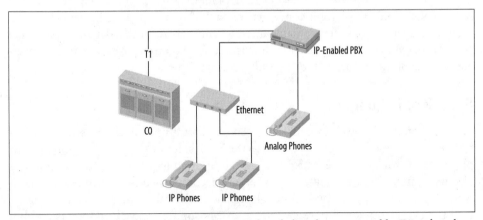

Figure 2-12. An IP-enabled voice switch supports IP-based, digital connections like T1 and analog connections

Key Issues: Voice over Data: Many Conversations, One Network

- VoIP can replace traditional telephony, but quality-of-service measures are required in order to make it as reliable as old-school gear.

- The OSI network model breaks down VoIP in terms of layers. The networking aspects run at the lower layers, and the application aspects run at the higher layers.

- VoIP media streams are delivered by connectionless UDP datagrams, and not TCP packets. This is because, in telephony and other real-time media applications, there's no point in error correction. VoIP administrators would rather strive for full error *abatement*. This means designing an IP network to carry voice, not just data.

- Most IP phones allow simple calls to be made directly to each other, dialed by IP address, without the need for a VoIP PBX server as an intermediary. The job of the server, among other things, is to provide a human-friendly addressing scheme and other features that the phones alone can't provide.

- Traditional telephony networking is characterized by client/server or mainframe-like tendencies. VoIP networks are characterized by distributed or fat-client tendencies.

- Most IP endpoints sit at the proverbial "edge" of the network, where PCs and printers also reside.

- Pure IP voice systems don't use any legacy interfacing or protocols—such as POTS or T1. Rather, they support only VoIP protocols and offload the media conversion required for such interfacing to other devices.

- IP-enabled, or hybrid IP, voice systems offer server-based interfacing for legacy links while also providing VoIP signaling, usually in one server chassis.

CHAPTER 3

Linux as a PBX

Evaluating VoIP for enterprise or for your home phone setup means a lot of experimentation, and you'll need to build a test server with which to hone your VoIP skills. That test server should be something you can get a lot out of without spending a bundle or committing to a specific vendor's commercial VoIP platform before you've done your homework. Free telephony software lets you do that homework.

Free Telephony Software

If you were learning engine repair instead of VoIP, you probably wouldn't use a Ferrari for your experiments. You would want something more forgiving and easier to work on, like a nice Dodge Omni. Luckily, there's Asterisk PBX software—the very open, roomy-under-the-hood telephony server. Like a Dodge Omni, Asterisk is easy to work on, support is a snap to find, and experimenting is cheap. In fact, Asterisk is free (although its development is supported by Digium, Inc., *http://www.digium.com*). So is its source code.

But like a Ferrari, Asterisk is very powerful. Asterisk supports several Voice over IP communication protocols: H.323, SIP, IAX, and others (see Chapter 7 for more on these). Using these protocols, it can support just about any IP telephone, as well as traditional analog and digital telephones. Asterisk has some industrial-strength features like call-queuing, conference calling, voice mail, and caller ID.

Using Asterisk, you can build something as simple as an answering machine that sends its recorded messages to your email address (as we'll do in Chapter 14) or something as sophisticated as a thousand-subscriber corporate communications system with least-cost call routing and advanced call accounting.

Not all PBX solutions bring such a wealth of features. By definition, a PBX is just a private call-routing exchange. In traditional telephony, advanced features such as voice mail and autoattendant are often provided by separate, outboard devices. Figure 3-1 shows a summary of Asterisk's functions.

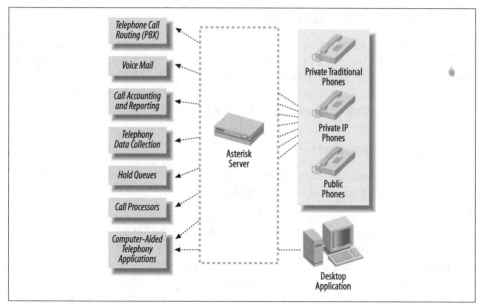

Figure 3-1. The functions of the Asterisk PBX software

With Asterisk and other freely available tools, you can build all kinds of telephony applications. The included Asterisk Gateway Interface allows you to develop computer-aided telephony tools using PHP, Perl, Java, or C, and the Asterisk Management API allows you to build socket-based monitoring and automation applications for your PBX. To bind telephony applications to data, Asterisk has a built-in database that is similar to the Windows registry.

You can teach Asterisk to do interactive voice response (IVR) and text-to-speech. Imagine an application that interfaces with home-control equipment so you can turn on and off the lights in your house using your cell phone. Other possibilities include logging your phone traffic to a web site or database, running your own caller ID blacklist to cut down on unwanted calls, recording calls, or creating data collection tools for your coworkers or customers.

You could use Asterisk's expansive PBX functionality to build an intercom calling system between offices or between rooms in your home. With a little help from wireless Ethernet, you could put an IP telephone out in your shed. Asterisk could even be the nerve center of an elaborate James Bond–style phone-tap system.

Of course, managing enterprise telecommunications is Asterisk's intended purpose, and it does an excellent, if exhaustive, job. Although it offers more features than can be easily documented, we'll make sure you understand enough of Asterisk to facilitate your own continued experimentation with the VoIP technology family.

Other Free Telephony Software

Other open source software, including the SIP Express Router, Open H.323 (covered further in Chapter 7), VOCAL, and ReSIProcate, offer task-specific, developer-oriented tools that can help you learn Voice over IP, too. But Asterisk is currently the most well-rounded and mature open source IP telephony system available. That's why it's been chosen to illustrate many of the examples used in this book.

Asterisk's Requirements

Asterisk is distributed like most open-source software—you download the source code and compile it yourself. Though Asterisk will run on FreeBSD, Solaris, and Mac OS X, it's easiest to compile using Linux.

Asterisk can run on many flavors of Linux, as long as the kernel version is 2.4.x or 2.6.x. Red Hat Linux 7.3, 8, and 9, which use these kernel versions, are all capable of running Asterisk, and this chapter assumes Red Hat Linux Version 9 unless noted. (Once Asterisk is compiled and installed, its VoIP uses are identical regardless of platform.) When installing Linux, be sure to include the kernel sources, Bison, and OpenSSL packages, which are all required by Asterisk. Most distributions include a copy of each, and almost all distributions are available for download from *http://www.linuxiso.org*.

If you're in doubt about which developer packages you need installed to build Asterisk, just install all the developer tools. That way, you're sure to have the required packages like the kernel headers, Bison, and so on.

A Pentium III PC with 128 MB of RAM, an Ethernet interface, and a few hundred MB of hard drive space is enough to support a very basic Asterisk configuration with a voice mail application and several SIP telephones connected by Ethernet. Just as an engine uses more horsepower to make a car accelerate faster, the PC you choose will need more processor power and memory in order to support more conversations, particularly if those conversations are between different types of phones (say, a SIP phone and an analog phone on the public telephone system).

For production Asterisk servers, Digium recommends a Pentium III/ 800 MHz with 512 MB RAM. Servers that will use legacy telephony interface cards will also need a revision 2.2 PCI bus.

If you are planning on connecting your Asterisk server to the telephone company, you'll also need to install a special PCI interface card—which would rule out using a laptop, since they don't sport PCI slots. We'll get into more interface card details later in this chapter.

It's a good idea to give your Linux PC a static IP address, too. Though some administrators prefer dynamic addressing for just about everything on their networks, IP telephones work better if their PBX is always located at the same address. Using a dynamic (DHCP) address for a VoIP call server is like putting diesel in that Dodge Omni—it might work for a short time, but pretty soon, it will cause problems.

For the duration of the book, the Asterisk server's address will be assumed to be 10.1.1.10.

Installing Legacy Interface Cards

Using PCI interface cards and USB devices from Digium, VoiceTronix, Quicknet, and others, Asterisk can communicate with POTS, FXO/FXS, and T1/E1 phone lines:

- Standard analog telephones (Quicknet Internet PhoneJack, Digium TDM400P, VoiceTronix OpenSwitch)
- Regular analog telephone lines (POTS) from the phone company (Digium X100P and TDM400P VoiceTronix OpenLine)
- T1 and E1 telephone lines from the phone company (Digium T100P, E100P, TE405P, and TE410P)

Many of the examples in this book use IP telephones, which communicate with the Asterisk server using Ethernet and therefore don't need specialized interface hardware to access the PBX. A legacy interface card is not required in order to use Asterisk for VoIP—in fact, VoIP can supplant all traditional telephony technologies. But since the PSTN will be here for years to come, analog trunks and phones are still important complements to VoIP. Since most phone companies don't yet offer dialtone trunks over IP, you'll need a legacy interface to connect your Asterisk server to the phone company.

The X100P Card

Most of the examples in this book use a single analog telephone line (POTS line) to access the telephone company. In order to use this line, your Asterisk server will need to be equipped with a Digium X100P analog trunk interface card, which provides a connector to plug in a single telephone company line.

 Just prior to publication of this book, Digium discontinued the X100P card. However a TDM400P card with a single FXO interface will work exactly the same in the examples contained here. The TDM400P is described later. If you'd prefer the X100P, an eBay search yielded dozens of X100P cards available from other sources.

The POTS pass-through connector

The X100P card has a second connector, a pass-through that you can connect an analog telephone to. You can't use this telephone with the Asterisk system, but when it's connected to the second interface, you can use it to tell whether the phone line connected to the Asterisk system is active with a phone company dial-tone. The setup of a basic VoIP-enabled telephony network is shown in Figure 3-2.

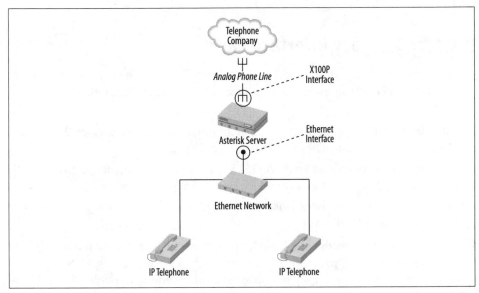

Figure 3-2. The basic Asterisk setup used throughout this book

Installing an X100P

To install the X100P, unplug the power from your PC, open the case, and snap the card into an empty PCI slot. Once it is installed, connect your telephone company wall jack to the primary jack of the X100P—the one marked with an etching of a telephone company plug. Optionally, connect a standard residential-style analog phone into the secondary jack, which is marked with an etched image of a tiny telephone. (If you don't have a spare analog phone, you can skip it.) Now hook your Linux system to your Ethernet network (wired or wireless), configure its IP settings (avoid DHCP), and you're ready to install the Asterisk software.

 The X100P is intended only for use in North America. Other system builders should consider the TDM400P card with FXO modules.

The TDM400P Card

The Digium TDM400P card (Figure 3-3) is a more powerful interface card, and it is not required for any of the projects in this book.

Like the X100P, the TDM400P is a PCI card that allows connections to the PSTN through a POTS line, as well as connections to an analog telephone using a regular two-wire phone cord. Unlike its simple cousin, the TDM400P has four modular ports, so you can use up to four POTS lines and/or analog phones.

These ports accept one of the two Digium proprietary hybrid interface modules, FXS signaling for analog trunks (POTS lines) and FXO signaling for analog phones. The card can host them in any combination. This card also has a Molex power connector, like a hard drive, for supplying analog phones with the line power they're accustomed to. Don't forget to connect it.

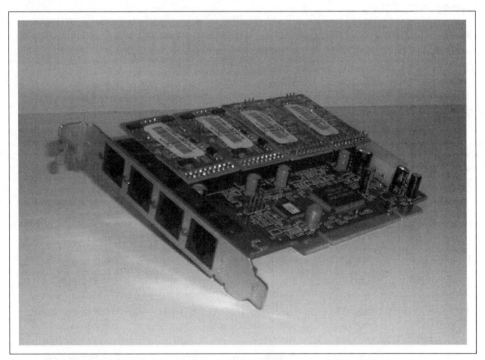

Figure 3-3. The TDM400P interface card, with two FXO and two FXS modules installed

Multiple TDM400P cards can be used in the same computer simultaneously. Since they have four ports per card, they are better than the X100P for high-density applications. Due to capacity issues on the PCI bus, however, you shouldn't use more than two TDM400P cards. A T1 interface is a great alternative, though: it allows you to connect two dozen phones at a time using a device called a channel bank. More on this later.

We've chosen the X100P for connecting our VoIP server to the PSTN because it's less costly than the TDM400P. Of course, you could substitute a TDM400P card with an FXO module in place of the X100P. Asterisk distinguishes little between a TDM400P connected to a POTS line and an X100P connected to a POTS line, so either card should work fine for most of our projects.

If you decide to use the TDM400P instead of the X100P for your VoIP test server, it needs a minimum of one FXO interface module, connected to a POTS line as if it were a single-line X100P card.

 The TDM400P has RJ45-looking sockets like an Ethernet card—but don't connect them to Ethernet, or you could damage the card, your PC, or your Ethernet switch!

Other Interfaces

Besides the X100P and the TDM400P, Digium offers some other specialized interface cards. Other vendors make interface devices that can be used with Asterisk, too. Table 3-1 describes some of the devices that are available, including the X100P and TDM400P.

Table 3-1. Some Asterisk-compatible VoIP interface devices

Device	Vendor	PC interface	Telephony interface	Purpose
X100P	Digium	PCI	RJ11-style	Connect an analog phone line (trunk)
TDM400P	Digium	PCI	Quad RJ45-style	Connect analog phone lines (trunks) or analog endpoints
TE400 Series	Digium	PCI	Quad RJ45-style / T1	Connect 1 to 4 T1 (E1) or PRI circuits
T100P and E100P	Digium	PCI	RJ45-style / T1	Connect a single T1 (E1) or PRI circuit
S100	Digium	Standalone device with 100BaseT Ethernet	RJ11-style	Connect a single analog endpoint
ATA-186	Cisco	Standalone device with 100BaseT Ethernet	RJ11-style	Connect a single analog endpoint
OpenLine 4	VoiceTronix	PCI	Quad RJ11-style	Connect 1 to 4 analog phone lines (trunks)
OpenSwitch 6	VoiceTronix	PCI	6 or 12 RJ45-style	Connect 1 to 12 analog trunks or endpoints

Table 3-1. Some Asterisk-compatible VoIP interface devices (continued)

Device	Vendor	PC interface	Telephony interface	Purpose
Internet LineJack	Quicknet	PCI	RJ11-style	Connect an analog phone line (trunk)
Ethernet Interface	Any that provides a Linux driver	PCI or ISA	RJ45-style 10/100 only	Connect to VoIP servers and endpoints via Ethernet

In addition to these, there are USB-based telephony devices. They can turn a USB-attached phone handset into a VoIP endpoint. These can be useful with softphones, though most serious enterprise implementers will avoid them because, like softphones, they are reliant upon the PC. For roughly the same money, you can purchase a low-end IP phone like the Grandstream Budgetone series.

Compiling and Installing Asterisk

RPM packages are available to simplify Asterisk's installation, but manual compiling is relatively easy. So we are going to download, compile, and install Asterisk the old-fashioned way. Real performance mechanics prefer a manual transmission over an automatic for better command and control, after all.

The development branch you'll download from is stable—though once you get comfortable with Asterisk, you'll want to jump out on the bleeding edge and try the developer releases, too. Each release tends to introduce something new and worthwhile, even if it's not in the stable branch yet.

The easiest place to download the Asterisk software is the CVS repository at Digium, the company responsible for Asterisk and the hardware components that work with it. To access the CVS repository, you'll need to be logged into your Linux computer at a shell prompt as root. Type these commands to run the CVS checkout routine and download the source code:

```
# cd /usr/src
# export CVSROOT=:pserver:anoncvs@cvs.digium.com:/usr/cvsroot
# cvs login
# cvs checkout zaptel libpri asterisk
```

Alternatively, you can specify a particular version of Asterisk:

```
# cvs checkout -r v1-0 asterisk
```

When prompted, use "anoncvs" as a password. If your Linux distribution doesn't use */usr/src*, then substitute the source path that's appropriate. The CVS client you're running here will create the */usr/src/asterisk* directory that contains all the Asterisk source code. Once the download completes, you are ready to begin compiling.

Install the Software Components of Asterisk

Asterisk consists of several software components for Linux. These packages are not all required, as some of them are drivers for Digium's interface cards. If you aren't planning to use Digium's card, you'll need to build only the last of the three, "Asterisk":

LIBPRI

A driver module that supports Digium T1/E1 interface cards so that ISDN and PRI trunks can be interfaced with Asterisk

ZAPTEL

A driver module that allows legacy telephone line interface cards to be used with Asterisk

Asterisk

A modular software daemon that provides telephony, management, and call-accounting features, including voice mail, SIP telephone support, dial-plan, and so on

If you're wondering about these technical terms, don't worry. As you experiment with Asterisk and learn more about VoIP, they'll become very familiar. For now, just compile and install all three of them.

Since you ran the CVS download, the source code for each of the Asterisk software components is sitting in its respective directory in */usr/src*. Let's compile each software component by issuing the following commands:

```
# cd zaptel
# make clean ; make install
# cd ../libpri
# make clean ; make install
# cd ../asterisk
# make clean ; make install
```

Again, *zaptel* and *libpri* need only be compiled if you're planning on using legacy interface cards. Many of the examples in this book make use of legacy devices, so it's probably a good idea to compile them all right now.

 You must compile Zaptel before you compile Asterisk, or else Zaptel features will be missing from the Asterisk build.

It should take 20 minutes at most to complete the whole build on an average PC. Once built, Asterisk is ready to use. But you can't drive this Ferrari without a license, and you can't really use Asterisk until you understand the basics of configuring it. So it's time for driving school. To get started, do this:

```
# make samples
```

Now, a basic sample set of Asterisk configuration files is set up in */etc/asterisk*.

More recent revisions of the Asterisk software provide more detailed examples in their sample config files than older revisions do.

For Those Who Prefer RPM Packages

Red Hat RPM packages for Asterisk and the driver modules can be obtained from *http://www.macvoip.com* or from *http://www.n2net.net*.

Loading the Interfacing Drivers

The software drivers that allow Asterisk to use telephone line interfaces (like the X100P) come in the form of kernel modules (files that add core functionality to the operating system), and we compiled them in the previous section. In order to make sure Asterisk will function correctly, you'll need to load the modules using these shell commands:

```
# modprobe zaptel
# modprobe wcfx    ## if you're using a TDM400P
# modprobe wcfxo   ## if you're using a X100P
# ztcfg -vvv       ## more verbosity with every 'v'
```

These commands enable the Digium X100P card in your computer (if you're using one) and create logical channels that Asterisk can use for voice traffic. The order of execution of these commands is important, because voice channels are numbered in the same order their interface cards are enabled. If you had two X100P cards installed or a TDM400P with two FXO modules, you would have two logical voice channels.

The Zaptel module doesn't need to be manually modprobed, as it is automatically loaded by the *wcfxs/wcfxo* modules, which do a dependency check to reveal Zaptel must be loaded first.

In the three preceding modprobe commands, the zaptel driver loads low-level support for any of Digium's cards, while the *wcfxo* and *wcfxs* modules load signaling support for analog telephone lines and phones, respectively. You'll use different combinations of kernel modules depending on what types of interfacing you're using: if you had an analog telephone connected directly to a TDM400P, for example, you would also need to modprobe the *wcfxs* module.

The ztcfg command tests the specific configuration of each voice channel as specified in the */etc/zaptel.conf* file. You'll need to visit this file with your text editor before you can use analog phone lines—and we'll do exactly that in Chapter 4. Non-Zaptel (i.e., non-Digium) is configured elsewhere.

Zaptel channel configuration is detailed in Chapter 17.

You aren't required to use Digium's interface cards in order to experiment with VoIP—but if you want your VoIP phones to be able to talk to the public telephone system, your Asterisk server will need to hook into a traditional telephone company line. See the section called "Installing Legacy Interface Cards" earlier in this chapter to learn how to install an interface card. Other kernel modules besides the ones shown here are relevant to Asterisk. Each supports a different model of interface hardware.

Here's a summary of the kernel modules that correspond to some of the interface cards described in the "Installing Legacy Interface Cards" section:

zaptel

Provides a foundation for using Digium's T1 and analog line interfaces. The Zaptel module is required in order to use other Digium interface modules.

wcfxs

Allows analog telephones to be connected to the Asterisk server like a traditional PBX.

wcfxo

Allows you to connect the public telephone system to your Asterisk server using a standard analog telephone line (like the one you probably have in your house).

wctdm

In new versions of Zaptel, *wcfxs* will be replaced by *wctdm*.

phonedev and ixj

Allow you to connect analog telephones to the Asterisk server using Quicknet's PhoneJack interface cards (these modules are provided with Quicknet's cards).

A basic Asterisk setup, like the one in Figure 3-2, uses VoIP phones on the LAN that can connect to the public telephone system through a single analog telephone line. Using an X100P card, you'll need to load only the *zaptel* and *wcfxo* modules. The right modules must be loaded in order to launch Asterisk if it is configured to use traditional voice interfaces. If you don't load the right modules first, the Asterisk server won't start and you'll get a message in Asterisk's standard error output about it.

It's very easy to accidentally enable Asterisk to use a channel that hasn't been set up by modprobing the right module or to install a new interface card and then forget to modprobe it before launching Asterisk. Both of these situations will result in an Asterisk that won't start up. One way avoid this problem is to have your startup scripts attempt to modprobe them all, regardless of whether the corresponding hardware is installed.

Starting and Stopping the Asterisk Server

The Asterisk binary is the program that you execute whenever you want to start the Asterisk server or connect to it for management purposes. The same program has two distinct modes of operation—server mode and client mode. The server is the instance of Asterisk that stays running all the time, handling calls, recording voice mail, greeting callers while users are away, and so on. The client is the instance of Asterisk that allows you to monitor and manipulate the server while it runs. The mode the program uses depends on how Asterisk is invoked at the command prompt or within a shell script.

To launch Asterisk in server mode, execute this command:

```
# asterisk &
```

To connect Asterisk in client mode on a machine already running in server mode, execute this command:

```
# asterisk -r
```

Once the Asterisk client is connected to the Asterisk server, you can use Asterisk's command-line interface to issues queries and commands about the telephony server. These include listing calls-in-progress, listing used and unused channels, and stopping the Asterisk server.

Several Asterisk CLI commands shut down the server:

- **restart now**
- **restart when convenient**
- **stop now**
- **stop when convenient**

The restart commands stop and then restart the Asterisk server process, which can be helpful in situations where the server's configuration has significantly changed and needs to be restarted. The stop commands just shut the Asterisk server process down. You'll have to execute the Asterisk program to get it running again.

The now and when convenient arguments tell Asterisk how quickly to shut down or restart. If you want to ignore the current calls and tasks in progress on the server, now is appropriate. If you want Asterisk to wait until all the calls and tasks are finished and there is no call activity at all, when convenient is appropriate. Generally, especially if you're planning to have any callers beside yourself on the system, get in the habit of using when convenient.

Of course, in extreme circumstances, like if the Asterisk CLI is unavailable or not responsive, you can stop the Asterisk server process from a shell like this. Get the process ID (PID) of the Asterisk server process:

```
# ps ax | grep asterisk
```

You'll see something like this:

```
424 std  S+     0:00.00 asterisk
```

The first bit of information is the PID, or process ID. Now, use that PID to kill the Asterisk server process:

```
# kill -9 425
```

The -9 is the same as using now at the Asterisk CLI. Any outstanding Asterisk threads, including calls in progress, filesystem requests such as logging, and administrative instructions, are immediately and abruptly ended so the server process can terminate. This technique is a last-ditch effort that should be avoided if possible.

Configuring for Automatic Startup at Boot Time

In Red Hat Linux, a tree of shell script files dictates the startup procedure for the operating system. After the kernel loads, user commands are executed in the sequence spelled out by the scripts located in /etc/rc.d. There are two accepted methods for adding Asterisk to the startup sequence: one that uses modifications to *rc.local* (the quick and dirty method), and another that involves a script wholly dedicated to Asterisk.

Using rc.local to launch Asterisk

To make Asterisk start automatically at boot time, modify one of the boot-time scripts, *rc.local*, to launch Asterisk in server mode. While you may find the second boot-time method more to your liking, this one shows how to launch Asterisk short and sweet—in four lines of shell commands.

Edit the script called /etc/rc.d/rc.local, adding commands to automatically start the Asterisk server. You can use Pico or your favorite Unix text editor to put these commands into the script:

```
export LD_ASSUME_KERNEL=2.4.1
modprobe zaptel    ## optional
modprobe wcfxo
ztcfg
/usr/sbin/asterisk -vvv &> /var/log/asterisk/pbx-tty.log &
```

The first line takes care of an incompatibility that Red Hat 9 imposes (you don't need this line for Red Hat 7.3 and 8), while the second and third lines load the Digium hardware kernel modules. The last line launches the Asterisk server using the -vvv switch for very verbose output (the more *v*s you use, the more detailed the logging is). The last line also redirects that output to a file called /var/log/asterisk/pbx-tty.log. Asterisk will now be launched as root by Red Hat's startup routine whenever the computer is booted.

Using the safe_asterisk script

The Asterisk distribution is equipped with a script that attempts to relaunch the Asterisk server if it crashes. While rare, the Asterisk software can have fatal exceptions—which result in the shutdown of Asterisk. The script can be found in */usr/sbin/safe_asterisk*. It passes all command-line arguments to the Asterisk instance, and it redirects all output to the shell, so you can further redirect the output like the previous argument. The *safe_asterisk* script does not restart Asterisk if you shut it down intentionally, say, using a `kill` command.

Setting up a Red Hat init script for Asterisk

The second method of adding Asterisk to your server's startup routing is to generate a Red Hat init script. This method produces a larger script that Linux newbies may have a hard time dissecting, but it also gives you more power. That is, the init script is useful at times other than boot time.

To generate Red Hat–friendly initialization scripts for Asterisk, you'll need to run a `make config` from the directory where the Asterisk source code resides:

```
# cd /usr/src/asterisk
# make config
```

You will see this output or something close to it:

```
if [ -d /etc/rc.d/init.d ]; then \
        install -m 755 init.asterisk /etc/rc.d/init.d/asterisk; \
        /sbin/chkconfig --add asterisk; \
elif [ -d /etc/init.d ]; then \
        install -m 755 init.asterisk /etc/init.d/asterisk; \
fi
```

Now, you can issue */etc/rc.d/init.d/asterisk* to control Asterisk with these command-line arguments:

start
: Launches Asterisk using the *safe_asterisk* script. Its output is directed to the console.

stop
: Attempts to shut Asterisk down by finding its PID and sending a kill signal.

restart
: Attempts to stop Asterisk with a kill signal and then restarts it using the *safe_asterisk* script.

reload
: Does the same thing as the `restart` option. This is different from the Asterisk CLI command reload, which keeps intact any in-progress calls on the system. This command is described in Chapter 17.

status
: Checks to see if Asterisk is running or stopped, and tells you.

If you use the Red Hat init script, make sure */etc/rc.d/rc.local* calls it instead of calling Asterisk directly as in the first method. Regardless of which method you choose, make sure you're loading the right modules (*wcfxo* or *wcfxs*) prior to the line in which Asterisk is launched.

Securing the Asterisk Instance

By default, Asterisk runs as root—the user account with total, unrestricted power. This is generally considered a bad idea, and it's the reason Apache, the web server, doesn't usually run as root. If Asterisk is compromised by an ill-willed network prowler, it's possible that the intruder could gain the power of root. That is, unless you make Asterisk run as a less godly user.

To do this, create a user called "asterisk" by issuing the Red Hat adduser command (other Unixes could use useradd instead):

```
# adduser -c "Asterisk PBX" -d /var/lib/asterisk asterisk
```

Next, you'll need to alter Asterisk's make file, located at */usr/src/asterisk/Makefile*. Using Pico or your favorite text editor, find the ASTVARRUNDIR constant in the file, and alter its definition to match what follows:

```
ASTVARRUNDIR=$(INSTALL_PREFIX)/var/run/asterisk
```

The directory referenced here needs to be writable by the user running Asterisk, and the directory normally used should be writable only by root. So by changing the setting, you're allowing Asterisk to use a directory that can be written by its own non-root user account. Now, recompile Asterisk using this sequence of commands:

```
# cd /usr/src/asterisk
# make clean ; make install
```

Once the recompile and install are done, you'll need to make sure the new user account has appropriate permission to several Asterisk-related directories, including the one you referenced in the altered Makefile:

```
# chown -R asterisk:asterisk /var/lib/asterisk
# chown -R asterisk:asterisk /var/log/asterisk
# chown -R asterisk:asterisk /var/run/asterisk
# chown -R asterisk:asterisk /var/spool/asterisk
# chown -R asterisk:asterisk /dev/zap
# chmod -R u=rwX,g=rX,o= /var/lib/asterisk
# chmod -R u=rwX,g=rX,o= /var/log/asterisk
# chmod -R u=rwX,g=rX,o= /var/run/asterisk
# chmod -R u=rwX,g=rX,o= /var/spool/asterisk
# chmod -R u=rwX,g=rX,o= /dev/zap
# chown -R root:asterisk /etc/asterisk
# chmod -R u=rwX,g=rX,o= /etc/asterisk
```

You can now launch the Asterisk server from the new user account or from root using the su command:

```
# su asterisk -c /usr/sbin/safe_asterisk
```

Finally, you'll need to adjust the *safe_asterisk* script so that it uses the new user account to launch Asterisk, rather than root. To do this, open */usr/sbin/safe_asterisk* in Pico or your favorite text editor, and add su asterisk -c before each instance of an asterisk command. Be sure to leave the commands unchanged, aside from prefixing them with the su command.

Once these steps are taken, Asterisk will have only as much power as you grant the "asterisk" user. Would-be attackers may be able to crash Asterisk, but in so doing, they won't be able to gain access to root's credentials.

Asterisk on Mac OS X

Besides Linux, Asterisk will run on some commercial Unixes, including the most popular of those, Mac OS X. A binary (already-compiled) distribution is available from *http://www.macvoip.com*. The configuration examples described earlier won't work on OS X, because it's a totally different flavor of Unix than Red Hat Linux. If you are looking for a good reference for learning the Mac's flavor of Unix, try *Mac OS X Panther for Unix Geeks* (O'Reilly).

As of this writing, drivers for legacy interface cards have not been ported to OS X, but with IP phones, the Mac port is fully functional. Indeed, with Applescript, iCal, and iTunes at every Mac user's disposal, there are big possibilities for fun Asterisk hacks. Again, all the examples in this book are geared to Red Hat Linux systems.

Developers interested in porting the hardware to the Mac should consider the work of the Asterisk on BSD project, for which Digium operates an email support list. Mac OS X has a version of BSD Unix called Darwin at its core, so it should be easier to port Asterisk-related code to the Mac from the existing BSD code rather than from the Linux code.

Project 3.1. Test an IP Phone with Asterisk

What you need for this project:

- Asterisk
- Grandstream IP phone (or similar)
- LAN

You've done some Asterisk compiling and customization using the Linux shell. But once Asterisk is installed, you'd probably prefer to interact with it from a more appropriate device—a telephone. In this project, we'll use the SIP telephone we configured in Projects 2.1 and 2.2 in order to access the de facto autoattendant greeting and access a brief demonstration of an IAX (Inter-Asterisk Exchange) trunk over the Internet. Sound like too much? Don't worry; most of this is already configured with Asterisk out of the box. The toughest part for a VoIP beginner will be making sure Asterisk is able to answer SIP calls—and that's pretty easy.

You won't need a regular phone line for this project—just a SIP phone, Asterisk, and an Internet connection.

SIP (Session Initiation Protocol) is one of several standards that allow IP voice endpoints and application servers such as Asterisk (and ReSIProcator and SIP Express Router) to establish, monitor, and tear down media sessions across the network. Asterisk uses SIP to facilitate calls on behalf of SIP-based IP phones like the Budgetone 101, the Cisco SIP IP Phone 7960, or the Avaya 4602.

Verify Asterisk's SIP server is running

To get started, Asterisk will need its SIP server module running so that it can listen for SIP calls. By default, Asterisk's SIP server module listens on UDP port 5060, the commonly used port number for SIP. If you use the SIP phone (10.1.1.103) to dial the Asterisk server (10.1.1.10) by IP address, you should get a 404 message on the phone's display: 404 is a SIP error code that means "Not Found"—just like the 404 message used on the Web. If you get this response from the Asterisk server, it means the SIP module is listening and has responded to you.

Now, in order to go from dialing only by IP address to dialing by extension, the IP phone must be pointed to the SIP server.

Set the IP phone to use a SIP server

The IP phone, whose address is 10.1.1.103, must be set to use a SIP server if it is to make calls other than the IP-to-IP type described in Chapter 2. In the test network, the IP phone should refer to the IP address of the Asterisk server, 10.1.1.10, as its SIP server. (For the details on SIP servers, refer to Chapter 7.)

Configure the SIP user ID setting as 103, too. For the DTMF Mode option, select SIP Info. Then apply the config changes and reboot the IP phone. The same configuration options are supported by other makes of SIP phone, too. The configuration page for a Budgetone phone that has been configured to use a local SIP server is shown in Figure 3-4.

Allow the IP phone to place calls via Asterisk

Until you authorize a SIP phone to communicate with Asterisk using Asterisk's SIP configuration file, you will always receive SIP error messages when trying to dial to (or through) the Asterisk server. This is Asterisk's way of ignoring what it sees as an unauthorized endpoint. Unlike traditional PBXs, which tend to give network access to any phone connected on an active port, SIP servers tend to enforce some security—usually in the form of authentication.

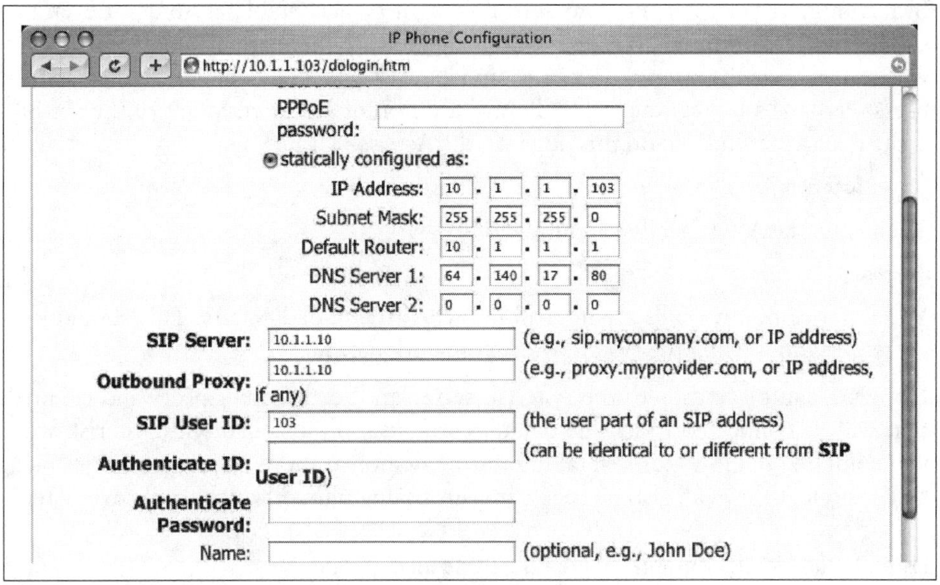

Figure 3-4. A Grandstream Budgetone that has been configured to use a SIP server at 10.1.1.10

Let's tell the Asterisk server to stop ignoring requests from your IP phone. Asterisk, the softPBX, refers to IP phones and other SIP devices as *peers*. SIP peers are defined in Asterisk's configuration file, */etc/asterisk/sip.conf*. To enable the phone as configured in Figure 3-4, add the following to the end of this file:

```
[defaultsip]
type=friend
context=default
username=103
fromuser=SIP Phone
callerid=103
host=10.1.1.103
nat=no
canreinvite=yes
dtfmode=info
disallow=all
allow=ulaw
```

The preceding section adds the 10.1.1.103 IP phone that matches the configuration of the earlier Grandstream Budgetone. Don't worry about the meaning of this syntax—these settings are covered in more detail later.

Restart or reload?

There are two ways to enable the configuration change you've just made. One is to restart Asterisk:

```
# /etc/rc.d/init.d/asterisk restart
```

Bear in mind that restarting your softPBX might be acceptable at home or even in a small-office environment, but you'd better make sure there are no calls in progress if you restart it in any production environment, lest you draw the ire of angry phone users. Perhaps a better way to handle the addition of a new endpoint to the softPBX is the reload method. To do this, log into the Asterisk CLI using:

```
# asterisk -r
```

Then issue the Asterisk reload command:

```
pbx*CLI> reload
```

You'll notice that no calls are interrupted when the reload occurs. This should keep everyone who is using the system while you work happy.

Now you can place calls to the Asterisk server and to the other peers and channels that will be connected to it. The default configuration installed with Asterisk when you compiled it allows for several interesting demonstrations of its capabilities using a SIP phone. (They can also be tried using an analog phone, which we'll cover later.)

Listen to the Asterisk automated greeting using an IP phone

In its default configuration, Asterisk has an autoattendant that can route calls using an automated attendant. To try it out, take the IP phone off hook and dial 2. Then dial Send. You will hear a friendly voice saying, "Asterisk is an open source, fully featured PBX and IVR platform..."

Listen to a voice over Internet greeting

While listening to the automated attendant greeting, dial 500. This will cause the Asterisk server to greet you; connect you to a server at Digium, Inc., using the Internet; and allow you to listen to another automated greeting—the one being played back by a production Asterisk PBX at Digium's office. This connection does not use the PSTN at all, but rather a Voice over IP "trunk" that is set up on the fly by Asterisk.

> The Voice over Internet demo requires UDP port 4569. If you're using a firewall or NAT device, be sure it permits outbound traffic on this port. Most home-grade firewalls will permit this type of traffic by default. The UDP port does *not* need to be inwardly mapped or proxied.

The connection to Digium is established using IAX, not SIP. So the Asterisk server is managing two different kinds of channels simultaneously in order to facilitate this call. Listen to the sound quality. Do you notice any difference between the quality of the autoattendant on your Asterisk server and the one on Digium's? The difference in quality should be negligible, if even noticeable, especially over a fast Internet connection.

You can also perform an echo test by dialing 600 and accessing Asterisk's built-in voice mail service by dialing 8500. These are covered in greater detail later.

Installing Mpg123

The application responsible for providing music and messages for holding callers is called Mpg123, but don't confuse it with the Mpg321 application that ships with Red Hat Linux. Mpg321 doesn't work with Asterisk, so you must replace it with Mpg123.

To do this, start by shutting down the Asterisk server process if it's running. Then, make sure there are no Mpg321 or Mpg123 processes running:

```
# ps ax | grep mpg123
900 std  S+     0:00.00 mpg123

# kill 900
# ps ax | grep mpg321
1024 std  S+     0:00.00 mpg321

# kill 1024
```

You may not actually have both applications running, or even installed, but it's best to make sure neither is running before you proceed.

 As of this writing, Mpg123 has a remotely exploitable security bug that is documented at *http://www.mpg123.de*. Be sure to read the developers' description of the issue and proceed with caution.

By default, Red Hat has symbolic links for the Mpg321 package that make it appear synonymously with Mpg123. These must be removed in order to install the correct package:

```
# rm /usr/bin/mpg123
# rm /usr/local/bin/mpg123
```

Next, download a Red Hat Linux–compatible build (RPM package) of Mpg123 from *http://www.mpg123.de* and install it with a command similar to this one, called from the directory where the downloaded file is saved:

```
# rpm -iv mpg123-0.59r-1.i386.rpm
```

Along with Mpg123, Asterisk uses the configuration file called */etc/asterisk/musiconhold.conf* to define various "classes" of music-on-hold. Each class can be used in different situations or contexts depending on how the Asterisk administrator opts to handle each hold scenario. Mpg123 isn't required to deliver prerecorded sounds; Asterisk can do that on its own using files in the GSM-encoded format (and other telephony codec formats). What Mpg123 does is allow MP3 files to be played back for holding callers to listen to while they wait. On-hold music is covered in detail in Chapters 12 and 17.

Monitoring Asterisk

There are several ways to monitor Asterisk. Most notably, the Asterisk CLI console application (*asterisk –r*) offers a real-time console log. When you launched Asterisk with the -v option, this was enabled. The more v's, the more detail goes into the console log. The same is true of the logfiles that Asterisk puts out.

Asterisk's logfiles

In addition to standard output and standard error, which you can redirect using the shell, Asterisk has some important logfiles. They are stored in */var/log/asterisk* by default. If you want them to be stored elsewhere, edit */etc/asterisk.conf* as such. The three ASCII logfiles enabled upon installation are:

event_log
> Stores Asterisk system events, when triggered according to the Asterisk configuration

messages
> Stores debugging, error, and warning messages generated by most Asterisk modules

cdr-csv
> Stores the Call Detail Record, or CDR, which records the channels, actions, and durations of each call placed through the softPBX

Chapter 10 covers Asterisk's logfiles in much greater detail.

Astman

The Asterisk Manager is a text-based socket API that allows management applications to monitor and control the Asterisk server. One such application is Astman, which is included in the Asterisk distribution. Astman allows you to watch a list of calls in progress and allows you to redirect calls and disconnect them.

Before you can use Astman, though, you need to make sure Asterisk's Management API is running, by way of the */etc/asterisk/manager.conf* config file. Here's a sample:

```
;
; Sample astman config
;
[general]
enabled = yes
port = 5038
bindaddr = 0.0.0.0

[ted]
secret = hansolo
allow = 0.0.0.0/0.0.0.0
read = system,call,log,verbose,command,agent
write = system,call,log,verbose,command,agent
```

This sample config starts the Asterisk Manager listening on TCP port 5038, and allows all hosts (allow=0.0.0.0...) to connect. If your Asterisk server is accessible from the Internet or untrusted hosts, you'll want to make the allow directive more strict. The user called "ted" will be authenticated by the password "hansolo" and his administrative permissions are described by the read and write directives. You could set up other users with different credentials.

> *manager.conf*'s settings are for the Asterisk Manager API, not just the Astman program. Other applications that you use with the API will be subject to these settings, too.

Once you've set up your *manager.conf* file, restart Asterisk and issue this command on the Asterisk server:

```
astman localhost
```

Log into Astman using the username and password you established in the *manager.conf* file (see Figure 3-5).

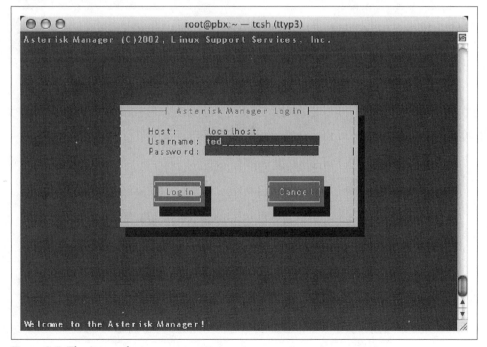

Figure 3-5. The Astman login prompt

Channels

Once you log in to Astman, you are shown a channel list. Channels are logical pathways for voice connections at the application layer, just as TCP and UDP provide logical pathways for data transfer and the transport layer. Whenever an endpoint contacts the Asterisk server, a channel is established that remains open for the duration of the connection. If one endpoint calls another endpoint via the Asterisk soft-PBX, two channels are established—one to each endpoint. If one endpoint calls another endpoint that is hosted by a completely separate Asterisk server, two channels on each server are established, meaning that, between the two servers, it required four channels to connect a single call. Astman can monitor the channels on only a single Asterisk server, though.

Proprietary IP telephony environments work similarly. Cisco, Avaya, and Nortel soft-PBXs handle logical voice channels, but none really states it as such in their management applications.

There is one line for each voice channel currently in use on the Asterisk server. In Figure 3-6, you can see that there is a call in progress to extension 500 using SIP, and an outbound call to a VoIP peer on the Internet using the IAX (Inter-Asterisk Exchange) protocol. In fact, the screen grab in Figure 3-6 was taken during the Internet-based VoIP demonstration outlined in Project 3.1.

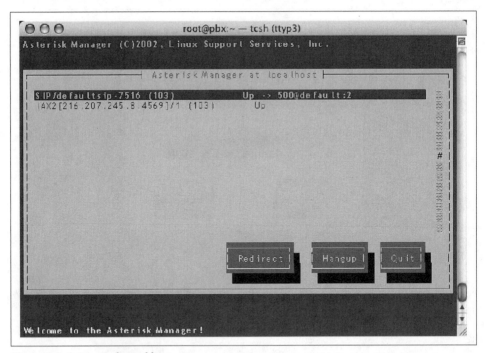

Figure 3-6. Astman's channel list

Most IP PBX vendors provide an application similar to Astman. Avaya's CallStatus, shown in Figure 3-7, is an administrator application that shows activity on Avaya's small-office softPBX (called IP Office), similar to Astman's function with Asterisk. Every call that is currently active on the Avaya system, whether placed by an inside user or received from the PSTN, is shown on CallStatus. Astman functions the same way, although it provides less detail about each call and has one channel per line rather than one call per line.

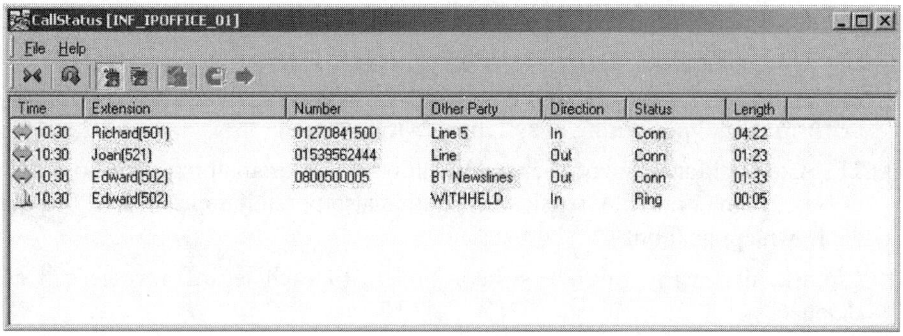

Figure 3-7. Avaya's proprietary CallStatus application is more advanced than Astman—but works only with Avaya's PBX

If you want to develop a more advanced version of Astman or create your own management or CTI (computer-telephony integration) applications, then the Asterisk Management API is the way to go. It's a text-based protocol that provides you with the ability to monitor the system, direct calls in progress, originate calls, and add or remove extensions.

The API can be used in conjunction with other applications to automate telephony functions. One example would be an attendant console application that somebody in a front-desk receptionist role could use to direct a call center. Another might be an autodialer for a personal information management application like Microsoft Outlook or Lotus Notes. It's also possible to link web-based applications with Asterisk using the Management API, so you could create an HTML frontend for a voice chat system or create a web-based answering machine.

The Management API has a number of quirks that are identified in Digium's bug-tracking system located at *http://bugs.digium.com*. It would be a good idea to review the information here before embarking on any ambitious API projects. Refer to the appendixes for a list of Management API commands.

Key Issues: Linux as a PBX

- Asterisk is an open source PBX in software that runs on Linux, BSD, and Mac OS X. It's a great test environment for evaluating VoIP; it's also a solid solution for production systems.

- Because Asterisk was written in C on Linux from the ground up, with many POSIX conventions at its heart, a duty-ready version of Asterisk for Windows doesn't exist.

- Digium, Quicknet, and others make interface cards that allow Linux softPBX servers to connect to the PSTN using POTS and T1/E1 lines.

- The Asterisk CLI is an administrative interface that allows programming of extensions and monitoring of system activity.

- The Asterisk Management socket API allows the Astman application to monitor calls in progress on an Asterisk softPBX. It's also possible to use the API to build your own applications.

- Asterisk, like many softPBX systems, refers to each leg of a voice call as a *channel*.

- Asterisk is commercially maintained by Digium, Inc., and supported by Digium and other consulting firms.

## CHAPTER 4

Circuit-Switched Telephony

Conventional telephone networks, whether public (PSTN) or private, bear several things in common. First, the phones used to make calls across them almost always use one- or two-pair physical connections. Second, the call-management device nearest the end user, be it a key system or a PBX, usually provides a dedicated, single-purpose circuit for each phone. The voice applications delivered by legacy systems are rigidly tied to the lower layers of the network. For instance, you can't get plain old telephone service from a cable company or a satellite provider because they can't provision copper telephone lines to your premises. Finally, the capacity of the data links used to carry traditional telephone calls rarely increases over time. It remains fixed, forever tied to the quantity of cable pathways between one point and the next.

These traits are common among legacy voice setups, whether they consist of heavy-duty TDM-bus PBX systems or just a few analog phones connected to the PSTN. Incidentally, VoIP doesn't exhibit these traits, but since your transition to VoIP may be incremental, it's important to understand the "goods and bads" of traditional circuit-switched telephony.

Regulation and Organization of the PSTN

The organization of the telephone network is a complex international affair. If you're dealing with telephony of anything deeper than a superficial level, you need to understand the players.

The FCC

In the United States, the public telephone system falls under each state's regulatory jurisdiction. This means that, while the FCC sets the rules for interstate service, it's up to the individual states' communications agencies to enforce and further define local service standards for the PSTN (see Figure 4-1).

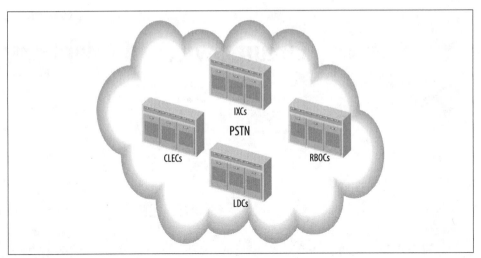

Figure 4-1. There are several kinds of network carriers on the PSTN

The FCC communicates with these bodies in order to make sure that the PSTN serves the public interest and is beneficial to consumers, while allowing an atmosphere of economic growth that is conducive to the involvement of private business. Though the PSTN's technologies are governed by the ITU, an international body, state governments and the FCC maintain lawful use of the PSTN in the U.S.A.

The International Telecommunications Union

The ITU-T (Telecommunications Standardization Sector) is the ITU's working group charged with governance of global standards for telephony practices and protocols. The ITU-T publishes recommendations that dictate the technologies used in public and private telephony. While the ITU-T has published some packet-based recommendations, like H.323, the majority of its work has focused on legacy, circuit-switched telephony networking. Its standards are categorized into groups by function. Each group is abbreviated using a letter. Here are the most relevant ones:

E   Overall network operation, telephone service, service operation, and user interface

G   Transmission systems and media

H   Audiovisual and multimedia systems

I   Integrated services digital network (ISDN)

O   Specifications of measuring equipment

P   Telephone transmission quality, telephone installations, local line networks

Q   Call switching and signaling

V   Data communication over the telephone network

X   Data networks and open system communications

Y   Global information infrastructure and Internet protocol issues

So Q.931 is a signaling protocol recommendation, while E.164 is a recommendation for a numbering scheme used for international dialing. V.90 is a modem communication recommendation, while H.248 is a media gateway signaling standard. A library of these standards is maintained online by the ITU-T. The service providers who are most directly affected by, and who most frequently use, all of these recommendations are the companies that operate the PSTN: RBOCs, CLECs, IXCs, and LD carriers.

RBOCs and CLECs

Regional Bell operating companies, or RBOCs, are the largest of the local phone service providers. These are the companies that tend to own most of the cabling infrastructure in areas that are routinely as big as an entire state. They are sometimes referred to as incumbent carriers because most of them got their start as a part of a huge, national corporation called AT&T that was once in charge of the national telephone system. In the early 1980s, that company was disbanded and broken into many smaller regional Bell operating companies.

 RBOCs are also called ILEC, or incumbent local exchange carriers—because they were around before CLECs were introduced by deregulation.

Competitive local exchange carriers, or CLECs, are local telephone companies operating competitively within each RBOC region. Unlike RBOCs, CLECs don't tend to own much cabling or switching infrastructure. Instead, they usually share facilities with the RBOCs. CLECs rose to notoriety in the late 1990s, after the Telecommunications Act of 1996 was passed by Congress, allowing regional competition for dial-tone services.

RBOCs and CLECs are also referred to simply as LECs, or local exchange carriers. Both are monitored and regulated by the FCC.

LD carriers

Long-distance carriers are service providers that provide network pathways between RBOCs' and CLECs' switches in different regions known as LATAs (local access transport areas). LD carriers route telephone calls for subscribers of the local telephone companies while billing the subscribers directly or as a part of the local phone companies' billing routines. They carry voice traffic across the country and throughout the world, so that the PSTN can extend beyond its RBOC roots. Due to deregulation of the LD carrier business in the past few years, it's now possible for RBOCs and LD carriers to offer the same services, and many do.

IXCs

Interexchange carriers are network operators that provide connections between PSTN network carriers. A network connection between a dedicated LD service provider and a regional Bell operating company may pass through an IXC. Many LD carriers are considered IXCs, too.

ISPs

Internet service providers, though not a part of the PSTN, are often operated by the same companies that provide PSTN services. Lots of large LD carriers, RBOCs, and CLECs are ISPs, too.

Components of the PSTN

The Public Switched Telephone Network and its signaling counterpart, SS7, connect, monitor, bill, and disconnect calls. At the edge of the PSTN are large, mainframe-like switches call exchanges, or central offices (COs). The role of each CO switch is to connect calls between channels on that switch and, when necessary, to connect calls to channels on other switches in the PSTN.

Calls travel along temporary pathways through the voice network. Though temporary, these pathways are end-to-end circuits, the root of the catchall moniker for legacy call management: *circuit switching*.

The Central Office

The CO is the building where the local exchange switch resides. A CO's switch may serve telephone service subscribers in a very narrow geographic area—such as a single large building. Or the CO's switch may serve subscribers for miles around. The CO's scope of service depends on the density of subscribers in its neighborhood and on the capacity of the switch it houses. As with many electronic services, capacity and efficiency have increased over time, so late-model CO switches are generally able to provide more channels and greater utilization ceilings than older ones.

Conventionally, you can tell which central office subscribers belong to by looking at the first three digits of their seven-digit telephone number. This three-digit section is called the prefix. One prefix is usually set aside for one central office, or for groups of small, geographically close COs. This tradition is disappearing, however. Newer signaling protocols and networking standards have made the relationship between the CO and the prefix less of a requirement and more of a historical artifact.

PSTN Distribution Frames

Distribution frames surround the CO. They are high-density cross-connection points where multiple subscribers' loops are tapped into the feed cables (that is, 50 to 800 pairs in a single cable) supplied by the CO. Usually, all connections to a distribution frame are copper. Distribution frames allow the telephone company to use high-density copper cabling that is less susceptible to breakage to feed groups of subscribers on the edge with a connection to the CO.

Those subscribers usually have a low-density (2–25 pairs) copper cable running from their buildings to the distribution frame. The frame is the aggregation point for many customers in a small area. One customer can use a CO loop that was once used by another customer due to the cross-connect terminals located inside the distribution frame. New buildings can be built in the area, and they can tap into the distribution frame rather than having to be connected directly, and usually over a longer distance, to the CO.

Main Distribution Frames

MDFs are distribution frames that have some smart switching or encoding equipment in them. Usually, this equipment has been placed there in order to facilitate the use of DSL (digital subscriber line) for access to the Internet by telephone company customers. MDFs can also be used to aggregate large bunches of customer copper loops into super-high-capacity, often optical, links that connect to the CO instead of the high-density copper feeds to the CO used in regular distribution frames (see Figure 4-2).

Switch-to-Switch Trunks

A link between two switches is called a trunk. A *switch-to-switch trunk* is one that carries calls between two switches on the PSTN. When a call from a particular switch's subscriber is destined for a subscriber on a different switch on the PSTN, one or more switch-to-switch trunks will be used to create a circuit for that call.

The trunks that connect a PBX to the local phone company are also considered switch-to-switch trunks, since the PBX itself is a switch. Large PSTN trunks, like those between very busy COs, may consist of synchronous optical networking (SONET) connections, or DS3 connections, which are described in greater detail later in this chapter (see Figure 4-3).

In-Band Signaling (DTMF)

DTMF tones are the sounds you hear when you press the numbered keys on a telephone keypad. The tones are used as signals to the switch, telling it which phone number you're trying to call or to access a telephony feature such as automatic call return (discussed more in Chapter 5). DTMF stands for dual-tone multifrequency

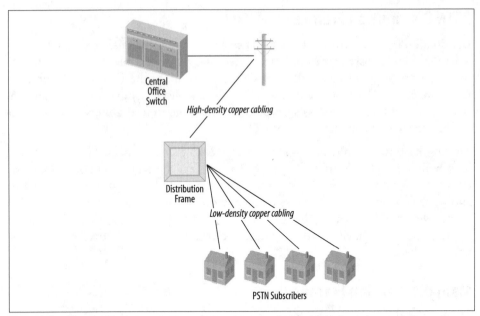

Figure 4-2. On the PSTN, the "last mile" of the loop is connected using copper wire pairs and distribution frames

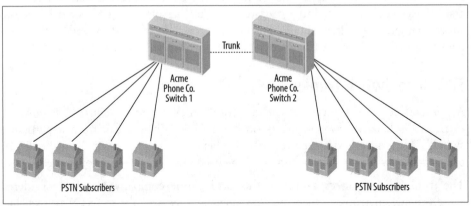

Figure 4-3. Connections that carry phone calls between phone switches on the PSTN or a private voice network are called trunks

signaling. Its tones consist of two pitches. DTMF is *in-band signaling* because the dual-tone signals occupy the same frequency band that is used to carry the voice sounds.

DTMF signaling replaced the electromechanical pulse-dial signaling that was prevalent as late as the late 1980s. The standard for DTMF is described by the International Telecommunication Union's Q.23 and Q.24 specs.

Standard tones

Sounds that you hear when the switch needs to notify the caller (or receiver) the progress of the call are called *standard tones*. These include the busy tone, the ring-back tone (the ringing the caller hears while waiting for his call to be connected), and the dial-tone. Standard tones are also defined by the ITU, in specs E.180 and Q.35.

City codes, area codes, and country codes

COs are grouped into logical, compact geographic areas that are large enough to support somewhere around five or six million phone numbers apiece. These groups, known commonly as area codes, are used to help the PSTN and SS7 route calls between subscribers in remote areas of the network. In this way, a caller from Boston can reach a caller in Tampa and so forth. Area codes are prepended onto the local CO portion of the phone number in order to place interarea phone calls. Country codes are similar prefixes, but for calls between different national PSTN designations, like the United States and Japan. City codes are a variation on area codes and are mainly used outside of North America. More information about phone numbers and international designations can be found in ITU recommendation E.164.

Out-of-Band Signaling and SS7

Common Signaling System 7, also called SS7 or C7, was developed by the ITU Telecommunications Standardization Sector in order to increase the efficiency of the public voice system. SS7 is a separate network whose duties are setting up, tearing down, monitoring, and routing calls on the PSTN.

SS7 is akin to TCP/IP in that it operates at several layers of the OSI model. And, like TCP/IP, SS7 is packet-based. It is the conceptual basis for VoIP call signaling, because it is a software-based system that operates independently of the voice transport itself (the PSTN).

SS7 works behind the scenes, so interacting with SS7 is something that the CO switch, not your phone or PBX, must do. SS7 is called an *out-of-band* signaling standard because, unlike DTMF, it doesn't use the same frequency band, or even the same transport, as the voice transmission. This isn't meant to infer that DTMF and SS7 are comparable—they aren't. DTMF has a very narrow purpose: sending dialed digits.

Out-of-band signaling is also called CCS, or common channel signaling. It's the technique used by all telecommunication vendors—including cellular phone service providers, long-distance companies, and local exchange carriers (LECs). All of these networks share one thing in common: a common bond in SS7. The ITU-T recommendation for SS7 is, coincidentally, Q.7 (see Figure 4-4).

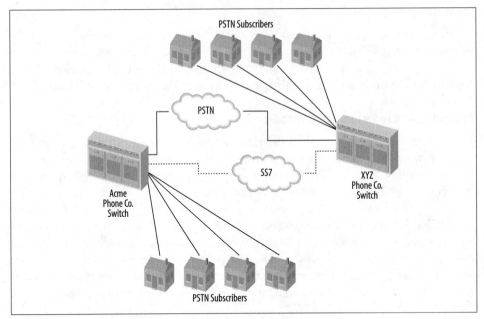

Figure 4-4. SS7 is accessed alongside the PSTN by CO switches and LD carrier switches in order to connect, disconnect, and bill calls

PRI/T1

The Primary Rate Interface is a digital access–signaling standard that allows 23 digitally encoded voice channels, called DS0s, to be carried across a T1 circuit between the CO and an enterprise PBX. These channels can be partitioned by an administrator using a DSU/CSU device, so that as few as a single one can be used for voice.

Channels not designated for voice tend to be used for data networking or not used at all. PRI service is sold by the telephone companies as a high-density dial-tone trunk solution. The 24th channel is called the *carrier channel* or the *signaling channel*, and it carries call-management signals over the T1. T1 is an integrated services digital network signaling standard and stems from ITU-T recommendation Q.931. For more about T1, see the later section titled "Time Division Multiplexing."

 Digital signaling standards differ from analog ones because they don't rely on electrical signals; instead, they use signals passed through a bit stream.

SONET and DS3

Synchronous optical network STS-1 (SONET) and DS3 (sometimes called T3 because it is a high density of T1-style paths) are methods of providing high-bandwidth-signaling pathways between switches on the PSTN. These links can be used for actual transmission of digitized voice, as well as SS7 signals, but aren't normally used for both simultaneously.

SONET network links can be used to deliver telephone service to PSTN subscribers, too. A common use for optical high-bandwidth dial-tone services, which are delivered using OC (optical carrier) circuits, is in call centers, where thousands of phone calls must be connected to a central group of phone operators.

DID

Direct inward dial service is a solution that allows inbound PSTN calls to the PBX to be routed directly to certain private extensions based on the phone number that the caller dialed, without the need for a call transfer or a human greeting to determine which extension each call should be transferred to. The way this works is quite simple.

When the DID trunk rings, the PBX answers and produces a dial-tone and a "wink" signal. This is an indicator that the PBX is ready to receive digits from the CO. The CO sends the digits that the caller dialed—usually just the last four digits—and the PBX decides how to connect the call based on them.

DID is most useful with PRI, where many DS0 channels can be assigned as DID channels. In this way, a single PRI trunk group can handle inbound calls to dozens or hundreds of phone numbers using 23 channels or less. An office with 80 desk phones could have 80 corresponding PSTN phone numbers and only a single PRI T1 for dial-tone trunking. Yet, the PBX knows which phone to ring when one of the 80 phone numbers is called because DID service tells the PBX which phone number the caller has dialed (see Figure 4-5).

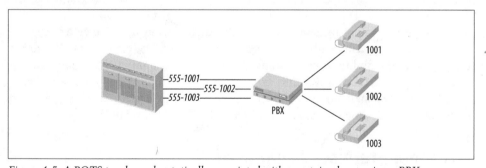

Figure 4-5. A POTS trunk can be statically associated with a certain phone using a PBX

Whereas each POTS line has an assigned phone number that can't be changed, as in Figure 4-5, the DID channels on a PRI can, in effect, have a different phone number every time they are used to connect an incoming call to the PBX, as in Figure 4-6. More information on DID services can be found in the ITU-T's Q.951 recommendations.

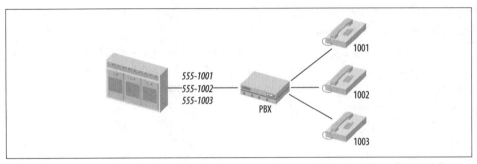

Figure 4-6. DID is a signaling standard that allows a call on an incoming trunk to dynamically signal which phone number is being called so the PBX can ring the appropriate phone

Hunt Groups

There's nothing more frustrating than a busy signal when you're trying to order a pizza on Super Bowl Sunday. That's why most pizza establishments use hunt groups. Hunt groups are groups of trunks that allow incoming phone calls to circumvent the busy trunk and "roll" to the next available trunk in the group. That way, the pizza place never misses a call, even when their phones are ringing off the hook.

Centrex

Centrex is POTS enhanced with business-grade telephony features like call conferencing, four-digit dialing, and per-call billing rather than per-minute billing. It was designed to curb the need for small businesses to invest in PBX equipment in order to get modern telephony features. A single Centrex customer can use many Centrex lines, collectively called a Centrex group. Within the group, each line can be called using four-digit dialing instead of the usual 7-digit dialing (i.e., the caller can omit the prefix when placing calls within her Centrex group). Some other PBX-like features include the ability to easily transfer calls between lines in the same group, or enable and disable call forwarding for a given line by dialing a special sequence of DTMF digits. Normally, users of Centrex have to dial an 8 or a 9 at the beginning of each call that is destined for a receiver outside their Centrex group.

Customer Premises Equipment

Any equipment installed by the telephone company for use in terminating voice calls or other connections is called customer premises equipment (CPE). In most small offices, CPE isn't used, and the occupant just plugs in a conventional analog phone or two. But in situations in which the telephone company itself manages the customer's PBX or manages a private voice trunk on the customer's behalf, then the PBX is called CPE. The softPBX described in Chapter 3 would be considered CPE if it were managed by the telephone company.

Switches

When you hear somebody refer to a switch that resides on the customer premises, that usually means a PBX or a KSU. See Chapter 1 for a refresher on PBXs and KSUs if needed.

The Demarc

The Demarc, or demarcation point, is the point at which telephone company–owned facilities terminate. Usually a cross-connection point or wiring terminal near the exterior of a building or outside it, the demarc is where the phone company's wiring connects to privately owned facilities. Typically, the demarc is where the phone company's troubleshooting stops, as well.

"Inside Wiring" and the Cable Plant

Inside wiring is the privately owned network of communications cabling and wiring terminals located on the customer side of the demarc. Inside wiring could be simple, two-pair wiring that connects a bunch of four-pin phone jacks (as in a house), or it could be a centralized distribution of cabling throughout the building—also called the *cable plant*.

 Any telephone equipment or data networking equipment that exists on the customer's side of the demarc and is maintained by the telephone company is called customer premises equipment (CPE).

CP Distribution Frames

Customer premises distribution frames are cross-connect blocks, terminals, or patch panels where endpoint locations are aggregated and centralized. Like the PSTN equivalent, CP distribution frames make it easier to move connections from one tenant to another in the same building or from one desk to another in the same office. These frames are most often used with twisted-pair cabling of Cat3 or Cat5 grades.

CP distribution frames are often accompanied by data networking gear like switches—devices that connect devices to the "edge" of the network. This is the proverbial "data closet," a place where telephone wiring and data networking equipment reside.

FXS and FXO Signaling

Foreign exchange station (FXS) refers to the signaling used by switch interfaces that serve analog telephones. An FXS interface is on the switch, with a phone connected to it.

Foreign exchange office (FXO) refers to the signaling used by telephones or PBX systems that are subscribers of the CO switch. These switches are either COs or PBXs. An FXO interface is on a phone (or PBX) that is connected to a port at the CO.

 If you remember one thing about FXO/FXS, remember this: FXO interfaces use FXS signaling to communicate with FXS interfaces and vice versa.

Loop Start Signaling

In order to establish an analog phone user's intent to begin a call, electromechanical signaling is used. Really, this is just a fancy way of saying that, at the outset of a call, a phone is lifted from the hook—the local loop is closed, and current begins to flow. This current becomes the sound of the dial-tone in your handset's transducer. The method of closing the loop at the hook switch to start the call is called *loop start*. It's by far the most common method. Less-common variations on this theme include *ground start*, *ear and mouth*, and several more. Usually, these methods are used with PBX trunks and not residential phones—which are almost always *loop start* devices.

Channel-Associated Signaling

A digital cousin of FXS and FXO uses an in-band or "robbed-bit" approach—this is called channel-associated signaling, or CAS. Digital telephones and trunks can use CAS. The phrase *robbed-bit* means that a portion of the digital bit stream normally used for sound transmission is allocated for signaling. Common PRI signaling is *not* CAS (see the earlier section "PRI/T1"). Unless you use channel banks extensively, you're unlikely to encounter CAS.

Other Legacy Standards

There are probably a half-dozen less-common analog and digital signaling techniques in use; a good description of each is available at *http://www.cisco.com/pcgi-bin/search/search.pl?searchPhrase=telephony+signaling*. Earlier in the chapter (see the section "PRI/T1"), we covered one of the most important of these standards—PRI.

Project 4.1. Create a Trunk Channel with a POTS Line

What you need for this project:

- Asterisk
- X100P interface card
- LAN

To build a softPBX that can communicate with the PSTN, you'll need at least one trunk channel to communicate with the FXS interface in the CO switch. This channel will provide you with the dial-tone from your local phone company so that calls to and from the PSTN can be handled by the Asterisk server. There's not much to setting up an FXO channel. The quickest way to add one to an Asterisk system is to install an X100P, TDM400P, or similar FXO line card.

First, plug the trunk into the X100P card by connecting an RJ11 standard phone patch cord from the wall jack of an active telephone company POTS line into the X100P card (or TDM400P card) in the appropriate port. On the X100P, this port is the one marked with an etching of a telephone wall jack. On the TDM400P, it's the port that corresponds to the slot number of the FXO module you've installed. For more information on the TDM400P and X100P cards, see Chapter 3.

Once the card is in a PCI slot on the Asterisk server and the phone line is connected, you've got to define the FXO trunk connection as a channel that is usable by Asterisk. Once defined, you can reference the channel within your Asterisk call-routing scheme. The POTS line can serve as the full-time gateway for all PSTN calls and all telephones in your home or office. Or the POTS line can just be a connection mechanism so that the Asterisk server can answer incoming calls on the POTS line if they aren't answered by a person within a certain number of rings. But first—the FXO connection must become a named Asterisk channel.

 Make sure the right Linux kernel modules are loaded once your FXO interface card is installed—or Asterisk won't be able to use it. See Chapter 3 for more information about using *modprobe* and *ztcfg* to load these modules.

Each voice channel in Asterisk has a number. This number consistently represents the same channel throughout all of Asterisk's configuration files and in its logging output. The numbering of voice channels—especially those that require a dedicated piece of interface hardware in the server—is determined by the order in which their drivers are loaded and the order in which they are identified in the PC's PCI bus. Figuring out which card is which—say when you have three or four X100P cards, each with its own POTS line—can be a bit of trial and error. In this project, we're using only one card and one line, so it should be a breeze.

The voice channel we're going to create will be called Zap/1-1. Asterisk follows a similar convention when naming all voice channels, even if they aren't analog phone line channels. The channel name is divided into two pieces. The first piece, Zap/1, refers to the physical Zaptel interface channel (which is either an FXO/FXS interface or a PRI channel), while the second piece, -1, refers to the line appearance (more on line appearances later).

Assuming you haven't touched the Asterisk configuration files since running make samples, you'll have to do only two quick config changes to fire up your POTS line. The first change will happen in */etc/zaptel.conf*. Add the following lines to the end of the file:

```
fxsks=1
loadzone=us
defaultzone=us
```

The first line tells the Zaptel configuration program, *ztcfg*, to set the X100P card to use FXS "Kewlstart" signaling—a variation of conventional FXS loop-start signaling. The reason the number 1 is referenced is because there's only one Digium card installed—so it's card number one, and its channel will be 1 as well. If there were two cards installed side by side, that first line would say fxsks=1-2 instead. If there were more than one channel per card (such as the TDM400P), then a single channel number is used for each channel on that card (i.e., fxsks=1-4 for a card with four lines attached). The following two lines localize the FXS signaling functionality of the X100P interface with loadzone and defaultzone. Other valid zones include fr, de, and uk.

Now, you may be asking yourself, "Why am I configuring the FXO interface card to use FXS signaling?" The answer is simple: in order to communicate with an FXS device (the interface at the CO), the local interface must use FXS signaling. Remember that only FXS devices can receive signals meaningfully from FXO devices, and vice versa.

 To alleviate confusion over the FXS/FXO kernel module naming, *wcfxs* has been deprecated in favor of *wctdm* in releases of Asterisk later than 1.0.5.

The next change you need to make is in the */etc/asterisk/zapata.conf* file. The sample configuration should be completely commented out (a semicolon [;] at the beginning of each line). If not, comment them all out by inserting the semicolons. Then, add the following lines to the end of the file:

```
context=default
signalling=fxs_ks
usecallerid=yes
echocancel=yes
callgroup=1
pickupgroup=1
immediate=no
callerid="* X100P POTS"<(123) 456-7890>
channel=>1
```

The first line tells Asterisk what set of assumptions to make (i.e. what "context" to choose) when handling incoming calls on the POTS line. The second line tells Asterisk (not *ztcfg*) what type of signaling the X100P has been set to use. The following lines turn on a few traditional telephony features—caller ID, echo cancellation, and other stuff that's covered in more detail later. The last line assigns the settings spelled out above it to channel 1. The assignment of these inherited settings uses the => assignment operator rather than just an equals sign (=). The Asterisk configuration parser doesn't distinguish between them; the convention is merely for ease of readability.

One more quick change you'll need to make to Asterisk's sample configs: change the Zap/g2 definition for $TRUNK in *extensions.conf* to Zap/1. (This step may not be necessary with earlier versions of the sample config.) This will allow outbound dialing to be directed to the correct channel, Zap/1, the one that represents the connection to the PSTN.

Now, since you've added a new hardware interface, Asterisk must be restarted. Once you've done that, try calling the POTS line you've connected to the X100P using a second phone line or your cell phone. After a few rings, assuming you've not changed it in the configuration, Asterisk will answer and you should hear the familiar demo greeting that you heard in Project 3.1. If you examine Asterisk's console output during this demo, you'll see something like this:

```
-- Starting simple switch on 'Zap/1-1'
-- Executing Wait("Zap/1-1", "1") in new stack
-- Executing Answer("Zap/1-1", "1") in new stack
...
-- Playing 'demo-abouttotry'
-- Executing Dial("Zap/1-1", IAX@/guest@misery.digium.com/s@default) in new stack
...
```

Through the console output, you can trace every step Asterisk took to recognize, answer, and process the incoming analog call from the PSTN and connect it using the IAX protocol to a remote server across the Internet. Note that, although this chapter is about legacy, circuit-switched telephony, we're using IAX to get our feet wet with VoIP. Plus, the IAX demo is so easy to run with Asterisk out of the box, it's a great way to demonstrate how a VoIP signaling protocol can be used with legacy signaling in a single PBX.

 IAX stands for Inter-Asterisk Exchange. It's a signaling protocol that uses UDP to connect voice calls. Although IAX is not a mandate of a standards body like the ITU or IETF, it has a number of cool characteristics—like the ability to easily work through firewalls—that distinguish it from other VoIP signaling protocols. IAX is detailed in later chapters.

This is the first project in which we've used a traditional analog device, the phone calling in through the PSTN, with Asterisk. While Asterisk itself is software, traditional PBX systems have long relied upon software-like algorithms in order to allow many analog and digital phones to communicate across a single, shared switching system. Chief among these algorithms is a technique called *time division multiplexing*, and almost all legacy switches use it to support those old-school phones.

Time Division Multiplexing

Traditional PBX systems use a digital bus to carry sound information between interfaces where phones and/or trunks are connected. The signals flowing across this bus are digitized audio that travel in an aggregate form—that is, one bus can carry many separate signals within a single bit stream. The transmission technique is called time division multiplexing (TDM). To understand how all modern telephony solutions work, including VoIP, a basic understanding of TDM is important.

Multiplexing means combining many signals onto a single transport mechanism, such as a T1 or a PBX bus. This bus can be a connection between two points—like a point-to-point T1 circuit, or it can be a large group of digital phones, like a PBX's bus.

Time division is the method of combining, and later dividing, the signals, with the purpose of yielding greater efficiency over the data link, be it a T1 circuit or a PBX backplane. Each signal is given a time slice, a small piece of the total bandwidth of the bus. At a very high rate, a TDM bus transmits a fixed sequence of time slices that are equal in duration. Each time slice contains a digitally sampled representation of the original analog waveform signal.

Each endpoint pulls a particular time slice out of the aggregate TDM bit stream and reassembles it, in real time, into a single cohesive digital signal. Each piece of the time slice used to reassemble that single signal is called a *frame*, just as pieces of the bit stream on an Ethernet data link or T1 are called frames.

Pulse Code Modulation and DS0 Channels

A sampling rate of about 8 KHz is sufficient to adequately record the human voice using analog instruments. When analog sound information is recorded at this rate, its amplitude, or power, is sampled 8,000 times per second. This is only the first step in putting the sound into digital format. It's still analog, because it is still measured on a scale with infinite, analog resolution. Imagine the sampled sound levels are 8,000 points on a line graph. Now draw curves to connect them. This is an analog waveform signal (see Figure 4-7).

In order to digitize that analog signal, it must be quantized, or put into a scale with limited resolution. Quantizing means taking each of those 8000 sample points and assigning each one to the nearest value on a finite, graduated scale. There are two major scales in use—Alaw and μlaw. μlaw is prevalent in North America, Japan, and Hong Kong, while Alaw is common elsewhere. There are 256 levels on both scales. Eight thousand times per second, an analog sample's strength, or amplitude, is rounded to the closest of the 256 levels (see Figure 4-8).

Once digitally quantized, the signal must now be encoded. Since an 8-bit expression can describe 256 discrete values, and the rate of sampling is 8 KHz, the digital sound

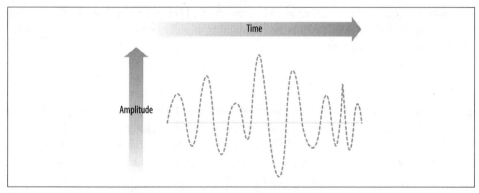

Figure 4-7. An analog sample of part of a waveform signal

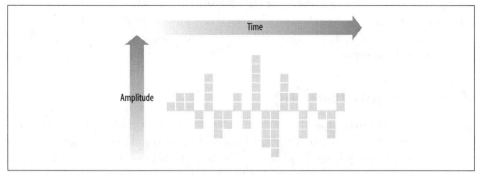

Figure 4-8. A digitized sample of the same part of the waveform signal in Figure 4-7

signal is transmitted at a data rate of 64 kbps, which is 8 bits times 8,000 samples per second. Lined up one after the other, 8,000 times per second, each 8-bit word forms a stream of binary data that travels the TDM bus in a single time slice. The common name for a 64 kbps voice channel is *DS0*, and the signaling and encoding technique is called PCM, or pulse code modulation.

DS0 channels are the building blocks of ISDN and T1 voice circuits, too. Twenty-four of them together comprise a T1. This is why T1s have a maximum bandwidth of 1.5 Mbps, which is the same as 24 times 64 kbps.

Channel Banks

Channel banks are devices that *multiplex* (combine) separate digital voice signals into a single T1 circuit and *demux* (split) it back into its individual signals again. In a nutshell, a channel bank combines up to 24 separate phone lines into a single digital link.

Channel banks are not DSU/CSUs, which are data transmission devices, though some DSU/CSU devices can perform channel bank functions like connection of one or more single analog phone lines to a T1. For example, certain AdTran DSU/CSUs can hold FXO/FXS interface cards for connecting phones, PSTN lines, or PBXs. One purpose of a channel bank might be connecting a group of digital phones to a PBX through that PBX's T1 interface.

BRI

Basic Rate Interface (BRI) is an access signaling standard for multiplexing two voice channels on a single digital circuit. A third channel is used for call control. This is the service loosely referred to as an ISDN line, though ISDN itself means far more than just BRI. In fact, the PRI standard is also a child of ISDN.

BRI uses two channels of voice (called B channels), each occupying 64 kbps of bandwidth, while the third channel (called the D channel) occupies 16 kbps for call signaling. The B channels are purely for transmission of payload—that's voice or data, while the third channel, the D channel, is used to signal connects and disconnects on the channel. Generally, digitizing for BRI is performed using a PCM algorithm, like a T1 or a digital TDM phone.

Point-to-Point Trunking

Besides the trunks that supply a PBX with a dial-tone from the telephone company CO, it is common to have trunk connections between multiple PBXs. These types of connections, which are often high-density copper cables if the PBXs are within a few hundred meters, are called *point-to-point trunks* or *private trunks*. Often, these trunks are established in order to forego the PSTN, and any associated fees, for calls between disparate business locations, as in Figure 4-9.

Most PSTN traditionalists insist that only one-pair analog phone lines that link two switches can be called *trunks*. But, as networks have progressed, so has the definition of *trunk*. Trunks can be T1s or radio links; what makes them trunks is that they connect two switches. Even in Ethernet switches, ports that connect VLAN data (more on this in Chapter 9) between two locations are called *trunk ports*.

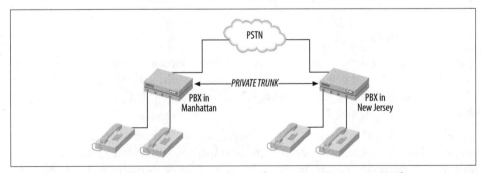

Figure 4-9. A private trunk connects two switches on the same private voice network

When privately owned cable can't be run between locations due to excessive distance, telco-owned facilities, like T1s, T3s, or wireless solutions can be used for private trunking. BRI-ISDN is often used for low-density switch-to-switch trunking across relatively short distances, but its high cost is a discouraging factor. BRI-ISDN allows two voice calls to be carried at once.

T1 circuits are far more abundant than BRI-ISDN in trunking situations, because they offer 12 times the capacity. Even though the cabling used to carry the T1 circuit is the phone company's property, and the phone company charges a fee for the use of that pathway, the T1 circuit used to connect two switches can still be called a private trunk because the voice calls it carries are not considered PSTN traffic.

The ITU's recommendation for T1 technology describes both the transport and data link layers. The physical layer for a T1 circuit tends to be two pairs of copper wiring. The phone company may connect the T1 through its local access transport areas (LATAs) using fiber, but the part of the circuit that connects to the customers' locations is almost always copper.

Even higher-density T1-like circuits are available for private trunking if the need exists. DS3 circuits multiplex many times the capacity of a T1. You can find out about even higher-capacity telephone circuits by studying the ITU-T's recommendations or by asking a local telephone company's presales engineer.

Aside from ISDN lines, T1, and DS3, there are four ways private voice trunks can be carried between locations: copper, fiber optics, radio waves, and free-space optics. The two most dominant, in traditional telephony, are copper and fiber-optic cabling, but wireless technologies are becoming more popular for trunking because of their low cost.

Copper Cabling

Copper is the material that is found in most local area data and voice network wiring. It's a flexible, resilient material that is highly conductive of electricity. It carries electrical signals via direct current over short distances, say from the CO to your house or from an Ethernet switch to a desktop PC or IP phone. Copper is the most common type of material used to wire the connections between analog and digital phones and their switches.

Copper is excellent and cheap for short-distance transmission. But over long hauls, it suffers from attenuation. This is why Ethernet twisted-pair data links have a length limit.

Fiber-Optic Cabling

Fiber-optic cables, or fibers, are very, very long strands of glass that are almost optically perfect. That is, when light beams shine into one end of a fiber-optic strand, they tend to emit from the other end without losing any power. Switching the light source on and off allows you to transmit a digital bit stream over the fiber. Attenuation is still a factor for fiber optics as with copper, but the length limits imposed are much higher—miles instead of meters.

Fiber itself is rather expensive and more fragile than copper, making it a poor choice for short-distance connectivity. But there are some local area applications where fiber beats copper. In areas with very high electrical noise, fiber will work where copper often fails, because fiber cabling carries optical, not electrical, signals.

Because fiber doesn't use electricity for signaling, it doesn't require a physical loop. This means that it can still operate with one strand of fiber, rather than two, intact, while a copper segment needs two strands intact in order to function at all.

Radio

Radio waves can be used to carry voice and data, though in modern enterprise networks, their use for data is far more common. 802.11b and 802.11g are common standards for deploying Ethernet wirelessly using radio. They offer short-distance (typically twenty-five to a few hundred meters, depending on the antenna and base) connectivity for many client devices using a single radio base station. These standards can also be used for simple Ethernet bridging.

Other types of radio can be used to connect, or trunk, between locations in a metro area network (MAN). These radio technologies use high-power microwave transmitter/receiver stations with a very narrow focus to transmit data across relatively great distances—as much as 15 or 20 miles in enterprise networks. If you've ever spotted the dish-shaped antenna that are sometimes used to connect skyscrapers in a downtown skyline, then you've seen how this technology facilitates point-to-point trunking.

> Because radio bridges are packet-based Ethernet devices, integrating them into a traditional voice network as a trunking solution isn't possible. To use a network link to trunk voice traffic, the voice devices on either end of the link must first support VoIP.

Radio is great for ease of deployment. Many manufacturers of radio datacom devices support license-free radio spectrum. The products themselves tend to be Ethernet bridges, making them largely configuration-free. But radio is less reliable than copper or fiber. Microwave radio needs a clear "line of sight" between connected locations, which often rules it out in mountainous regions or terrains with very tall trees. While antenna towers can solve this problem, they are often cost prohibitive, and an eyesore, too.

Some service providers sell a service that they call wireless T1. Don't confuse this with actual T1 service from a phone company. Wireless T1 doesn't have anything to do with T1, it's usually no good for VoIP, and it doesn't use T1 signaling. In fact, wireless T1 is usually just a name ISPs give to a radio Ethernet bridge that provides a subscriber's data network access at roughly the same speed as a real T1 circuit.

Free-Space Optics

A relatively new development, free-space optical (FSO) uses light beams transmitted over the open air, like a television remote control, to connect remote locations. Like radio, FSO requires a line of sight and is subject to the same terrain challenges. Like microwave radio antenna, FSO transmitters and receivers must be carefully kept in alignment.

But unlike radio products, FSO links are capable of much higher data throughput approaching 3 gbps. FSO's range is limited to about a mile. Where radio is susceptible to RF interference, FSO is susceptible to light interference—especially from the sun.

More information about metro area networking is available in "Metro Ethernet," available at O'Reilly's Safari Bookshelf (*http://safari.oreilly.com*), and in O'Reilly's *802.11 Wireless Networks: The Definitive Guide*.

Legacy Endpoints

In conventional telephony, every endpoint is an analog transducer set with a dial-pad (touch-tone) or a rotary-dial pulse wheel. Rotary-dial phones are a dying breed, of course, as the dominance of DTMF has marginalized them to the degree that many telephone companies don't support them any more. Both rotary and touch-tone phones share support for some variant of FXS signaling—an electrical protocol described earlier—in order to communicate with the CO switch.

FXS/Analog Endpoints

While most legacy switches support analog endpoints through onboard analog ports, VoIP systems require additional interfacing in order to use them. SoftPBX vendors support analog endpoints with FXS signaling in the following fashions:

Analog telephone adapter
   A device that connects an analog telephone to a softPBX via Ethernet. ATAs usually have one RJ11 port, for the phone, and one RJ45 port, for the Ethernet switch. The ATA accomplishes all the digitizing and packet encoding needed to use the analog phone in a VoIP environment. Generally, pulse-dial phones aren't supported.

Outboard analog media gateway
> A device that serves the same purpose as an ATA, but allows more than one analog phone to be used. Some analog media gateways are highly expandable and can support dozens of analog endpoints. Analog media gateways may also have some built-in call-routing functionality.

Onboard analog media interface
> An interface port on the softPBX server itself that allows for direct connection of a single analog phone.

Of course, analog endpoints are more than just phones. Sometimes they are modems, fax machines, answering machines, burglary or fire alarm systems, or those automated dialers with prerecorded telemarketing messages that sometimes call and interrupt your dinner. Generally, analog endpoints can be connected to a VoIP network using one of the three mechanisms listed.

But many data transmission protocols used by faxes and modems are broken by the voice digitization process used on VoIP networks. Quantization, as described in the earlier section "Time Division Multiplexing," has a knack for screwing up the analog sound wave signals used by modems and faxes to represent data. But don't fret—there's a solution to this problem, and it's covered in Chapter 14.

Digital Endpoints

Digital phones are functionally similar to analog phones, but internally quite different. Their connection to the local switch is a digital connection, carried on two pairs of copper instead of a single loop. Digital endpoints are integrated into the TDM bus, so that the signal carried to them is completely digitized. They have integrated circuitry onboard that can decode that signal and encode one to send back through the TDM bus. Digital phones tend to be more expensive, and more feature-packed, than their analog counterparts. Their signaling standards also tend to be proprietary interpretations of ITU recommendations. For this reason, it can be a challenge using one vendor's digital phone on another vendor's PBX.

Using (or not using) digital endpoints with a VoIP server

While it is possible to use digital endpoints in a VoIP environment, the specifics of supporting TDM phones (such as the one shown in Figure 4-10), channel banks, and proprietary TDM-based switches are outside the scope of this book. There are, of course, ways of interfacing digital phones directly with a TDM-aware VoIP switch, but why bother? All you'd really be doing is re-creating a TDM bus—which is one of the legacy technologies we're trying to get away from, right?

A better approach would be to connect the old-school PBX to a VoIP server using a single interface to trunk between them. Then you can migrate each user from a digital/legacy phone to an IP phone over time, and you don't have to heave-ho all those

Figure 4-10. Traditional digital business phones like this one are a part of the PBX system's TDM bus

digital phones right away. A good source for information about time-division phone standards is the book *Digital Telephony* (Wiley).

Don't feel as if you have to reinvent the wheel to accommodate VoIP. When the time comes, you'll replace your digital phones and all your TDM equipment, too. But a single trunk between the VoIP switch and the PBX switch can be a godsend in a migration to VoIP, because it can carry calls between an old conventional PBX and a new softPBX.

Speakerphone

A phone than can be used without holding the receiver next to one's head is called a *speakerphone*. It has a speaker that is loud enough to clearly hear the remote party, and it has a microphone that can adequately pick up the local user's voice. Most speakerphones have a handset onboard; in fact, most business phones *are* speaker-phones. There are two kinds: full-duplex and half-duplex.

A speakerphone—especially one of the analog persuasion—is naturally prone to acoustic feedback, usually exhibited as an annoying high-pitched squeal or hiss emitted by the speaker. This is caused by the microphone in the phone picking up the sound coming from the speaker and creating a feedback loop, or a signal of a continually amplified and destabilized waveform. The result is that awful hiss. To get around this problem, half-duplex speaker phones have a limiter or gate circuit that cuts off the speaker when the local user is talking. The only drawback is that sometimes both parties can talk at the same time without realizing it, because the speaker is cut off. If you've ever used a half-duplex speakerphone, you know how aggravating this effect can be.

Full-duplex speakerphones solve the feedback loop problem by using DSP (digital signal processing) circuitry to equalize the signal against frequencies of sound that are prone to feeding back or to modify the sound signal in a way that makes it immune to feedback but is undetectable by the human ear.

Lines and line appearances

With legacy phone systems, there are two ways to refer to voice channels on each phone. These channels, the circuits across which voice information travels, are either *lines* or *line appearances*. What's the difference?

A line is a physical connection to the PBX—a single voice circuit that is capable of carrying one concurrent phone call. A line appearance, conversely, *is* the one concurrent phone call.

Consider the PSTN feature called three-way calling. This feature allows a single phone user to call two other PSTN subscribers simultaneously, switch back and forth between the two conversations, and even bridge the conversations into a three-party conference call. Each of the conversations "appears" to be on a separate line and is therefore called a line appearance, since only one physical line is in use at the caller's phone.

The same is true of analog or traditional phones connected to a PBX. From a PBX subscriber's phone, you may dial extension 101, put that call on hold, dial extension 102, put that call on hold, dial extension 103, put that call on hold, and then switch back and forth among them. Three line appearances, one phone, one line to that phone from the PBX.

In key systems, line appearances and lines tend to be one and the same, as different line appearance buttons on key system phones correspond to individual POTS lines that are connected to the key system.

Ring groups and private hunt groups

In some environments, the person closest to a phone is most qualified to answer a call. Ring groups allow two or more phones to ring simultaneously when a call is incoming. Once answered on any phone in the ring group, the ringing stops, and subsequent calls will ring all the phones in the group again.

Private hunt groups are more advanced ring groups. Like the PSTN hunt groups, private hunt groups allow calls destined for a phone that is currently unavailable to roll to another channel—usually another phone. Depending on why the phone is unavailable to receive a call, the hunt group may have a different type:

- Sequential hunt groups seek the next available channel from a list.
- RNA (ring no answer) hunt groups roll the call to the next channel when there's no answer on the first phone, in addition to when that first phone is in a call.
- Simultaneous hunt groups may ring all phones simultaneously, like a ring group, but only when a "lead phone" is busy on a call.

Most PBXs support some combination of ring groups and hunt groups.

The group of analog phones in most homes is *not* a ring group, but rather an electrical aggregate: a parallel cluster of analog PSTN endpoints. In an all-analog residential setup, all phones ring at the same time, like a ring group. But unlike a ring group, the phone that answers the call doesn't become exclusively connected to that call.

Project 4.2. Simulate a Simple Key System with a Ring Group

What you need for this project:

- Asterisk
- X100P interface card
- Cisco SIP 7960 IP phone (or similar)
- Grandstream Budgetone 101 IP phone (or similar)
- LAN

Using the POTS line that was connected to the Asterisk server in Project 4.1 and any type of phone (analog or IP), it's pretty easy to emulate a key system. In Project 5.1, we'll assign extension numbers to two IP phones, allow them to dial out using a POTS trunk, and put them together as a ring group for all incoming calls.

In this project, we'll use a Cisco SIP 7960 (phone A) and a Grandstream Budgetone 101 (phone B) to participate in the ring group and the X100P for the PSTN interface. You don't have to use these particular phones if you know how to configure the IP phone you have—as long as the IP phone you're using supports SIP.

A key system is not as sophisticated as a PBX—it doesn't do station-to-station calling, and it usually doesn't have extension numbers for each phone. Instead, a key system aggregates one or more PSTN lines so that a larger group of phones, usually no more than 16, can all make use of them. Actual call switching is still handled by the CO switch.

A simple key system can be built around a group of IP endpoints and a basic gateway device, like an appropriately configured Cisco media gateway. Something as sophisticated as Asterisk isn't really necessary, but since it's free, it's a great place to start. The completed project is shown in Figure 4-11.

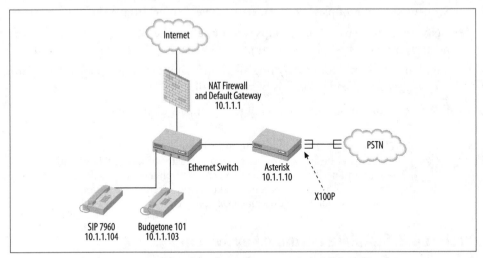

Figure 4-11. When Project 4.2 is complete, the VoIP test network will look like this

In this setup, all incoming calls on the POTS line will ring all IP phones on the private Ethernet network simultaneously and then connect each incoming call with the IP phone that answers first. If you're using two 7960s rather than one 7960 and one Budgetone, then you can skip Step 2 below, and just repeat Step 1 with the second 7960's configuration in mind. If you're using two Budgetones, skip Step 1.

Step 1: Unlock and configure phone A (Cisco 7960)

Before you can go anywhere with Asterisk and a Cisco 7960 SIP phone, you've got to make sure the phone is running SIP firmware and not Cisco-proprietary SCCP firmware. Though Asterisk offers rudimentary support for the SCCP protocol through a software module, SIP is far more accessible to VoIP beginners—not to mention an open, community-based standard.

The quickest way to tell if your 7960 is loaded with SIP firmware is to power it up and watch the upper-right corner of the LCD display. If, after booting up, the word *SIP* appears there, then you're in business. If not, it's time to get your hands on another phone. Incidentally, Cisco prohibits the loading of any firmware on its IP phones because each phone is covered by a service agreement.

 You'll need Cisco SIP firmware Version 7.1 to do most SIP projects in this book if you're doing them with a 7960. Check your firmware version by pressing the Settings button, followed by 5, followed by 3. Look for Application Load ID Version "POS3-07-1-00" or higher.

Once the Cisco 7960 is booted up, unlock it by pressing the Settings button, then 9. Then use the keypad to dial the word *cisco*. That's the 7960's default password for unlocking its configuration. Next, press the Accept softkey. Now, the phone's configuration is changeable.

First, assign an IP address by pressing Settings, dialing 3 for Network Configuration, and dialing 5 for IP Address. Enter 10.1.1.104. If the 7960 doesn't allow you to enter an IP address, be sure to disable DHCP by scrolling down to option 25 and pressing the No softkey.

Find the Default Router 1 setting and enter 10.1.1.1 (or whatever the address of your gateway is on the 10.x network). Change the phone's subnet mask to 255.255.255.0. Be sure to press the Save softkey after each setting change.

Now, from the main configuration menu on the 7960, choose SIP settings, and then choose Line 1 settings. For the Name and Shortname settings, enter 104. Change the Proxy Address setting to 10.1.1.10—i.e., the Asterisk server. Check that the Proxy Port setting is 5060—SIP's commonly used port number. When settings are complete, they should resemble this list:

    IP Address: 10.1.1.104
    Subnet Mask: 255.255.255.0
    SIP Proxy: 10.1.1.10
    Username: 104

Once all these changes are complete and you're back to the 7960's main screen (the one with the Cisco copyright notice), you'll see 104 in the upper-right corner, along with an icon that looks like a telephone. This icon is an indicator that the IP phone has initiated communication with the SIP registrar—that's one of many SIP terms that means softPBX. The Asterisk server is intentionally ignoring SIP requests from 10.1.1.104, so this indicator is nothing to worry about. It will be gone after the Asterisk server is told how to handle this particular IP phone.

Finally, connect the 7960 to your Ethernet switch.

Step 2: Configure phone B (Budgetone 101)

You can apply the following configuration to a second 7960, if you prefer, or set it up on a Budgetone 100 series:

    IP Address: 10.1.1.103
    Subnet Mask: 255.255.255.0
    SIP Proxy: 10.1.1.10
    Username: 103

Use the Budgetone's web-based configuration page to change these settings, as you did in Project 3.1. Then connect the Budgetone to your Ethernet switch.

Step 3: Define SIP peers for the IP phones

Open */etc/asterisk/sip.conf*. This is the file where SIP channels, which Asterisk calls *peers*, are defined. In this case, we'll need to define two—one for each IP phone. In this file, you'll see this block at the top of the file:

```
[general]
port = 5060              ; Port to bind to
bindaddr = 0.0.0.0       ; Address to bind to
context = default        ; Default for incoming calls
```

The SIP configuration file is structured so that every section *other* than the [general] section defines one SIP peer apiece. The sample configuration file provided with Asterisk has some suggested configurations, all commented out. Add the following to the end of the file to set up the two IP phones:

```
[103]
type=friend
context=default
username=103
callerid=103
host=10.1.1.103
nat=no
canreinvite=no
port=5060
dtmfmode=info  ; this is required for Budgetone 101 phones

[104]
context=default
type=friend
username=104
callerid=104
host=10.1.1.104
nat=no
canreinvite=no
port=5060
```

The bracketed headings, [103] and [104], correspond to the two SIP phones whose SIP usernames you configured, respectively. They apply the settings that follow to a SIP channel that is reserved for the endpoint claiming a particular username—either 103 or 104. Now, to make them ring when somebody calls in on the POTS line.

Step 4: Create a ring group for the two IP phones

In order to disable the demonstration greetings and configuration that ships with Asterisk out of the box, you'll need to open */etc/extensions.conf* and find the section that begins with [default]. Find the directive include => demo and comment it out by placing a semicolon (;) at the beginning of the line. Then, add this entry right below it, still in the [default] section:

```
include => keysystem
```

Now, move up the file to just above the [default] section and add the following bit of Asterisk configs:

```
[keysystem]
exten => s,1,Answer( )                    ; Answer the line
exten => s,2,Dial(SIP/104&SIP/103,40,r) ; ring our SIP phones
```

These two entries, placed in the new [keysystem] section, establish the two tasks the softPBX must do in order to ring the two IP phones when the POTS line rings. First, the Answer( ) application command tells the softPBX that, by default, all incoming calls will be answered by the softPBX.

> The Answer( ) command isn't required first. If you omit Answer( ) and just use the Dial( ), the POTS line will be answered only if one of the phones in the Dial command is picked up. This way, the caller's phone company won't bill them if nobody answers. However, if you want to play music on hold or some recorded message while the caller is waiting for the call to be answered, an Answer( ) is needed.

The second command, Dial, causes Asterisk to ring both SIP phones—103 and 104—simultaneously. The one to answer the call will cause the ringing to stop and the other will no longer be able to join the call merely by answering. Note the ampersand (&), which delineates multiple channels for simultaneous ringing.

The SIP/104 and SIP/103 notations refer to SIP channels, just as Zap/1-1 referred to a Zaptel analog channel in the earlier project. These SIP voice channels were established in Step 3.

To try out the ring group, execute a "restart now" at the Asterisk CLI, and then use a cell phone or an extra phone line to call in to the POTS line that's connected to Zap/1-1. Both IP phones will ring while displaying the incoming caller ID information, and the first phone to pick up will field the incoming call.

Step 5: Allow the SIP phones to dial out

As it stands now, an incoming call on the POTS line would indeed ring on the 103 and 104 phones simultaneously. But neither phone can currently place an outbound call through the POTS line on channel Zap/1-1. To do this, you'll need to include another context in *extensions.conf*. In the [default] section, add the following:

```
include => trunkld
```

This change allows the call routing for outbound calls, as contained in the [trunkld] section of the file, to be applied so that Zap1 can be used to originate PSTN calls. Now, after an Asterisk CLI "reload," your SIP phones will be able to dial 9 followed by a valid 11-digit PSTN number in order to place outbound calls.

All of these configurations are essentially just modifications to the softPBX's call-routing scheme. Historically, and in VoIP, this scheme is referred to as a *dial-plan*.

To truly simulate a key system, you'd need to allow each phone to select which trunk it wants to use when dialing out—but since this example uses only one trunk, there's not much to choose.

Dial-Plan and PBX Design

The dial-plan is a set of rules that governs the call-routing behavior of a PBX. When a business user picks up his telephone and dials a number, the PBX refers to the rules in the dial-plan in order to determine how best to connect that call. While dial-plans have no standardized syntax or established set of best practices, they are always expected to describe:

- How phones are uniquely identified on the network—usually through extension numbers
- How to connect calls based on the caller's DTMF digits—i.e., which channels to open to route the call
- How to connect or restrict calls based on their origin
- What effects the origin of a call has on its priority
- How to group phones together for common applications, such as group ring or hunt groups

Dial-plans needn't be confined to a single PBX within an enterprise. A large network can have dozens of PBXs that are all subject to a common dial-plan. Dial-plans aren't exclusive to legacy telephony equipment either. SoftPBX systems support and require a dial-plan, too.

The specifics of configuring a dial-plan vary from platform to platform and from PBX to PBX. One legacy AT&T switch may have a very different configuration approach from an old Panasonic switch. This inconsistency continues among Voice over IP call-switching systems, too. In this book, we've concentrated mainly on Asterisk up until this point. But it's important to know that even though the principle of dial-plan design and capabilities are more or less universal, the platform-specific implementations of dial-plan can be vastly different in structure.

In Project 4.2, we built a very simple dial-plan that assigned extension numbers to two IP phones, allowed them to dial out using a POTS trunk, and put them together as a ring group for all incoming calls. This dial-plan was intended to emulate a conventional key system—a simple and relatively easy thing to do on any VoIP call-routing device whose interfaces include Ethernet and FXO. An enterprise-class Avaya MultiVantage server or a simple Cisco media gateway could behave like a simple key system, but that's not how most are set up.

With a traditional PBX, phones are called *extensions*, connections to the phone company are called *dial-tone trunks*, and connections to other private PBXs are called *private trunks*. Usually this terminology is applied the same way in a VoIP network. Many of the same considerations have to be made when creating a dial-plan:

- How many phones will there be in the network? Does the extension number allow for enough digits for all of them?
- Will phone users dial a special digit (8 or 9) at the beginning of PSTN-bound calls, or will the phone system just figure out how to route their calls?
- Will calling between private switches be possible?
- Will extension numbers be based on a mitigating or symbolic convention—such as the organizational location of the phone—i.e., which office location it's at, or a desire to match a particular user's DID phone number? (See the section "DID" earlier in this chapter.)
- Will each user's voice mail box number match her extension number?
- Will calls routed from a particular user to the PSTN always travel the same trunk, or will the trunks be selected randomly when calls are placed?

Some of these mundane questions can have a big social and economic impact. Using a long extension number (five or six digits) could be unpopular with users but necessary because of the size or growth potential of the voice network. Routing outbound calls to the PSTN on a user-per-line basis, in combination with caller ID, can make it easier for people to get hold of the right person when they return missed calls, but at the cost of complexity.

In circuit-switched telephony, these issues have much greater economic significance than with VoIP. As each dedicated legacy interface is replaced with an extensible, flexible protocol running over a converged network, the cost of hardware maintenance drops. Maintaining individual trunk ports, individual extension cards, and individual phone lines decreases. But in the meantime, a legacy or hybrid/IP-enabled environment's dial-plan must reflect cost-effectively on the physical interfacing available to it. Stated another way—it's especially important not to waste capacity in an analog/TDM environment, where capacity is more expensive.

That means if a particular user is on the phone constantly making outbound calls— say a collections person—it might make sense to assign a trunk to that user specifically. But you wouldn't want to give a dedicated trunk to the guest phone in the reception area, because it's more economical for this phone to share trunks with other phones.

Extension Numbering

A number of strategies can be used to assign extension numbers.

Extensions based on DID

In organizations using DID, extension numbers often match the DID number given to each phone user. That is, if a user's DID phone number is 335-1464, then his extension would be 1464. This approach is popular in organizations that use DID for every user.

Extensions based on geography

In organizations that have a geographically diverse network of office locations and PBX switches, giving figures of the extension number a geographical significance can be helpful. For example, in a four-digit extension, one could say that the first two digits represent the office location. Here's a scheme based on this idea:

- Extensions 1100–1199: Detroit Sales Office
- Extensions 1200–1299: Southfield Sales Office
- Extensions 2000–2999: Kalamazoo Customer Service Call Center

In this scheme, 100 extension numbers are available in Detroit and Southfield, while the Kalamazoo office has 1000 available extension numbers. It's up to the dial-plan to enforce proper call routing with regard to this scheme.

Extensions based on department

Like the geographic approach, this type of scheme takes groups of extension numbers and assigns them to an organizational unit—not a location, but a department:

- Extensions 1000–1199: Marketing
- Extensions 1200–1299: Accounting
- Extensions 1300–1599: Engineering
- Extensions 1600–1649: Information Technology

It's possible to combine geographic and departmental schemes, dividing up the extension numbers in an even more detailed manner.

There's no right or wrong way to establish a call-routing scheme unless its complexity or economy make it unmanageable—then it's wrong, of course. Remember, whatever scheme you, as a voice system designer, cook up, you must also implement and support it. Simplicity is always elegant.

Extensions based on type of device

In a PBX, you'll sometimes address trunks and outboard devices like voice mail servers using extension numbers, too. This is because some PBX systems require you to do so. You might pick a block of extension numbers that are very unlike those used to call real human-answered extensions, to prevent accidental dialing by a user. Some PBX systems can restrict users from calling an extension used to signify a trunk or voice mail port, but most don't. Fortunately, VoIP is not as geared around extension numbers as legacy PBX systems are.

Extensions "just because"

Some users may request a particular extension because it's the extension number they had on their former employer's system or because it's easy to remember. Other extension numbers may be baggage from generations of phone system upgrades. There may even be a hodgepodge of geography or legacy issues and special number requests cluttering your dial-plan today, before you've even had a chance to add VoIP to the dial-plan mix. Fortunately, VoIP networks can be eminently more programmable than the old PBX system.

Dial-Tone Trunks

Dial-tone trunks are the PSTN pathways of a PBX. Outbound or PSTN-bound traffic flows to the dial-tone trunks, and inward traffic flows across them to the PBX. Traditional dial-tone trunks can be POTS, Centrex, BRI channels, or PRI DS0 channels.

Chapter 12 discusses the impact of voice applications on the selection of PSTN trunk technologies and sizing.

It could make sense to have a very large dial-tone trunk group to make sure the rest of the world never gets a busy signal when calling in to your phone system. But the cost of phone lines—or T1 circuits—adds up quickly. There's an easy way to figure an adequate number of dial-tone trunks for your PBX, whether it's a conventional TDM system or a softPBX. Use the Erlang traffic-engineering method, described here, to balance the cost of trunks against your system's capacity requirements.

An Erlang is a unit of voice traffic. Each Erlang is equal to the average number of calls to a system in an hour times the average duration of the calls in seconds, divided by the number of seconds in an hour—3,600. The Erlang formula, therefore, is e=cs/3600. For a system with 120 calls per hour at 250 seconds apiece, the formula is applied like this:

```
(120 cph x 250 seconds ) / 3,600 = 8.3 Erlangs
```

You can use this formula to calculate the Erlang rating of your phone system. Remember, the Erlang is a unit of voice traffic, not a unit of physical capacity. You'll need to use something called the Erlang B table in order to correlate your system's Erlang rating with a physical capacity number—a trunk count. So, in a nutshell, figure out the Erlang rating and look it up in Table 4-1 to figure out how many trunks your system requires.

Table 4-1. Erlang B table

Trunks required	Blocking 1%, Erlangs	Blocking 5%, Erlangs
1	.01	.05
2	.15	.4
3	.5	.9
4	.9	1.5
5	1.4	2.2
6	1.9	3.0
7	2.5	3.7
8	3	4.5
9	3.8	5.4
10	4.5	6.2
11	5.2	7.1
12	5.9	8.0
13	6.6	8.8
14	7.4	9.7
15	8.1	10.6
16	8.9	11.5
17	9.7	12.5
18	10.4	13.4
19	11.2	14.3
20	12	15.3
21	12.8	16.2
22	13.7	17.1
23	14.5	18.1
24	15.3	19

By looking up the number of Erlangs in the Erlang B table, you can tell how many trunks are required to satisfy several levels of probability for blocking due to busy state.

So for a system with an Erlang rating of 8.3, 15 or 16 dial-tone trunks is enough to support a 1% probability of blocking, while the same system would need only 12 or 13 trunks to support a 5% probability of blocking.

The blocking formula can be used to size private trunk groups, too, but tolerance for blocking on private trunks may be higher. Because blocking of private trunks can often be overcome using the PSTN as a go-between when the private trunks are blocked, a higher probability percentage can be applied to save money. Ultimately, the size of each trunk group depends on how tolerant your telephony applications, and users, are of busy signals.

Think about your use of the telephone at home. Now, let's apply the Erlang formula. Figuring on two calls per hour with an average length of 120 seconds, we get .06 Erlangs:

```
( 2 cph x 120 seconds ) / 3,600 = .06 Erlangs
```

By finding that Erlang rating in the Erlang B table, you can see that this fictitious residential telephone setup will have about a 5% probability of blocking with one POTS line. So, 1 out of every 20 calls will be blocked with a busy signal.

Key Issues: Circuit-Switched Telephony

In this chapter, the structure of traditional voice networks has been dissected. In a nutshell:

- The PSTN provides global interconnectivity and calling features for its subscribers.
- SS7 is the signaling network that runs alongside the PSTN.
- Businesses use PBXs to build their own private voice networks that interface with the PSTN using trunks.
- PBXs use a data-multiplexing technology called TDM to provide connections to digital phones and other switches.
- A single voice call across the PSTN or a PBX requires 64 kbps of bandwidth.
- Analog, one-pair phones use FXS electrical signaling and transmit voice as an analog waveform signal.
- Digital, two-pair phones use digital signaling and transmit voice as a slice in the TDM bit stream.
- Extension numbering and dial-plan are key steps in PBX design.
- The Erlang B table can help you decide how many trunks a phone system needs.

It's important to understand the structure of the PSTN, PBXs, and the connections between them. But the network and its devices—the structure—are merely a means to a much more meaningful end: telephony as a set of applications. Human interaction is the deliverable of telephony, and the network can't deliver without programmed applications running on it.

In the next chapter, we'll examine some of the most important legacy telephony applications: intercom, call-waiting, hold, call transfer, conference call, caller ID, call park, and others.

CHAPTER 5
Enterprise Telephony Applications

In the previous chapter, the technologies of legacy voice network systems were discussed. Some might find that subject fascinating enough to have spent more than a chapter on it. In fact, there are volumes on the subject, and the ITU web site (*http://www.itu.int*) is filled with papers that describe it all in painfully unsparing verbosity.

But it's *telephony*, the application functionality within the voice network, that is the fun part. Telephony accommodates and assists human interaction in a very real, personal way, which is why it's such an engaging subject. Unlike written forms of communication, such as email or instant messaging, telephony's distinguishing traits are its use of sound and its immediate, real-time nature. It's a much more fundamental mode of interaction than the written form—because when we use telephony, we *talk*, the same thing we do when we're together.

Telephony can use live, immediate speech or speech that's recorded, stored, and played back later, depending upon the needs of the application—and it can be largely automated using well-defined standards. In fact, computer-integrated telephony applications have even been programmed to recognize and respond to human voice commands.

Telephony is an extensible, evolving enabler of human communication. Telephony applications such as character terminals have been created for the deaf, and applications such as speaking caller identification devices and Braille terminals have been created for the blind. Developers have expanded call-switching systems to include video features, too.

What all this means is: when you consider the breadth, variety, and social significance of telephony applications in day-to-day life, the details of endpoints and call signaling seem downright mundane. If you want to build a good voice network that enhances the lives of its users, don't lose sight of applications and deliverables because of a technical fascination with clients, servers, data links, and protocols. The users of the system don't care about that stuff anyway. They're a means—not an end.

Application Terminology

Many PBX vendors refer to their telephony functions using words like *actions* or *routes* or *forwarding plans*. Others, the author included, prefer another word: *applications*. This word is appropriate for a modern, soft-driven telephony world. The words *function* and *action* invoke an algorithmic meaning, while *application* is much more broad and programmatic. Traditional TDM systems and VoIP systems have a similar distinction: TDM is monolithic, rigid; VoIP is programmable, flexible.

But whatever the semantics, things the network can do with a phone call are referred to in this book as applications. They've been organized here in five groups: basic call handling, administrative, messaging, advanced call handling, and CTI. What's covered in this chapter is by no means an exhaustive study of telephony apps. Indeed, while I was busy writing this book, new apps were being developed.

Applications themselves are invented out of necessity and driven by market forces more than by the creativity of technologists. When the market demand for certain functionality reaches a boiling point, a vendor or a group of vendors produce a solution to the demand. Sometimes the solution is submitted to the ITU, ANSI, or IETF for recommendation as a standard.

Not all applications are standards-based. Not every feature in a shiny new Avaya Communication Manager PBX is enabled by SIP or some other open IETF or ITU recommendation. In fact, most PBX vendors, even the ones making commercial-grade VoIP servers, deliver a majority of features using their own proprietary methods. Telephony equipment manufacturers are constantly competing on the basis of features, and often the open standards don't support the features the market is looking for. The standards have to be augmented, and this takes time. The 802.3af standard (power over Ethernet) was ratified in 2003, several years after Cisco had begun shipping its own proprietary solution to the need it addressed. The standards bodies may not have a published recommendation for a certain feature until years after the feature itself is made available in proprietary form.

So, while Cisco, Avaya, and Nortel all have rudimentary SIP call-signaling support *functions*, they all offer a much broader selection of telephony *applications* through their own closed, proprietary, or "flavored standard" signaling protocols and software. At the same time, no telephony platform supports 100% of the features of SIP or H.323.

Basic Call Handling

Call-handling features are the essential traffic-processing duties of a PBX and the group of applications that most resembles the services available to a POTS subscriber. Connecting, disconnecting, and call transfer are the building blocks of all call-handling applications.

Intercom Call

Intercom calling is the oldest of all telephony applications. In fact, it was the intercom calling application that lead to the invention of the telephone. This one telephony application came even before the network did. When Alexander Graham Bell's assistant heard the inventor's voice coming through a transducer, quite by accident, the first intercom call was placed.

An intercom call is merely a conversation between two private endpoints. Most intercom calls are two-way conversations, meaning both parties can talk and hear each other talk. In PBX terms, intercom calls are calls between extensions. The setup in Project 4.2 allows you to place intercom calls between the two IP phones.

Mute, Hold, Call Transfer, and Multiparty Conference

Certain features have become more or less staples of a PBX environment. *Mute* allows the local party to temporarily stop sending sound to the remote party. *Hold* is a form of mute that allows the local party to leave the phone set for an extended period of time, unattended, and then return to resume the conversation with the remote party. *Call transfer* allows the local party to hand the call in progress off to another extension, so it can be resumed there. *Multiparty conference* allows the local party to set up conference calls from her phone.

Blind and consultative transfer

A *blind call transfer* is one in which the transferring party hangs up immediately following the transfer, having no knowledge of the availability of the party to whom the call was transferred. A *consultative transfer* allows the transferring party to speak with the receiving party prior to making the transfer or merely verify the receiver's availability before completing the transfer.

Conference

Multiparty conference, or just *conference*, allows more than two parties to have a call, all able to hear one another at the same time. With PBXs, these features are accomplished by proprietary programming. In VoIP environments, packet-based signaling systems like SIP and MEGACO (media gateway control protocol) work with DSP (digital signal processing) servers called *conference mixers* to facilitate this kind of functionality. In the PSTN, SS7 does the job.

Meet-me

Meet-me conferences are ad hoc conferences in which users can voluntarily join a conference in progress by dialing a code on their phones. This code can be permanently set in the PBX on behalf of a certain user, or it can be established on the fly by the host of the conference. Users attempting to join the conference may be required to enter a password or PIN code. Not all PBX systems support this kind of conference.

Caller ID

Caller identification, also called *calling party identification*, allows the recipient of a phone call to know which endpoint is calling both before and after answering. Caller ID signals can be sent using in-band or out-of-band signaling. Some endpoints are able to display the ID information on a built-in display. Some endpoints require an outboard display device. With CTI (computer-telephony integration) programming, caller ID information can be displayed as a part of a PC or web application. SoftPBXs can use caller ID data as criteria for call routing.

Administrative Applications

In addition to call-handling applications, we need applications for performing house-keeping. If nothing else, the traditional telephony providers understood accounting and billing!

Call Accounting

The all-important billing system is a required element of a service provider's soft-PBX. But there's more to call accounting than merely charging customers for the amount of service they use or the number of calls they place. Call accounting includes gauging to whom, when, and to where calls were placed so that utilization of the PBX can be fairly accounted for within the enterprise. Call accounting allows system managers to enforce voluntary system policies, too: abuse of telephone privileges can be documented and so on.

Project 5.1. Analyze Call Detail Records and Call Accounting

What you need for this project:

* Asterisk
* Excel or similar spreadsheet application

Most commercial softPBX systems provide a detailed logging mechanism for keeping track of when and to whom calls were made and received. Asterisk provides this, too. In */var/log/asterisk/cdr-csv/Master.csv*, a flat text log of all call activity is retained. It's a snap to import this into Excel or your favorite spreadsheet for analysis. You can download the file using FTP from your server, or you can run the following command to email it to you. Keep in mind, large logfiles may not work well with this trick:

```
# more /var/log/asterisk/cdr-csv/Master.csv | mail me@mydomain.com
```

Of course, replace *me@mydomain.com* with your email address. If your Linux server has sendmail or a similar SMTP agent running (most do), you'll receive the log in your email. You can then copy and paste it into Excel, as in Figure 5-1. Place the cursor on column A, row 1 before pasting.

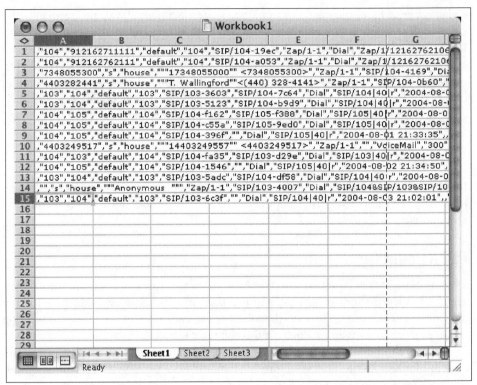

Figure 5-1. A portion of the Asterisk ASCII CDR logfile, copied and pasted into the Macintosh version of Excel; the Windows version will work just as well to handle these logfiles

Once you paste the text or open the file, select column A by clicking the A column heading. Then use Excel's Text to Columns function, on its Data menu. This launches a wizard that helps you organize the text file into columns. You'll see a preview of the text you pasted in the bottom portion of the window that appears. Leave the Delimited radio button selected and then click Next.

Select Comma as a delimiting character, make sure there are no other delimiters selected (as in Figure 5-2), and click Finish. Now, you're ready to label the column headings according to their purposes. Insert a blank row at the top of the spreadsheet, and you can label them as follows.

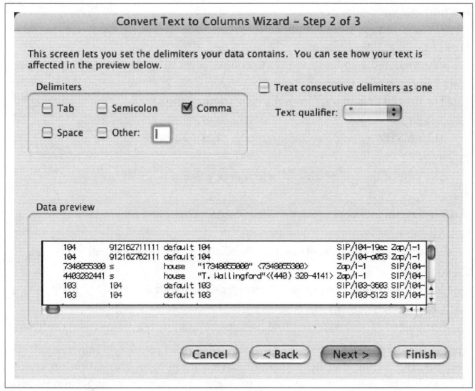

Figure 5-2. The second step of the Text to Columns Wizard breaks up the CDR log text into meaningful cells of data

Asterisk CDR default fields

Account Code
A tag that can be used in billing and analysis

Source
The unique identifier of the endpoint placing the call

Destination
The unique identifier of the endpoint receiving the call

*Context_*
The dial-plan context of the call (more on this later)

Caller ID
The calling party identification signals supplied by the calling endpoint

Incoming Channel
The voice channel that routes to the caller

Destination Channel
The voice channel that routes to the receiver

*Application_*
> The software function handling the call

*Last Data Sent to Application_*
> Information the application uses to connect the call

Start Time
> The time of first contact from the caller to the softPBX

Answered Time
> The time the receiving endpoint answered, if applicable

End Time
> The time of the end of the call, whether or not it was answered

Duration
> The length, in seconds, from the first contact to the end time

Billable Duration
> The length, in seconds, of connected, billable time during the call

Disposition
> The last known status of the call during this application

*AMA Flags_*
> Automated Machine Accounting flags, used by some telephony billing software

The underscored field names record very Asterisk-proprietary information. For example, the Application field may not have a meaningful correlation on another softPBX because not all softPBXs refer to telephony functions as applications. More detailed descriptions of these Asterisk terminologies can be found in Chapter 17.

The idea here is that once the CDR is imported into Excel—or another data analysis tool—you can interpret it in interesting ways. Suppose you want to figure out which customer places the most calls to your technical support department. You can count occurrences of that customer's caller ID in the CDR. Or if your teenage daughter is receiving a dozen calls a day, you can bill her accurately for them!

With CDR records in Excel, Crystal Reports, or even a homegrown Perl program, a savvy telephony administrator can:

- Determine which channels are used the most and the least
- Determine which endpoints are called most often
- Calculate the percentage of outgoing calls that are out of your area code
- Create a list of calls, broken down by endpoint
- Create an invoice for a paying subscriber to the softPBX

 Asterisk's CDR records can also be stored in PostgreSQL or even syslog. Open source software modules have been developed to simplify the use of a third-party database for call accounting.

Console

Most PBX environments support a central, "big picture" application called an *attendant console*. This application can monitor the status of calls in progress throughout the PBX or within a certain scope of users, tell who's on the phone and who's not, and sometimes even keep track of who's in the office and who's out of the office. The console application may or may not have a phone directly connected. In some systems, the console itself is an endpoint with a large, specialized display built in. In others, it's a PC application or terminal connected to a serial interface on the PBX.

In Asterisk, there isn't a dedicated console application, but some people have used Astman as a substitute for a homegrown console application. Cisco, Avaya, and Nortel provide PC-based or web-based console applications for their softPBX systems.

In-Out, DND, and Call Forward

Many traditional PBX systems offer endpoints with an *in-out* function—that is, the endpoint can be used to alert the console when an office occupant is in or out of their office. That way, the human attendant, who uses the console, will better know how to handle that user's calls.

Do not disturb, or DND, is another PBX endpoint function that allows each user to silence the ringer on her phone, placing it in a state of silence so incoming calls don't disrupt the user's activities.

Call forward gives PBX endpoints the ability to temporarily relay all incoming calls to another endpoint, either on the same system or, if the PBX supports it, on the PSTN.

On the public switched telephone network, SS7 facilitates signaling of all advanced calling features, called CLAS (custom local area calling services). These include call-waiting, a form of hold with two line appearances, and three-way calling, a three-party conference call.

Call Logs and Missed-Call Indications

Some PBX endpoints and caller ID–equipped analog phones can store a log of calls placed, received, and missed. This feature tends to be something built into the endpoint, though the same information is almost always stored in the PBX's call accounting system.

Messaging Applications

Telephony messaging applications are one-way in nature, distinguishing them from call-handling applications. That is, a message is sent from one party to another, and unlike a two-way phone call, the receiving party cannot respond in the same medium without initiating a return call. For example, a voice mail message is received. The receiving party must then make a call back to the sender in order to respond, if a response is necessary.

Overhead Paging

Overhead paging is when an endpoint broadcasts the voice of the caller over a large group of endpoints simultaneously. Most legacy overhead paging applications use public address speakers or ceiling speakers wired to a power amplifier. This PA is connected to an analog port on the PBX—the "paging channel"—and the PBX sees the amplifier as an analog endpoint. Some PBX systems permit the use of TDM and analog telephones to be used as broadcast endpoints, too.

Barging

Barging is similar to overhead paging, but instead of broadcasting the caller's voice over a large group of endpoints, the caller can send his voice to a single extension. It's called barging because this feature tends to allow callers to interrupt calls in progress. Some PBX dialects call this feature *announce*.

Voice Mail

Voice mail is one of the most beloved telephony applications. Essentially, it records voice messages on behalf of people who aren't available to answer their calls. Practically everybody uses a form of voice mail, from the old-fashioned cassette tape answering machine all the way up to sophisticated multisubscriber voice mail services.

Some switches have voice mail built in, but most offload the voice mail functions to a dedicated server, usually a PC running a flavor of Unix, Windows, or even OS/2. Messages are stored as digitized sound files when the caller is prompted to speak. Later, the intended recipient plays back the voice mail messages from a phone—either the phone attached to the same switch as the voice mail server or perhaps from a phone on the PSTN that has dialed access to the voice mail server.

Message notification (pager, email, etc.)

Some voice mail servers add the ability to notify each user when she has messages waiting. This may mean signaling the PBX to light an indication lamp on that user's phone, or it may mean causing that user's phone to produce a stuttering, abnormal dial-tone when the receiver is off the hook. In either event, the idea is to notify the user that messages are awaiting an ear.

Some network-aware voice mail servers can send an email to a user when he has messages waiting, or even email the message itself, in the form of a WAV file, for example. Most voice mail servers support some kind of dialed notification as well—

meaning numeric or alphanumeric paging or an automatic agent that calls the user's cell phone to tell, via a recorded voice, that new messages are waiting to be heard. The user can then listen to the messages if he wants.

With some tweaking and a few additional software packages, a Linux-based Asterisk server can be set up to replicate all of these scenarios. In its default configuration, Asterisk can be used to automatically notify voice mail subscribers by email when they've received a call—using sendmail or qmail.

Advanced Call-Handling Applications

Here are some call-handling applications that, while a little more advanced, your users may well insist on having.

Call Parking and Orbit

Call parking is a form of hold and transfer offered on some PBXs. It allows calls to be placed on hold at ad hoc virtual extensions (i.e., parked) until the intended recipient of the transfer can be located. She then dials the virtual extension and is connected with the holding caller.

Orbit adds a bit of functionality to call parking by causing the holding call to eventually ring back at the phone where it was parked to begin with (orbiting) or perhaps ring another phone in a programmed sequential group.

Project 5.2. Set Up Call Parking

Where do calls go when they get parked? Parking zones. In reality, these zones are just extensions that Asterisk sets up and tears down using its built-in parking application. The extensions themselves cannot be used for any other purpose—so, if you define a particular range of extension numbers to be used for parking, those numbers are off-limits as "real" extensions.

By default, Asterisk gives you 20 parking zones, extensions 701-720. The way you park a call in progress is to transfer it to extension 700, and it will automatically be parked in the lowest available parking zone, starting with 701.

To enable call parking, you'll just need to make sure:

- The parkedcalls context is included in the default context in the extensions.conf configuration file: `include => parkedcalls`.
- For later versions of Asterisk, make sure the sample file */etc/asterisk/features.conf* exists and hasn't been modified since you generated it with make samples.
- For earlier versions of Asterisk (0.4 and earlier), make sure the sample */etc/asterisk/parking.conf* exists and hasn't been modified since you generated it with make samples.

The contents of *features.conf* (or *parking.conf*) should look like this:

```
[general]
parkext => 700
parkpos => 701-720
context => parkedcalls
```

Don't forget to issue a "restart now" at the Asterisk CLI.

Try calling into your PSTN interface from a second telephone company line or from your cell phone. When your ring group from Project 4.2 rings, answer the call on one of the SIP phones, then perform a transfer to extension 700. On the Cisco 7960, this is done by pressing the More softkey, then the Transfer softkey, then dialing 700, then pressing the Dial softkey. A spoken voice will tell you which park extension is going to hold the call. Then, press the Transfer softkey again to complete the park. The call with your cell phone will cease on the 7960, but the softPBX is keeping the call on hold, parked at extension 701. Subsequent calls would be parked at 702, 703, and so on.

Now, go get yourself a cup of coffee or a refreshing ice water. When you return, assuming your cell phone hasn't run out of battery power, that call will still be parked. To unpark it and resume it, call 701 from one of your SIP phones.

Put that call in orbit

To add an orbit-style timer to the parking zone, so that the holding call automatically rings back to the parker, add the following to *features.conf*:

```
parkingtime => 120
```

Now, parked calls will orbit back to the parker after exactly 2 minutes. This will help if, in the course of getting that cup of coffee, you forgot you parked a call.

Automatic Call Return

ACR is a function generally ascribed to the PSTN, especially when caller ID isn't available. By dialing a DTMF code on the phone, *69 in many areas, a PSTN subscriber can automatically return the call of the last person to have called his line. (This is a function of SS7's CLAS feature set.) Some PBXs support a similar function.

Hunt Groups and Ring Groups

Hunt groups are logical, sequential groups of endpoints that ring individually when the phone before them in the sequence is busy or unavailable. In a group of three phones, phone B will ring only when phone A is busy, and phone C will ring only when phones A and B are both busy.

Ring groups, conversely, are groups of phones that ring simultaneously until one of them is answered, connecting the call. Then the rest of the phones stop ringing until

---

the next call comes to the ring group. An example of a ring group configuration using Asterisk is given in Chapter 4.

Project 5.3. Set Up a Private Hunt Group

What you need for this project:

- Asterisk
- At least two SIP phones (soft or hard)
- LAN

To configure Asterisk for a sequential hunt group of private phones, let's start with the extension configuration (*extensions.conf*) for the IP phones used in the last couple of examples:

```
exten => 103,1,Dial(SIP/103,40,r)
exten => 104,1,Dial(SIP/104,40,r)
```

Consider that extensions 103 and 104 both ring a SIP phone for 40 seconds each, individually. A hunt group would ring them both, one at a time for a certain duration, in sequence. Here's a bit of dial-plan that establishes a sequential ring group at extension number 100 consisting of SIP phones 103 and 104:

```
exten => 100,1,Dial(SIP/103,10,r)
exten => 100,2,Dial(SIP/104,10,r)
```

So, 100's first priority is to attempt to connect the call on SIP phone 103. It rings that phone for 10 seconds before giving up and going on to its next priority, which is to ring SIP phone 104. It's the second property of the exten declaration that assigns a priority—check out Chapter 17 for the details on the exten declaration in Asterisk.

Add cell phone bridging to this hunt group

It's not just private phones that can be the destination for a sequential hunt group. You can add outside dialing instructions, too. The following configuration adds a third sequence in the hunt group, the dialing of a PSTN phone number. This could be a cell phone, which would have the effect of making you very hard *not* to get hold of when somebody dials your 100 extension:

```
exten => 100,3,Dial(Zap/1/12165241234,40,r)
```

For more tips on using PSTN trunks for nifty bridging applications, check out Chapter 12.

Hold Queues

When more people are calling into a system than there are attendants to answer their calls, those inbound callers must be put on hold until an attendant becomes available to deal with them. The *hold queue* is a call-ordering application that keeps people waiting in line in the same order in which the phone system originally fielded their calls. This way, people who called in after you don't get on the phone with an attendant before you do. Hold queues are most common in call centers where lots of people are answering a very high volume of calls simultaneously. If you've ever called the electric company or insurance company to ask a question about your bill, then you've almost certainly made your way through a hold queue.

Directories

A *directory* application in telephony is the same as a directory in the phone book sense or in the network object sense. If you've ever dealt with LDAP, Novell NDS, or Microsoft Active Directory, then the concept of an online directory should be familiar. In the case of telephony, the directory provides subscribers and inbound callers alike with information about how to reach people on the phone system. This information is presented either on an endpoint's display, in a web-based or PC application, or through audible interactive voice reponse prompts in a phone connected to the host system.

Most telephone system directories allow you to search for a contact's extension by "spelling" her name using the alpha-equivalent dial-pad technique or by choosing from a list of departments in order to narrow a large directory down to a particular contact in several steps. Once found, most directories can dial the contact's extension or phone number and connect the user by transferring the call.

 Chapter 13 spells out the way directories work in telephony applications and shows you how to create an interactive directory on an IP phone's display.

Presence

Presence isn't so much an application as a genre of applications designed to make it easier to find a particular phone user. This may mean ringing his desk phone and cell phone simultaneously or sequentially, or it may mean simply allowing him to notify an attendant when he's available to receive calls. It may also mean publishing information about his availability, or presence, in a directory application. This information might be the user's current location or possibly a list of people from whom he is willing to receive phone calls at the moment.

This is the rough extent of presence applications in traditional telephony. Presence is implemented much more completely in VoIP than in PBX settings, thanks to standards that more capably deal with presence issues. One such standard is SIM-PLE, or SIP Instant Messaging Presence Leveraging Exensions.

Bridging

When a call is *bridged*, it is simultaneously connected to more than one endpoint. This could mean it is bridged between two users on separate phones or bridged between two phones on behalf of a single user. Why would a single user ever want to field the same call on two phones at once? People don't call themselves, after all.

Well, really, it's quite useful. Beginning a call on the desk phone, bridging it to a cell phone, walking out the building and driving away while continuing the call on that cell phone, then returning to the office and resuming the call on the desk phone, all without any disconnection or interruption of the call—that's an obvious "killer app" for bridging. But there are other uses for it—recording calls, for example, requires bridging, because both the person talking and the recording device need a connection to the call. This function is fairly rare among traditional PBX systems, but many who have it swear by it.

CTI Applications

CTI (computer telephony integration) applications combine the telephony functions of a PBX with database or personal productivity functions on a personal computer or web site.

Automated Attendants

Automated attendant applications answer calls using a recorded voice and then respond to caller commands that have been issued through DTMF tones—usually from the buttons pressed on the caller's dial-pad. Automated attendants can transfer calls to private phone system users in the absence of a human console attendant.

Interactive voice response and data collection

IVR is the technology underlying the autoattendant. Because IVR enables customized voice greetings and response handling, it lets callers have much more sophisticated interaction with the system. Callers may be able to enter their responses by dialing digits or, on more advanced systems with voice recognition, by speaking their responses.

Privacy management

One common use for IVR and caller ID is privacy management. It's possible to use CTI applications to recognize and attend to certain kinds of incoming calls automatically, based on their caller ID information. One approach is to have the application greet the caller, record the caller's name, and then ring an internal phone to ask the local party whether she wishes to speak with the person on the recording. If the local party indicates "no," the call is disconnected, but if she indicates "yes," the call is allowed to proceed as normal with both parties connected. This application is offered by most local telephone companies.

 Asterisk has a built-in privacy management application that uses IVR to capture the caller's identity when he hasn't provided a caller ID. This application, `PrivacyManager( )`, is called from the Asterisk dial-plan configuration in */etc/asterisk/extensions.conf*.

Call Centers

Call centers are the telecom equivalent of data centers: a concentration of application resources dedicated to a concentration of users. In English, a whole bunch of phones and telephony processes running in a relatively confined space. Call centers can be vast gardens of telephony apps, complex dial-plans, and high-capacity call-switching equipment.

Insurance claims processing departments are call centers, filled with armies of telephone operators, CTI applications, and friendly IVR bots that take information from customers and facilitate an appropriate response over the phone. When you call an 800 number on the TV shopping channel, you're calling a call center, and it's likely filled with dozens, or hundreds, of telephone operators, normally configured in large inbound hunt groups. Telemarketers and market research companies use call centers in reverse: they primarily make outbound calls; theirs are *outbound* call centers.

Supporting a call center may be one or more PBX systems and special-purpose IVR servers running customized software that has been built for specific customer-service or knowledge-processing applications. Often they include a voice mail system and a well-integrated email system, too.

Key Issues: Telephony Applications

Telephony apps are programmed functions for the voice network. In Asterisk, sometimes commands used in the dial-plan configuration files, including *extensions.conf*, are called, merely, applications. This is somewhat of a misnomer, since the dial-plan commands themselves are just functional pieces of greater overall telephony apps.

Indeed, it's even correct to say that the dial-plan itself is a telephony application, albeit a more complex, more global one. Usually, at least in this book, a *telephony application* refers to a group of functions working together to produce a desirable result, just as all of the applications discussed in this chapter can be achieved by piecing together different software and hardware tools in the VoIP family.

Here are the most important points about telephony applications to remember when replicating them in a VoIP network:

- Telephony applications are distinguished from PC applications because they are centered on direct, real-time spoken communication between people. This is such a natural and organic process that failure or performance problems are never tolerated as they may be with PC apps.
- Most telephony applications stem from the dial-plan—i.e., users access applications as a result of rules established in the dial-plan.
- Intercom calling, conferencing, voice mail, and automated attendant are telephony apps.
- Administrative features such as call logging are telephony apps, too, because they are functionality resulting from the dial-plan configuration.
- Call centers are offices with a large concentration of phones, telephony applications, and very large inbound hunt groups or outbound operators.

CHAPTER 6

Replacing the Voice Circuit with VoIP

In circuit-switched voice networks, every time a call is placed, the network establishes a dedicated pathway from the calling endpoint through the network to the receiving endpoint. This pathway is called a *loop*, and it always uses the same amount of network resources (bandwidth) for the duration of the call, even if nobody is speaking. Fortunately, VoIP provides a way of leveraging packet-switched networks in order to make more economic use of available bandwidth.

This chapter describes the software and hardware elements of the voice loop as it exists in Voice over IP so you can get the most out of it. Indeed, a VoIP admin can do more to improve the quality and economics of his network by "tweaking the loop" than he can by fiddling with any other aspect of the network.

The "Dumb" Transport

In a VoIP network, each loop, or pathway, from caller to receiver is virtualized and controlled using software. So, during times of silence, for instance, the call's pathway doesn't need to utilize a full amount of bandwidth, and the shared resources of the network may be better utilized by another call—or perhaps by another application altogether.

LAN and WAN data links, each capable of carrying TCP/IP, are just systems for moving bits, and the low layers don't distinguish between voice and data traffic, because it's all just TCP/IP packets. This dumb-transport notion allows for changes to VoIP's internal workings for voice digitization and calling features, just as it allows database application developers, for example, to add new functions to their systems. Future changes to the voice network, especially those that increase bandwidth economy, are simplified because the data link and the voice carrier aren't codependent.

Compare that to the trusty old voice T1, which is a rigidly defined data link designed with DS0 voice channels in mind: each channel is fixed at 64 kbps, and there can be 24 simultaneous phone calls per T1.

T1s are prevalent mainly in North America. In many other parts of the world, high-density digital voice links are called E1 and allow 31 simultaneous phone calls.

Of course, all of VoIP's economy and flexibility come at a price: sophistication of infrastructure. Many traditional telephone technicians aren't hip to what TCP/IP brings to the table: disaster survivability, a uniform addressing scheme, data link independence, and better integration with directory services. So, a lot of so-called "phone guys" bash IP telephony, dismiss it as a fad, or even claim that it doesn't work. If you've done the first several projects in this book, then you know how incorrect those claims are—because you've already used a simple VoIP network successfully.

Voice Channels

A VoIP softswitch has two main functions: call management (or switching), which is covered in the next chapter, and voice transmission.

Voice transmission—the packaging, transmittal, receiving, and reconstruction of digitized voice data—occurs inside virtualized pathways across the TCP/IP network. Many softPBX systems, Asterisk included, call them *channels*. The word *channel* does mean different things to different vendors, though. Keep that in mind as you read VoIP documentation. It also means different things at different layers. RTP protocol has *media channels*, which are streams of sound or video data, while a path across the network for call signaling is also sometimes called a *channel*.

For this chapter, we'll give the word *channel* a wide definition: the complete virtualized transport that takes the mouth-to-ear analog signal and transports it over a great distance using networked software.

There are several steps in the process of transmitting voice sounds over a channel: sampling, digitizing, encoding, transport, decoding, and playback. Usually, each step occurs once per packet of voice data. Complex applications like conference call, surveillance, or overhead paging may handle these steps in a unique way, but for this chapter, we'll concentrate on how the process works for a standard, two-party, point-to-point phone call.

Sampling and Digitizing

Digital-to-analog conversion (DAC) and analog-to-digital conversion are the processes that convert sound from the format in which it is heard—analog sound waves—into the format that VoIP uses to carry it—digital streams—and back again. These processes are necessary in order for inherently analog devices—namely, human ears—to use digital sound signals. In the world of traditional telephony, the process is fairly simple, because variations in DAC techniques are driven by requirements of different data links and devices and by regional standards variations.

The DAC processes employed in Voice over IP aren't tied to the data link, so they can vary greatly: different DAC, digitizing, and compression techniques are used in different circumstances. Sometimes the data link's properties, like bandwidth capacity and latency, are factors in the selection of these techniques, but not always. DAC is required in all telephony environments, even where VoIP isn't used, because just about every traditional telephony system employs digital carriers and analog sound reproduction devices like speakers and transducers.

DAC includes of quantizing or digital "sampling" of sounds, filtering for bandwidth preservation, and signal compression for bandwidth efficiency. Pulse code modulation (PCM) is the most common sampling technique used to turn audible sounds into digital signals. We'll deal with DAC subjects in greater detail later.

The 64 kbps channel

To connect a phone call, a traditional telephone, whether analog or digital, requires a loop with enough quality for 64 kilobits per second of digital throughput. In fact, 64 kbps is the fixed line speed of any POTS line. Analog and (most) digital telephone systems offer similar sound clarity because they operate at the same sampling frequency, 8,000 Hz. This frequency, when combined with a sampling resolution of 8 bits, requires 64 kbps of bandwidth—hence the 64 kbps line speed.

It should suffice for now that each concurrent voice conversation—analog or digital—requires a link capable of a speed or bandwidth* of 64 kbps. As you'll discover, the 64 kbps channel is a baseline unit for dealing with sizing issues on your VoIP network. It's important to look at bandwidth conservation methods in relationship to 64 kbps—the "standard unit" of voice bandwidth.

Encoding

Framing

Framing is the real-time process of dividing a stream of digital sound information into manageable, equal-sized hunks for transport over the network. Consider Figure 6-1, which shows a representation of 60 milliseconds of digitized sound. It's divided into three frames, each 20 milliseconds in duration. At this rate, it takes 50 frames to represent 1 second of digitized sound.

---

* *Bandwidth* is not an entirely accurate term to describe link speed, but the industry has accepted it as synonymous with link speed.

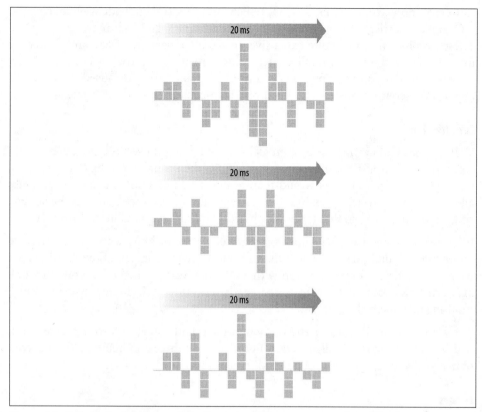

Figure 6-1. Framing is the process of dividing a digital stream into equal-sized hunks

Digital versus packet based

Unlike an analog phone line, the sound signals transmitted and received by an IP endpoint are digital. This makes them more akin to those carried over a traditional voice T1 or ISDN circuit. But unlike a digital phone company circuit, VoIP calls are also packet based. This means that the sound frames are carried across the network in units that are also used to carry other kinds of data—in VoIP's case, UDP datagrams.

Multiplexing

The PSTN offers a way of providing a far-higher call capacity than a single POTS line—24 simultaneous calls using two pairs of wire rather than one pair per simultaneous call. This high-density technology is called T1. It's often used to provide links between PBX systems. For instance, one could use a T1 circuit to link PBXs in separate buildings at disparate locations. T1 provides a far more economical way of allowing many simultaneous calls between users at opposing locations than does POTS. The technique is called *multiplexing*. Even denser multiplexed voice circuits can carry more voice channels: DS3, which supports 672 individual channels, and OC (optical carrier) circuits are also used to multiplex and link between switches. These circuits tend to be quite expensive. In voice applications, DS3s and higher are most likely to show up in call center environments or as trunks between PSTN switches. In data applications, DS3s and OC circuits are often used by ISPs and application service providers that need very high-capacity Internet connectivity.

Compression

VoIP provides an even more economical way of linking those PBXs together. If 100 calls are to occur at the same time using a PBX, then roughly 6 mbps of composite bandwidth is required. This would require five T1s. But VoIP encoding techniques allow for significant compression of the sound sample used to represent the spoken voice in the network, so that far fewer physical links are required in this instance.

It's possible to reduce a 64 kbps voice call down to 44 kbps without a noticeable reduction in sound quality—a feat that, setting aside the concept of overhead (which we'll cover later), is quite common with VoIP compression methods. Now, that link between PBXs uses only 4 mbps, and needs only three T1 circuits instead of five, resulting in a much cheaper trunk.

The algorithms VoIP uses to encode sound data, and sometimes to decrease bandwidth requirements, are called *codecs*. In order to get three T1 circuits to do the work of five in a voice application, bandwidth-conserving codecs are used.

Codecs

Codecs, short for coder/decoders, are algorithms for packaging multimedia data in order to stream it, or transport it in real time, over the network. There are dozens of codecs for audio and video. We'll be talking about audio codecs, since they are most common on VoIP networks.

Most of the codecs in use on VoIP networks were defined by ITU-T recommendations in of the G variety (transmission systems and media). A few are well-suited to a very high fidelity application like music streaming, but most are suitable only to spoken word. They're the ones we'll be concentrating on.

Telephony audio codecs break down into two groups: those that are based on pulse code modulation and those that restructure the digital representation of PCM into a more portable format. So the two groups of telephony codecs are PCM codecs, which are the basic 64 kbps codecs, and *vocoders*, which are the codecs that go a step beyond the essential PCM algorithm. Here are the codecs you'll see most often:

G.711

    This codec is a 64 kbps encoding/decoding algorithm that uses straightforward 8-bit PCM digitization for 8 kHz linear audio monaural signals. In other words, that's a toll-grade telephone audio signal. G.711 is the least processor-intensive codec, and it's the encoding scheme used by most traditional digital telephony circuits, like T1s. G.711 does not provide any compression.

> μLaw and ALaw are two variations of the PCM digitizing technique used in the G.711 codec. One uses a logarithmic digitizing scale to grade amplitude levels, while the other uses a linear one. They are not compatible and must be transcoded if the caller is using one and the receiver is using the other. This isn't likely to be a problem unless you have callers accessing your VoIP network from other continents. μLaw is used throughout North America and parts of the Far East, while ALaw is prevalent elsewhere.

G.721, G.723, G.726, G.728, and G.729A

These codecs enable significantly more economic use of the network, permitting high-quality sound reproduction at a bitrate of 8 to 32 kbps. Unlike G.711, this group of codecs uses ADPCM or CELP algorithms to reduce bandwidth requirements. Adaptive differential pulse code modulation (ADPCM) conserves bandwidth by measuring the deviation of each sample from a predicted point rather than from zero, allowing fewer bits to be required to represent the historically 8-bit PCM scales. CELP, or code excited linear prediction, uses a newer variation of this approach.

G.722

This codec is called a wideband codec because it uses double the sampling rate (16 kHz rather than 8). The effect is much higher sound quality than the other VoIP codecs. Other than that, it's identical to G.711.

GSM

The global systems for mobile codec offers a 13 kbps bit stream that has its roots in the mobile phone industry. Like many of the ITU-recommended codecs, GSM uses a form of CELP to achieve sample scale compression but is much less processor intensive.

iLBC

The Internet low-bitrate codec is a free, proprietary audio codec that provides similar bandwidth consumption and processing intensity to G.729A, but with better resilience to packet loss. iLBC is available from *http://www.ilbcfreeware.org*.

Speex

The Speex codec supports sampling rates of 8 to 32 kHz and a variable packet rate. Speex also allows the bitrate to change in midstream without a new call setup. This can be useful in bursty congestion situations, but is unlikely to matter much to enterprise networks that have quality-of-service measures and more reliability than the Internet. Speex is free, and open source implementations exist.

Each of the codecs has some pros and cons. G.711 is great on data links where there's plenty of capacity and very little latency, like Ethernet. It's also highly resilient to errors. But you wouldn't want to use it on a 56 k frame relay link because there would not be enough bandwidth. Conversely, the codecs that provide compression do so at a loss, or degradation in quality, to the sound. That's why some call them "lossy" codecs.

Codec packet rates

Besides the bits that represent data, all data packets carry bits used for routing and sometimes for error correction. These "overhead bits" have no direct benefit to voice applications, other than allowing the lower levels to function—things like Ethernet headers, IP routing headers, other information necessary for transport of the packet. When longer durations of sound are carried by each packet, these overhead items don't have to be transmitted as often, because fewer packets are required to transport the same sound. The net result of decreasing overhead is that the application uses the network more efficiently. Reducing overhead is crucial, just as it is in a business plan, because overhead, while necessary, provides no direct benefit to the application (or to the business).

The knee-jerk way to lower overhead in a VoIP network is to reduce the number of packets per second used to transmit the sound. But this increases the impact of network errors on the voice call. So there needs to be some balance between what's acceptable overhead and what's acceptable resiliency to errors. This is where a diversity of available codecs can help. Different codecs have different packet rates and overhead ratios—which gives VoIP system builders a way to fine-tune their network's voice bandwidth economy.

The *packet rate* is the number of packets required per second (pps) of sound transmitted. Again, different audio codecs use different rates. The gap between transmitted packets is called the *packet interval*, and it is expressed in *converse* proportion to the packet rate. The shorter the packet interval, the more packets are required per second. Some of the codecs, especially those that use very advanced CELP algorithms, can require a longer duration of audio at a time (say, 30 ms rather than 20 ms) in order to encode and decode. The packet interval has the most obvious effect on overhead. The shorter it is, the more overhead is required to transmit the sound. The longer it is, the less overhead is required.

 A G.711 call, which normally fits neatly on a 64 kbps channel, won't fit so snugly into a 64 kbps IP WAN connection. This is because it is wrapped in RTP and UDP packets, which are necessary overhead. Remember to include UDP and RTP overhead when calculating the capacities of your IP network connections.

But with longer packet intervals comes increased lag (see Figure 6-2). The longer the interval, the longer the lag will be between the time the sound is spoken and the time it is encoded, transported, decoded, and played back for the listener. An IP packet isn't transmitted until it is completely constructed, so a VoIP sound frame can't travel across the network until it's completely encoded. A 30 ms sound frame takes a third longer to encode than a 20 ms one, and inflicts 10 ms more lag, too. As with all networked apps, lag is bad. It's *especially* bad in VoIP.

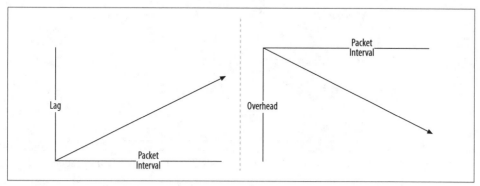

Figure 6-2. Longer packet intervals cause lag, but decrease overhead

Long packet intervals have another drawback: the greater the duration of sound carried by each packet, the greater the chance that a listener will notice a negative effect on the sound if a packet is dropped due to congestion or a network error. Dropping a packet carrying 20 ms of sound is almost imperceptible with the G.711 codec, but dropping a 60 ms packet is quite obtrusive. Since VoIP sound frames are carried in "unreliable" UDP datagrams, dropped packets aren't retransmitted. Even if TCP packets were used instead of UDP, error awareness and retransmission would take so long that, by the time the retransmitted packet arrived at the receiving phone, it would be hopelessly out of sequence.

Consider that 8,000 samples per second are required for a basic voice signal at 8 bits per sample. Now, assuming a 20 ms packet interval (1/50th of a second), you can see that it takes a minimum of 1,280 bits of G.711 data in each packet to adequately carry the sound:

> 64,000 bits per second / 50 = 1,280 bits per packet

Mathematically, increasing the sound data in each packet means reduction of packet overhead. Figure 6-3 illustrates a very simplified cross-section of a VoIP packet carrying 20 ms of G.711 data.

Following the previous example, increasing the packet interval to 30 ms (1/33rd of a second) results in a reduction in the number of packets required per second, raising

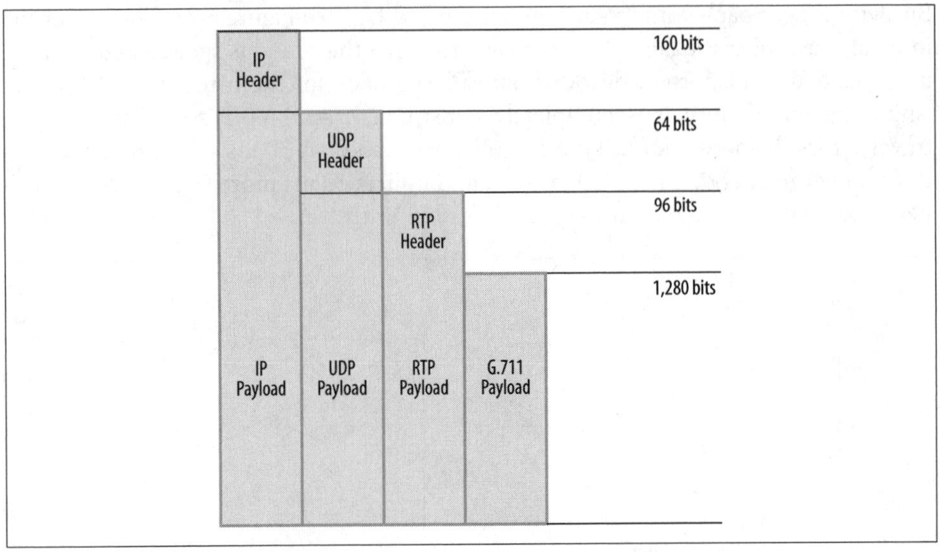

Figure 6-3. TCP/IP adds overhead to a VoIP channel; this IP packet carries 20 ms of sound

the bit count per packet and reducing the amount of overhead required to transmit the sound:

64,000 bits per second / 33 = 1,940 bits per packet

Generally, on Ethernet-to-Ethernet calls, the use of G.711 with a 20 ms packet interval is encouraged, because a 100 mbps data link can support hundreds of simultaneous 64 kbps calls without congestion, and a dropped packet at 20 ms interval is almost imperceptible.

On calls that cross low-bandwidth links, it's up to the administrator to balance between latency, possible reductions in sound quality incurred by using a compression codec, and network congestion.

Up until this point, we've been talking about the overhead of each packet merely as it relates to the amount of voice payload it carries, and with good reason: codec selection and framing are something over which you, as an administrator, have the most control.

But packet overhead is affected by the network and data link layers, too. Ethernet frames have a different size and different overhead than ATM cells or frame relay frames. Network overhead is addressed in the following section.

Different codecs have different bandwidth requirements. Table 6-1 shows the characteristics of the most popular VoIP codecs.

Table 6-1. *VoIP codec characteristics*

Codec	Algorithm	Bandwidth used for sound	Packet interval	Voice bits per packet	Processing intensity
G.711	PCM	64 kbps	20 ms	1,280 bits	Low
G.726	ADPCM	32 kbps	20 ms	640 bits	Medium
G.728	CELP	16 kbps	10 ms	160 bits	High
G.729A	CELP	8 kbps	10 ms	160 bits	High
GSM	RPE[a] or CELP	13 kbps	20 ms	160 bits	Depends upon algorithm used; CELP is higher

[a] Regular pulse excitement, yet another sound encoding/compression technique.

Transport

The T1 carrier versus VoIP

The T1's 24 DS0 channels are each never-ending streams of digitized voice information. In reality though, the T1 circuit itself is one big stream of binary digits that uses TDM to divide the T1 into those 24 DS0 channels. Each is assigned a time slice of the big stream, and each time slice is further divided into frames, as shown in Figure 6-1. All voice T1s use the same amount of bandwidth no matter how many calls are in progress—roughly 1.54 mbps. Trunking with T1s is very stable and predictable as a result.

VoIP frees the system builder from requirements traditionally imposed on the lower OSI layers of the network. In a T1, the transport and data link layers are defined together as a bundled carrier, and you have to use the G.711 PCM codec on all the channels, yielding 24 simultaneous voice channels in the available bandwidth. VoIP lets you pick and choose the codec, packet interval, and transport technologies you want and thus gives you ultimate control. Using the G.729A codec and a T1, you could conceivably trunk hundreds of calls at once.

Unlike G.711 traffic on a T1, VoIP's "carrier" is TCP/IP. So VoIP can traverse Ethernet, T1s, DSL lines, cable internet lines, POTS lines, frame relay networks, virtual private networks (VPNs), microwave radio, satellite connections, ATM, and just about any other link. If IP can go there, VoIP can go there—just with varying levels of quality.

All VoIP systems to date use equal-sized frames of the same fidelity during each call. You'll never have a sample rate of 8 kHz change to 12 kHz during a single session the way you could with a variable-rate MP3, for example.

Voice packet structure

The layered appearance of a VoIP packet is similar to that of other types of networked applications that run within the TCP/IP protocol: the lower layers encapsulate the higher layers recursively.

The lowest layer, shown leftmost in Figure 6-3, is the Internet Protocol (IP) packet header. It contains routing information so that the packet can be handled correctly by the devices responsible for carrying it across the network. It also contains a flag that indicates which protocol of the TCP/IP suite this packet is carrying: TCP, UDP, or something else. Voice packets are almost always UDP. Among other things, the IP header may include a Type of Service flag that allows routers and switches to treat it with a certain priority based on its sensitivity to delay. At a minimum, the IP packet header is 160 bits in length.

The payload of the IP packet is the UDP packet, whose header is 64 bits long. Its first 32 bits contain the source and destination ports of the UDP traffic it carries in its payload, along with 8 bits for optional error checking and 8 bits for describing the length of its payload in multiples of 8 bits.

The payload of the UDP packet is the RTP packet, whose header is 96 bits long. It contains information about the sequence and timing of the packet within the greater data stream.

Real-Time Transport Protocol

The Real-Time Transport Protocol (RTP) defines a simple way of sending and receiving encoded media streams[*] in connectionless sessions. It provides headers that afford VoIP systems an easy way of discriminating between multiple sessions on the same host. Remember that the codec merely describes how the digitized sample is encoded, compressed, and decoded. RTP is responsible for transporting the encoded sound data within a UDP datagram. RTP was designed for use outside the realm of telephony, too: streaming audio and video for entertainment and education are common with RTP.

RTP supports mixing several streams into a single session in order to support applications like conference calling. It doesn't, however, provide adequate controls for defining multiplexed voice pathways that are normally associated with telephony, like trunks. This is the responsibility of the softPBX and its signaling protocols.

---

[*] When certain signaling protocols, such as IAX, are used, RTP may not be employed to packetize the voice stream. IAX has its own built-in voice-packetization capability.

Control of RTP's media sessions, and collection of data relevant to those sessions, is accomplished by RTP's sister, RTCP (Real-Time Transport Control Protocol). Together, RTP and RTCP provide:

- Packetizing and transport of digitized, encoded voice or video signals, including unique identification of each RTP stream
- Multicast sessions for conferencing applications
- Basic performance feedback about the utilization of RTP media sessions

For the VoIP administrator, RTP is largely invisible. Most VoIP frameworks and system-building tools, including Asterisk and Open H.323, implement RTP so seamlessly that the administrator rarely has to worry about its inner workings. If you are interested in RTP, check out Internet RFC (Request for Comments) 1889, published by the IETF's Network Working Group.

As shown in Figure 6-3, the only part of each VoIP packet *not* considered overhead is the payload of the RTP packet, which is encoded sound data.

Ethernet

Ethernet is a physical and logical data link specification that makes provisions for error-correction on locally connected network devices. Ethernet packets, called *frames*, are typically less than 1,500 bytes, or about 12,000 bits. VoIP packets are very rarely larger than 250 bytes, or 2,000 bits. Not accounting for Ethernet overhead, the packet in Figure 6-3 is 1,600 bits long, a rather small packet.

Like RTP, UDP, and IP, Ethernet adds some bulk to each packet. The overhead Ethernet imposes is 176 bits for its header and 128 for its CRC "footer"—a bumper at the end of each Ethernet frame that provides an error-detection mechanism used by network interfaces on participating hosts. Figure 6-4 shows an Ethernet VoIP frame.

The total size of a G.711 Ethernet VoIP frame is 1,904 bits. At a standard packet interval of 20 ms and 50 pps, a voice call digitized using plain-vanilla PCM at a rate of 64 kbps consumes a healthy amount of overhead—specifically, 15.2 kbps of Ethernet overhead and 16 kbps of combined RTP, UDP, and IP overhead.

When you add in the payload to all that overhead, an Ethernet-transported voice channel using the G.711 codec requires 95.2 kbps of bandwidth.

While a G.729A voice channel requires only 8 kbps of bandwidth to frame the sound stream, the overhead of IP, UDP, RTP, and Ethernet adds 31.2 kbps, putting the total bandwidth consumption of a G.729A call at 39.2 kbps. Table 6-2 shows the total Ethernet bandwidth consumed by several of the most popular codecs.

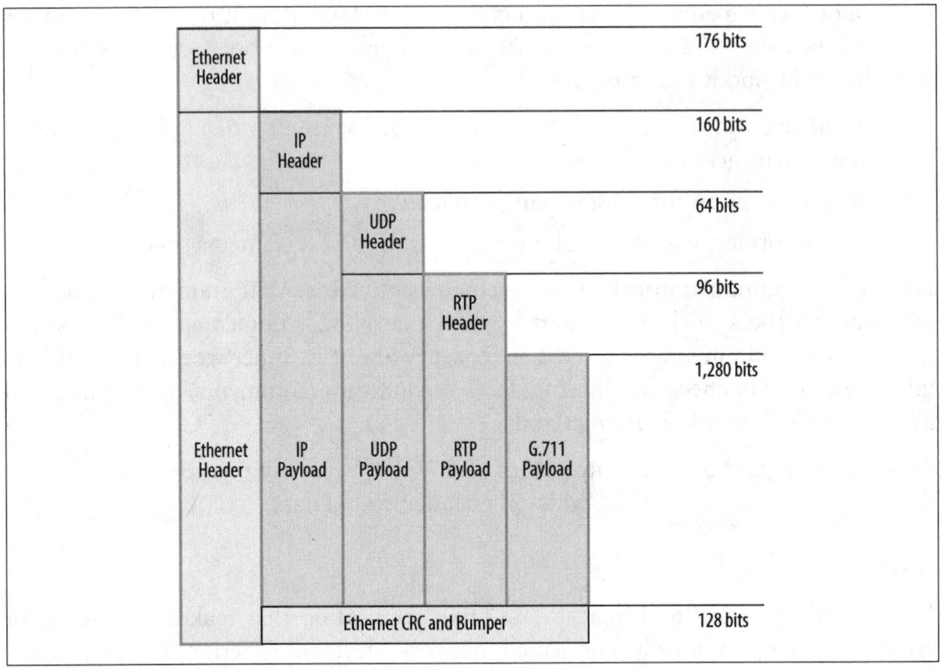

Figure 6-4. An Ethernet-encapsulated 20 ms VoIP packet

Table 6-2. VoIP codec bandwidth consumption

Codec	Encoded sound bandwidth	Ethernet overhead bandwidth	Total bandwidth
G.711	64 kbps	31.2 kbps	95.2 kbps
G.726	32 kbps	31.2 kbps	63.2 kbps
G.728	16 kbps	31.2 kbps	78.4 kbps[a]
G.729A	8 kbps	31.2 kbps	39.2 kbps
GSM	13 kbps	31.2 kbps	44.2 kbps

[a] G.728 uses four voice frames at 16 kbps per packet. This accounts for the deviation in overhead bandwidth.

Ethernet isn't the only data link suitable for carrying VoIP packets—ATM, frame-relay, point-to-point circuits, and other technologies can be used, and each introduces its own overhead factors.

Decoding and Playback

When a VoIP packet is received, it is decoded according to the codec employed to encode it. It is then played back on the analog hardware of the receiving endpoint—a speaker—while undergoing DAC, or digital-to-analog conversion. Decoding generally takes about as much processing power as encoding, depending on the codec employed.

G.729A and Asterisk

Digium's implementation of the G.729A codec for Asterisk is a licensed commercial version made by Voice Age (*http://www.voiceage.com*). The codec is patent-protected, so if you want to use G.729A endpoints with the Asterisk server, you must pay for a commercial license. Digium sells the licensed, and GPL-friendly, version of G.729A at the price of *around* $10 per simultaneous call. Using unlicensed versions of G.729A (which do exist) violates the GPL under which Asterisk is distributed, because the GPL requires end users to adhere to local patent law, by which G.792A is governed in several nations.

Most IP phones and ATAs support several codecs, as shown in Table 6-3. All support G.711 using both the μlaw and Alaw scales, and a majority support G.729A, though with variance in quality and completeness of their implementation. It's fair to say that G.711 and G.729A are the two most popular VoIP codecs in use today.

Table 6-3. Codecs supported by some leading VoIP endpoint devices

Phone / ATA	G.711	G.726	G.728	GSM	Speex	G.729A
Cisco 7960 IP Phone	Yes	No	No	No	No	Yes
Avaya 4602 IP Phone	Yes	No	No	No	No	Yes
Grandstream Budgetone 101 IP Phone	Yes	Yes	Yes	No	No	Yes
Digium IAXy i100 ATA	Yes	Yes	No	No	No	No
Grandstream Handytone ATA	Yes	Yes	Yes	No	No	Yes
Cisco ATA-186 ATA	Yes	No	No	No	No	Yes
3com 3102 IP Phone	Yes	Yes	No	No	No	Yes
SNOM 220 IP Phone	Yes	No	No	Yes	No	Yes
X-Lite Softphone	Yes	No	No	Yes	Yes	No

Things that degrade playback quality

Several factors can degrade the quality of audio transmitted over the network:

Jitter
> This effect occurs when gaps between packets occur at durations greater than the packet interval. The effect is missing or garbled speech. Jitter can be caused by network congestion or processing overloads on the encoding or decoding endpoints (this usually doesn't happen on dedicated devices, only on softphones). Some IP phones and softPBXs offer a jitter buffer to compensate for packets arriving out of sequence or at odd intervals, but jitter buffers introduce lag—and lag is bad.

Lag (also called latency)

The time it takes from the moment the caller speaks until the moment the listener hears what was spoken; the longer the lag, the more difficult the conversation becomes. On the PSTN, lag tends to be around 15 ms, or practically imperceptible. On Ethernet, packetization overhead, varying by packet interval, introduces at least 20–50 ms of lag. Over slow links like a frame-relay network, where larger packet intervals are required to conserve bandwidth, lag can be downright annoying. Lag as great as 120 ms is common on slow frame-relay links, for example. Using VoIP across a virtual private network can be even worse. Compound that by adding all the laggy cell phone networks in use today and you could be approaching a half-second or more of lag.

Packet loss

Excessive packet loss will kill a VoIP network. Packet loss is a fact of life, though, especially in Ethernet networks, and even the most well-intentioned network engineer will only be able to minimize, not completely eliminate, it. That being said, highly compressed codecs like G.729 are at the greatest risk of quality degradation due to packet loss, because their packet intervals tend to be larger; plus, their predictive algorithms break down quickly when robbed of voice frames. Packet loss is caused by network congestion, malfunctioning equipment, and processing overload.

Transcoding

In Project 3.1, a SIP endpoint was used to connect through the Asterisk server to a demonstration server on the Internet via the IAX signaling protocol. Though the voice channel from the SIP phone to the Asterisk server was encoded using G.711 law, the cross-Internet voice channel to the demonstration server was encoded using GSM, as shown in Figure 6-5.

Figure 6-5. A call path that uses two codecs

When a call path requires it to use more than one codec—different ones on the two endpoints of the call, as in the case of Figure 6-5—the softPBX or another specialized server device called a gateway must transform in real time, or *transcode* each leg of the call. Certain connectivity mediums don't provide enough bandwidth to facilitate G.711 from end to end. A 64 kbps circuit, for example, can't carry a G.711 call, because it requires more than 64 kbps when IP packet overhead is accounted for. So bandwidth-conserving codecs have their uses, but not all endpoints support every codec. Transcoding is the solution.

Transcoding is a processing-intensive task, so it's a good idea to minimize the number of codecs that you support as standards on your network. A few conference calls with three or four codecs apiece could be a real handful. Cisco and other commercial vendors recommend G.711 for local calls over Ethernet, and G.729A over low-bandwidth WAN connections. The softPBX will insert itself into the path of the call in order to negotiate the appropriate codec on each leg and then perform transcoding.

Call paths

Even though the softPBX is the central call-management and -signaling element on the VoIP network, it doesn't always sit in the call path. One of the purposes of SIP, and other signaling protocols, is to allow endpoints to discover what codecs their peers support, so that, when beginning a call, both endpoints can be using the same one. Another purpose of SIP is to allow for multiple pathways through the voice network based on the capabilities of each endpoint and the preferences of the administrator. These pathways are known as call paths.

For example, an IP phone can place a call through the softPBX, and the softPBX can act as a proxy for the sound signals, receiving them from that caller and sending them milliseconds later to the receiver. In this case, the softPBX may or may not be transcoding, but it is a point in the call path. You could call this softPBX call path or a proxied call path.

But an IP phone placing a call mustn't always have its call path cross through the softPBX. In fact, in most commercial VoIP softPBX implementations, this isn't the preferred method. Indeed, with Cisco's CallManager, it isn't even possible out of the box, with exceptions for a few centralized applications like conferencing, music-on-hold, and bridging. In these setups, the softPBX sets up the call using a signaling protocol, and then the phones themselves communicate the sound data directly to each other in UDP bursts. The big advantage of an independent call path is that there's less processing load incurred on the softPBX. One disadvantage is that it's impossible to run centralized applications that deal with the sound stream in the call, like, say, a clandestine call-recording application.

 Independent call paths are sometimes called direct paths. Call paths are the paths of the media channels used in the call—*not* the call-management signals. In most setups, the call signals tend always to pass through the PBX.

When transcoding is employed, the call path always crosses the softPBX or another gateway device that speaks all the necessary codecs. Fortunately, transcoding tends only to be used when a medium other than Ethernet is being used for connectivity, and a codec besides G.711 is employed for that leg of the call path.

So, the call path determination is affected by several issues:

- Do the endpoints involved in the call all support a common codec? Yes indicates an independent call path. No indicates a softPBX call path because transcoding is needed.

- Is that common codec sufficiently bandwidth conservative to be used across all data links involved? Yes indicates an independent call path. No indicates a softPBX call path because transcoding may be required for one or more endpoints.

- Is there a centralized application, such as a call recorder or conferencing server, that must be in the call path? Yes indicates a softPBX call path. No indicates an independent call path, in order to minimize processing load on the softPBX.

- Is there an IP firewall between the caller and the receiver that will interfere with the voice transmission? Yes indicates a softPBX call path that circumvents the firewall by the administrator's design. No indicates an independent call path. (Firewall problems are covered more deeply in Chapter 13.)

Commercial vendors support automatic selection of a call path to varying degrees. As indicated earlier, some don't support a softPBX call path at all, unless its purpose is to deliver conferencing applications and not necessarily for transcoding. Others support automated negotiation of the call path during call setup signaling.

Asterisk falls into the latter group. It allows either kind of path for SIP calls (except a conference call) according to the administrator's design. Project 6.1 describes how to enable an independent call path using a SIP feature called Reinvite.

Silence suppression and comfort noise generation

When nobody is speaking, there's a great opportunity to save bandwidth, because during periods of silence, no sound data needs to be transmitted over the network, right?

Several codecs have taken this idea to heart. GSM, G.723.1, and others support *silence suppression*, a technique that suspends the packet stream during periods of silence. In order to create a seamless experience for the person listening to that silence, silence suppression is usually accompanied by comfort noise generation, or a small amount of white noise. This white noise is created by the endpoint of the listener, rather than being transmitted to her over the network.

For a good demonstration of silence suppression, try the peer-to-peer softphone Skype (*http://www.skype.com*). Before making a call with Skype, shut down all other networked applications on your PC. Then, while placing a Skype call, watch as the traffic load to and from the PC all but stops during times of silence, but notice the white noise you hear. During these times, the white noise is simulated by your Skype client.

Project 6.1. Set Up Custom Codec Selection and Enable an Independent Call Path

What you need for this project:

- Asterisk
- Two or more hard or soft SIP phones
- LAN

Within Asterisk, it's possible to set preferences for each SIP channel as to which codecs should be used, or allowed. This can be a necessity if a particular channel is connected to the server using a low-bandwidth link. It would be prudent to limit that SIP channel to codecs that preserve bandwidth. Figure 6-6 shows how calls across a WAN might use G.729A, a low-bandwidth codec, while local Ethernet-based calls might use G.711.

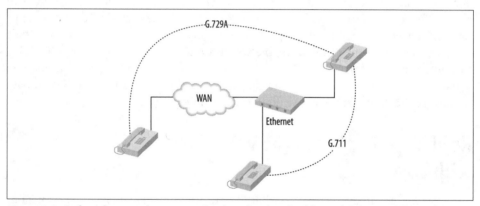

Figure 6-6. Codec selection schemes are often based on bandwidth availability on different data links

Codec selection is a function of the call setup routine, handled by SIP, H.323, or a proprietary signaling protocol, but there are varying degrees to which you can tune the process. Depending on vendor, the softPBX has some control over codec selection on a phone-by-phone basis. Some VoIP vendors call this *per-peer* selection.

Per-peer codec selection on Cisco media gateways

Cisco's media gateway devices, themselves just simple VoIP softPBXs, allow the definition of endpoint and trunk peers using SIP and H.323 signaling configurations. One of the parameters that can be assigned to each peer is a codec preference.

Per-peer codec selection on Asterisk

In Asterisk's *sip.conf* configuration, it's possible to set an order for codec preferences on a per-peer basis. Consider the following snippet of a *sip.conf* file:

```
[103]
type=friend
context=default
username=103
callerid=103 Budgetone Phone
canreinvite=no
disallow=all
allow=ulaw
allow=gsm
```

This definition, for a SIP peer called 103, allows only G.711 μlaw and GSM codecs to be employed. Asterisk prefers the codecs in the same order specified in each SIP peer definition.

What this sample SIP peer *cannot* do is have an independent call path. In order to allow SIP peers to operate with independent call paths, the canreinvite setting must be yes. This enables SIP phones to establish a direct media channel with the phone on the other end of the call, cutting the softPBX out of the proverbial loop. Enabling this option doesn't force the SIP phones to use an independent call path, though: in situations in which transcoding is needed or the softPBX is required in the call path because of the design of the network, Reinvite won't get used.

Key Issues: Replacing the Voice Circuit with VoIP

- The two fundamental duties of a softPBX are call signaling and voice transmission.
- Voice channels in a VoIP network are analogous to loops or circuits in a traditional voice network.
- Voice transmission occurs across virtual channels on an IP network, most frequently using UDP packets because they have less overhead than TCP packets.
- Codecs are algorithms that digitize and package sound and/or video for transport across the network.
- The most common codec for LAN use is G.711.
- G.711 uses one of two digitizing scales: μlaw or Alaw. μlaw is the standard in North America. Each requires 64 kbps of bandwidth.
- As a rule, TCP/IP and Ethernet add 32 kbps overhead to the bandwidth required by the codec.

- Selection of the codec for each call is accomplished by the signaling function of the softPBX or by direct negotiation between two endpoints at the beginning of that call.

- Transcoding is the real-time translation of one codec or another. This may be necessary when two endpoints cannot use the same codec or when a conferencing application dictates the use of more than one codec to support multiple endpoints.

- Different codecs offer different levels of bandwidth economy.

- The top causes of perceived quality problems in voice transmission are jitter, lag, and packet loss. Generally, the more bandwidth-conservative a codec is, the more prone to noticeable quality breakdowns caused by these issues.

- A call path is the chain of virtualized voice channels used to connect a call across the network. Each channel may employ a different codec.

- A call path that is established directly between two endpoints, without a soft-PBX, proxy, or transcoding element in between, is an independent call path.

- Independent call paths result in lower processing intensity because the softPBX doesn't have to handle the audio. Complex, highly compressed codecs and soft-PBX call paths result in higher processing intensity.

## CHAPTER 7

Replacing Call Signaling with VoIP

In Chapter 6, the process of transmitting voice sounds in virtual media channels, via codecs, was presented. In order for those media channels to be set up, monitored, and destroyed when needed, a PBX calls on signaling. Different methods are used for different kinds of endpoints and trunks. On the PSTN, the SS7 network handles signaling. On a POTS voice channel, the signaling is accomplished using analog FXS signaling.

SS7, FXS, and the dozens of other signaling technologies in use on the PSTN, though all signaling protocols, are outside the realm of VoIP. They could all be considered legacy technology, since just about all of their signaling functions have been replicated using several new, modern, open TCP/IP-centric standards. Even though SS7 is a packet-based protocol and there are attempts underway to make it compatible with VoIP softPBX systems (Asterisk included), its roots are in the PSTN, not the Internet.

This chapter describes the standards for call signaling in a softPBX-based VoIP network; it also describes the ways these standards compete with and complement one another.

VoIP Signaling Protocols

A signaling protocol is a common language spoken by telephones and call-management servers, the PSTN, and legacy PBX systems as they communicate to set up, monitor, and tear down calls.

The Voice over IP technology family provides several signaling protocols. Some commercial softPBX systems support one or two of them. Others, like Asterisk, support just about all of them. Most signaling protocols have a few things in common:

- Their purpose is to signal, record, and facilitate key events in a call: the beginning, the end, and when the users are attempting to use telephony features like transfers or conference calling.

- Though call signals are usually sent using UDP, they aren't viewed as real-time traffic like the media channels, which also use UDP.

- The traffic patterns they incur on the network are short and bursty, as opposed to media channels, which tend to be consistent and lengthy.

- They tend *not* to be supported simultaneously on any single IP phone. There are SIP phones, and there are H.323 phones. But there are none that do both.

- Most are available in free implementations such as Asterisk, GnoPhone (an open source softphone), OpenH323 (an open source softPBX), VOCAL, and others.

There are two primary contenders in VoIP call signaling: the Session Initiation Protocol, developed by an Internet Engineering Task Force working group, and H.323, developed by an International Telecommunications Union working group. Other signaling standards, like Cisco's Skinny Client Control Protocol (SCCP), Nortel's UNISTIM, and Digium's IAX, have been developed by private companies intent on bringing features to call signaling that don't exist in the published implementations.

VOCAL is an open source suite of VoIP software. It implements a softPBX with support for lots of codecs and telephony apps, such as Asterisk. But it also provides the building blocks for open source quality-of-service measures and load-balancing VoIP servers. More information on VOCAL can be found in O'Reilly's *Practical VoIP Using VOCAL*.

The result is a vast, sometimes confusing, assortment of signaling systems that have about a 75% feature overlap. So, with SCCP, MGCP, MEGACO/H.248, IAX, H.323, SIP, and a half-dozen others, is it really worth it to educate oneself about all of them?

Not really. Carrier-grade telephony technicians and engineers may have a reason to learn the ins and outs of MEGACO/H.248, for example, but only the most detail-obsessed enterprise integrator would bother with it. The people who work at Nortel and Cisco who design media gateways for a living need to know all of these protocols—but for the rest of us hobbyists, system administrators, and enterprise telecom managers, two have the biggest potential impact: H.323 and SIP.

The reason these are the two most important signaling protocols you'll use in your VoIP network is that they operate in the realm of the private. That is, H.323 and SIP operate on *your* side of the telephone company demarc. You can control them, leverage their features, monitor them, and exploit their benefits to your heart's content. This makes your use of them immune to potential regulation.

H.323 and SIP provide a common signaling dialect for endpoints, gateways, and PBXs. They signal for trunks and phones. Administrators of a VoIP network need to understand H.323 and SIP, because they really are the roots of VoIP and the harbingers of voice protocols to come.

This isn't always true of the other protocols, MEGACO/H.248 and MGCP, which are seen more often in the realm of the equipment vendors and the telephone company.

These protocols don't tend to be as practical when used as endpoint-signaling protocols as they are when used as trunk and gateway ones. When an administrator needs to alter the configuration of a MEGACO device, he may have a tendency to "set it and forget it," because changes to trunks and gateways are typically rare compared to endpoint changes. This isn't to say that there aren't any MGCP-based IP phones, because there are. But the realm of IP phone and softPBX signaling favors H.323 and SIP.

The day-to-day administration required for enterprise VoIP leans more heavily toward the endpoint-to-endpoint, not gateway-to-gateway, communications, just as in traditional telephony. As a result, most enterprise administrators will spend the bulk of their time working with H.323 or SIP devices.

H.323, SIP... How Do I Choose?

More often than not, if you're using a commercial VoIP platform to build your network, the choice will already be made for you. If you're using a Cisco CallManager softPBX, then you probably aren't going to be using SIP, because CallManager's SIP support is far from complete for even mundane telephony apps. (CallManager uses Cisco-proprietary SCCP signaling instead.)

Regardless of how well each vendor claims to support the standards, there will be certain standards not supported by every vendor. If you standardize on Avaya Communication Manager or Nortel Meridian, you aren't going to use Cisco SCCP. The point being: most commercial vendors offer full support for a single standard and partial support for a second standard. The aim of this strategy is to promote signaling features that are proprietary and unique in nature, while still allowing interaction with other vendors' softPBX systems by way of trunking. So, while Cisco's unique features have been implemented in its proprietary SCCP protocol, its softPBX still allows trunking to other devices by way of H.323 and SIP.

Sometimes, lack of support for a particular signaling standard poses a problem. In a 10-person office in which it's all right for every user to be equipped with the same model of phone, interoperability probably isn't a big need. But suppose that 10-person business is purchased by another business that uses a different signaling standard and management wants to integrate the phone systems. Now, interoperability suddenly matters.

If interoperability is more important to you than anything else, you're better off implementing Asterisk or Vocal, because they are built around published standards and support them with the overt intention of compliance, whereas the commercial vendors don't always support the complete spec of a given standard, only the parts that matter to them. Of course, even in PC applications and programming, everyone aspires to compatibility and then slowly sacrifices it in order to add features.

The best thing you can do is understand the key signaling standards and make an informed decision that weighs your technology preferences, your budget, and your organization's politics. If a little back-rubbing at the executive level has your management team really fired up about a commercial phone system, then all the justification in the world in favor of Asterisk probably won't change their minds. So you've got to be prepared to support the *standards,* rather than the *brands,* that make everybody in the organization happy: give the users the features they want, and get yourself the raise you want.

Putting aside which vendors support which standards, there are some real, substantial differences between the main signaling standards, as described in Table 7-1. The biggest distinguishing factor between signaling protocol families is in the kinds of call paths they were designed to handle. H.323 makes provisions for switch-to-switch, PSTN-to-switch, and endpoint-to-switch call paths, meaning it supports interfacing with traditional telephony systems, especially the PSTN. Comparatively, SIP is much more limited in network scope. It doesn't support any legacy (analog or TDM) components and was designed only with an end-to-end IP network in mind.

Table 7-1. VoIP signaling protocol families[a]

Protocol families	Intended signaling scenarios	Maintainer	Built-in legacy device support
H.323	Telephony, video	ITU-T	Yes, via gateways
SIP	Telephony, instant messaging, video	IETF	None recommended
IAX	Telephony	Digium, Inc.	None recommended
SCCP	Telephony (switch-to-endpoint)	Cisco Systems	None recommended
MEGACO/ H.248	Telephony, gateway control	IETF and ITU-T	Yes
MGCP	Telephony, gateway control	IETF	Yes

[a] There are other telephony signaling standards, but these are the ones most relevant to a VoIP network and also the most popular ones.

Then, there are protocols that are gateway and server oriented, such as MGCP and MEGACO. SS7, too, is a "backend" protocol. These are probably of the least concern to an *enterprise* administrator, while a carrier-class system builder or engineer might really prefer one or the other.

Signaling for non-telephony apps

One big advantage of SIP is its comfort in non-telephony applications, such as instant messaging, video, and white-boarding. Apple's iChat and Microsoft's MSN Messenger use SIP for these purposes. Earthlink's SIPShare application allows peer-to-peer media sharing (a la Kazaa) using SIP. Extensibility is an appealing characteristic of SIP. Lack of extensibility is the drawback of H.323.

H.323

H.323, currently in Version 2, is the ITU-T's recommendation for PBX-styled signaling that supports a packet-based network. H.323 doesn't have to be delivered entirely using an IP network. Certain subrecommendations of H.323 allow for traditional telephony networks to be integrated with seamless signaling across all participating devices. For example, the H.323 suite allows for signaling over POTS on the PSTN using recommendations in H.320 and H.324.

 H.323 was originally proposed as a solution for video conferencing using LANs. Later revisions saw it morph into a full-fledged PBX-replacement plan.

H.323 is implemented in Microsoft's NetMeeting conferencing software, as well as in Avaya's MultiVantage and Nortel's Meridian IP-PBX systems. While quite mature and well-documented by the ITU-T, H.323 has been implemented in vendor-specific "flavors" that are, unfortunately, not totally interoperable. This hasn't been a big hindrance to acceptance of H.323 for VoIP applications, however, since most integrators who choose this standard tend to choose a single vendor's implementation across their entire network.

Incompatibility of H.323 implementations is a problem only when trying to link opposing vendors' H.323 systems together. To get around this, many integrators link the systems using legacy trunking methods like T1, because opposing vendors' implementations of legacy telephony protocols are almost always totally interoperable.

H.323's message packets are compact, and H.323 signaling is very fast, especially compared to SIP, which uses comparatively long and wordy message packets. The design of H.323 is very much rooted in the design philosophy of the PSTN: brevity and availability are striven for. H.323 signals are short. The network is used as little as possible to carry call signaling and as much as possible to carry sound.

H.323 Gatekeeper

The *gatekeeper* is the host on the network that provides centralized call monitoring and signaling capabilities for H.323 endpoints (called terminals). The gatekeeper's scope may be a particular LAN segment or an entire continent, depending on the designer's intentions.

The scope of the network that a gatekeeper operates within, or the *reach* of that gatekeeper, is called a *zone*. IP endpoints may be on the table next to the gatekeeper or on a different continent, 15 router hops away, and all can be considered one zone. This isn't always practical, of course, just *possible*. There can be only one gatekeeper per zone and one zone per gatekeeper. It is common to refer to an H.323 gatekeeper as a softswitch or softPBX.

Registration

In order to be made available for telephony applications, each endpoint and/or gateway in a gatekeeper's zone must undergo a process called *registration*. This means that each H.323 endpoint must inform the gatekeeper what its unique identifying characteristics (i.e., phone number, IP address, etc.) are. This process can be authenticated, too.

Each H.323 endpoint, or terminal, must either be programmed with the IP address or resolvable hostname of the gatekeeper or be able to discover the gatekeeper's presence using IP multicast to 224.0.1.41:1718. This latter method is most conducive to redundancy, since multiple gatekeepers can be configured to respond to multicast locate requests like failover partners.

The registration process is defined by a unique setup protocol called RAS, which stands for Registration, Admission, and Status. This protocol governs only the registration process, not any call setups.

All components of an H.323 voice network must register if a gatekeeper is utilized—which is just about always. It would be difficult to do much more than some dedicated trunk-style channels without employing a gatekeeper.

According to the ITU-T recommendations, a compliant gatekeeper must provide:

- Address resolution via a standard called E.164, discussed later
- Registration authentication
- Bandwidth controls
- Zone management of registration and calling
- Call control signaling
- Call authorization

Endpoints follow a specific procedure when registering:

1. The endpoint sends an RRQ Registration Request message to the gatekeeper, which consists of the endpoint's H.225 socket (IP address and port number), E.164 address, and/or username alias to be used for caller ID purposes.

2. The gatekeeper stores all that information in memory for later use when authenticating the endpoint, along with a hash that is used to prevent fraudulent use of the phone's identity by MAC address spoofing.

3. The gatekeeper sends the requesting endpoint an RCF Registration Request Confirm message indicating that it is clear to place and/or receive calls on the network now.

H.323 Terminal

Each H.323 terminal—either a software endpoint or a hardphone—contains a stack of software elements that cover different aspects of the calling process:

- H.245, which provides capabilities negotiation in order to make sure there's a compatible application and codec running on the calling terminal and the receiving one.
- H.225, which provides billing and monitoring functions necessary for reliable call setup and carrier-grade call accounting.
- RTP, the IETF standard for transmitting packets of encoded real-time media.
- A selection of one or more audio codecs.
- Optionally, an H.323 terminal may offer T.120, a protocol for enabling interactive data applications such as white-boarding. A great example of this is Net-Meeting's shared drawing board feature.

H.323 Gateway

The purpose of a gateway is to interface the IP-based voice channels to legacy signaling and transport technologies such as FXO, FXS, E&M, ISDN BRI and PRI, T1, and DID. A great example of this device is Cisco's modular media gateways, which can be equipped with interfaces that can support all of these legacy technologies and more. An H.323 gateway is required only when an interface to a legacy voice network is necessary—which is usually the case.

H.323 gateways offer specialized convergence signaling protocols that support certain kinds of legacy circuits:

- H.320 supports packetized voice over ISDN circuits and T1.
- H.324 supports packetized voice over POTS using G.711.
- Analog ports on the gateway can support old-style analog POTS, too.

H.323 gateways must register with a gatekeeper for the zone they serve if calls are to be routed through their media interfaces.

Multipoint Control Units

An MCU is an H.323-specific device that has a single purpose: conferencing multiple H.323 media channels. MCUs are the official prescribed method of doing conference call applications in an H.323 VoIP network, though some proprietary extensions to H.323 exist in this role (hence H.323's less-than-stellar interoperability reputation). An MCU can be a dedicated server, or MCUs can be built into H.323 terminals.

Sometimes, MCUs are referred to as conference bridge servers or DSP farms, because it is via digital signal processing that all parties are able to hear the other parties on an MCU-hosted conference call. DSP is used to converge, or interleave, participants' audio signals into a new, combined audio signal that is sent back out to the participants. (Traditional telephony may also use DSP for conference mixing—this characteristic isn't unique to VoIP.)

The software element within the MCU that actually performs the DSP function in order to aggregate media channels into a conference call is known as an MP, or multipoint processing, element. This is a part of the MCU that is multicodec aware and can handle transcoding if necessary. MCUs are also RTP hosts, as RTP itself is designed to support mixer-controlled conferences.

The second component of an MCU is the MC, or multipoint controller—this is the part that is signaling-savvy and can negotiate with the endpoints using H.225 and H. 245 to get them added, and removed from, conference calls as needed.

The H.323 Call-Signaling Process

There are five general steps in the H.323 signaling process for each leg of a call path: setup/teardown, capabilities negotiation, open media channel, perform call, and release.

Setup/teardown

To initiate an H.323 call, H.225 is required for the setup step. During this step, each endpoint involved in the call is kept apprised of the status of the call setup, expressed in one of H.225's named states, the last of which doesn't happen until the end of the call, of course:

Proceeding
  This means that the calling endpoint is trying to establish a network connection with the called endpoint.

Alerting
  This means that the called endpoint is being notified that somebody is trying to reach it. In other words, the called endpoint is ringing, and the calling endpoint is receiving *ring back*, an indication of ringing on the remote end.

Connect
  This means that the called endpoint has accepted and a media channel can be established.

Release
  This means that one of the endpoints has signaled an end to the call. When release is indicated, the call is actually being torn down, not set up.

Capabilities negotiation

After setup, H.245 is enlisted to negotiate the application requirements of the call and select appropriate codecs. H.245 determines:

- What kinds of application media each terminal can support: audio, video, white board, and so on.
- What codecs each terminal is capable of and which it may prefer.
- How the media channel will be structured, and what packet interval will be used.
- Which terminal will be the master and which will be the slave for the duration of the call. Master and slave roles distinguish the client/server role assumptions for future signals during the call and are just a protocol formality.
- How best to notify the caller if negotiation fails. Usually the endpoint will display an error code while playing a busy signal. The busy signal is standard when a call cannot be connected on a VoIP network.

Open media channel

Once capabilities negotiation has succeeded, RTCP (RTP Control Protocol) springs into action to establish a UDP socket for the media channel. Then the RTP media channel is opened, and a stream of encoded UDP packets with RTP payload flows across the network using the negotiated codec and packet interval.

Perform call

As the call is in progress, RTCP, which runs alongside RTP (usually on separate, consecutive UDP ports that are selected during call setup), can keep tabs on the media channel, which remains intact via connectionless UDP for the duration. This continues until the call is finished.

Release

At the conclusion of the call, H.225 enters its release state, signaling an end to the media channel, an end to the H.245 application capabilities session, and an end to the call-accounting transaction on the gatekeeper. Depending on the endpoint, the caller and callee will hear a dial-tone or a busy signal.

In Figure 7-1, you can see how the RAS and H.225 are used with a gatekeeper to establish a call between two endpoints.

In this example, phone 3001 is attempting to call phone 3002. Once the user had dialed the call:

1. The caller sends an ARQ Admission Request message to the gatekeeper, identifying itself and the called party's E.164 addresses—that's the ITU's fancy term for phone number. This message is a part of the RAS protocol, and it can optionally

---

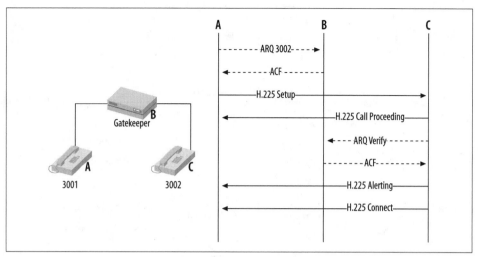

Figure 7-1. An H.323 call flow: 3001 is calling 3002

advise the gatekeeper of the bandwidth requirements of the call based on that endpoint's preferred codec.

2. The gatekeeper replies with an ACF, or ARQ Confirm. This tells the endpoint that the session requested by the caller is OK with the gatekeeper.

3. The caller sends an H.225 Setup message to the called endpoint.

4. The called endpoint sends a provisional H.225 Call Proceeding message back to the caller. It's provisional because it means the called endpoint has to verify the authenticity of the caller before proceeding any further.

5. The called endpoint sends a Called Party ARQ Admission Request message to the gatekeeper, asking whether the call is legitimate. At this point, the gatekeeper should still have a record of the original ARQ from the caller to match up with this Called Party ARQ.

6. If the gatekeeper has an ARQ match, it returns a Called Party ACF message to the called endpoint, clearing the way for that endpoint to ring.

7. Once the ACF is received by the called endpoint, it sends an H.225 Alerting message to the calling endpoint.

8. When and if the called party answers her ringing phone, the called endpoint sends an H.225 Connect message to the calling endpoint. This clears the way for H.245 capabilities negotiation to begin.

Figure 7-2 shows a gatekeeper-signaled call, which is identical to the preceding example except that the gatekeeper is a proxy for all of the signaling messages, not just the RAS ones.

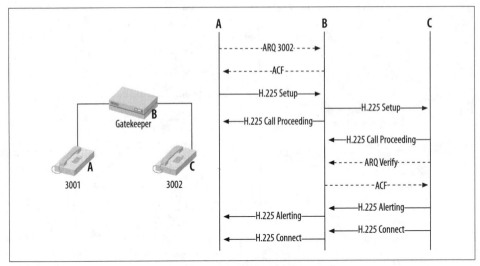

Figure 7-2. Gatekeeper-signaled call setup for a call from 3001 to 3002

The difference between gatekeeper-signaled and direct-signaled calls is the role of the gatekeeper in the H.225 session. If a gatekeeper involved, then the call is a gatekeeper-signaled call. Don't confuse this concept with the idea of call paths, which are the paths of the media channels—*not* the setup signaling, which is what we're talking about here. For a refresher on call paths, take a quick look through Chapter 6.

In Figure 7-3, a gateway is added to the signaling process, to facilitate a call from 3001 to a phone on the PSTN.

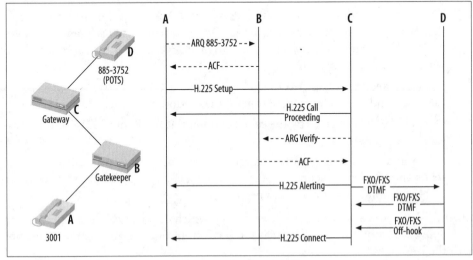

Figure 7-3. A gateway is an H.323 endpoint's portal to the PSTN

E.164 Address Scheme

E.164 is a convention for assigning telephone numbers to endpoints on a VoIP network. It's also the global convention for numbering of country codes, and is backward-compatible with older ITU recommendations for numbering. The phone numbers you dial on your analog phone when you call the pizza place and the barber shop—those are E.164-compliant numbers. Unlike older recommendations, however, E.164 allows endpoints on a VoIP network to dynamically register their E.164 address (number) from a list of available numbers stored in a database on the gatekeeper.

This database is an administrator-maintained list of Ethernet MAC hardware addresses, each of which corresponds to one or more assigned E.164 addresses. In this fashion, the administrator controls which endpoints are able to use a given E.164 address. In effect, the MAC address becomes the key to the endpoint's phone number, which makes telephone moves easier than they are with traditional telephony: no matter which port the phone moves to on the network, its E.164 address always follows.

Unfortunately, there are drawbacks to using the MAC address as a key for the E.164 address. First, MAC addresses follow a somewhat esoteric hexadecimal convention that isn't user friendly. Unless you're a serious geek, you aren't going to memorize many MAC addresses, so putting new IP phones into the gatekeeper can be a loathsome data-entry task. Second, the MAC address is usually hardcoded into the phone's firmware or network interface CMOS chip. This means it can never be changed.

Of course, MAC addresses are very much necessary: Ethernet wouldn't work without them, but there are better ways to handle alias assignment to a VoIP endpoint. Reliance on MAC addresses, and therefore intrinsic reliance on Ethernet, is one of the biggest drawbacks of H.323 when compared to SIP.

H.245

The H.245 session is established after the gatekeeper has cleared the way for the two endpoints to connect *and after the called party has picked up the phone.* In an instant, H.245 sets up a master/slave relationship between the endpoints, negotiates an agreeable RTP profile, including port selection, codec selection, frame size, and type of media (voice, video, or data). Then, the RTP/UDP datagrams immediately start to flow, and both parties can speak to and hear each other.

Fast-start

Fast-start is an optional parameter sent in the H.225 setup session when placing calls through H.323 gateways—usually to endpoints that don't support H.245, like those on the PSTN. If the gateway is able to confirm that the switching element (a CO switch or foreign PBX) on the called end supports Q.931, a fast-start–compatible signaling protocol, then H.225 can establish the media parameters immediately, foregoing the H.245 session after the called party has picked up the phone. In this event, the RTP session will be open by the time the called party answers. This is useful in minimizing the perception of lag in call startup when gateways are involved.

SoftPBX H.323 implementations

Because H.323 is such a distributed approach (gatekeepers, gateways, MCUs can all be on different hardware), it's tempting not to think of the standard as a way of achieving software-based PBX functionality. But in reality, most softPBX makers who've standardized on H.323 are putting gatekeepers, gateways, and MCUs all in one box and calling it a softPBX. This is an accurate description of what a fully functional H.323 network does—uses software to switch calls across multiple data links using a variety of physical interfaces, the same purpose as a softPBX.

Avaya, Nortel, and others that support H.323 offer most of these functions on a single Ethernet host—often a server blade that mounts inside their PBX chassis.

Cisco offers a consolidated approach in their media gateways. While they don't offer an MCU, Cisco's media gateways do provide H.323 gatekeeper and gateway functions and therefore are softPBXs in their own right. H.323 terminals can register with a Cisco media gateway if it's running a gatekeeper and place calls through it to the PSTN if it has a POTS or PRI interface with a phone company dial-tone plugged in.

 Telephony vendors' use of the term *media gateway* doesn't necessarily mean an H.323 gateway, a device for connecting VoIP networks to the PSTN. Cisco's media gateways, for example, are modular router/server hybrids that offer some softPBX functionality, often including an H.323 gateway, but not always.

Project 7.1. Build an H.323 Gatekeeper Using Open H.323

What you need for this project:

- Open H.323 software
- GnuGK (Gnu gatekeeper) software
- H.323 softphone such as MS NetMeeting
- LAN

Open H.323 is an open source implementation of the H.323 signaling protocol suite, managed by QuickNet Technologies, the same company that makes the Internet Phone Jack line of analog interface cards. Open H.323 is distributed in binary and source code forms for both Linux and Windows, though a crafty hacker should be able to get it running on a BSD-ish OS, too.

This project will allow a Microsoft NetMeeting H.323 softphone and an Open H.323 OhPhone softphone to place calls through an H.323 gatekeeper running on a Linux computer. In this example, we'll use Microsoft NetMeeting on Windows XP and OhPhone on Mac OS X.

Although Open H.323 provides a framework of tools for developing H.323 servers and endpoints, it also natively implements a complete H.323 gateway, MCU, and endpoint. Here's a partial list of software packages that accompany Open H.323:

OpenGK
> A simple H.323 gatekeeper server example

OhPhone
> An H.323 softphone for Linux and Windows (OhPhoneX is the Macintosh version)

OpenMCU
> An H.323 conference bridge server

PSTNgw
> An H.323 gateway server

Each of these requires the base distributions of Open H.323 and its prerequisite, PWLib, a project-specific class library.

Open H.323's requirements

A machine with specs like those required by Asterisk should be sufficient to run a small-scale H.323 gatekeeper. A Pentium III clocked at 600 mHz will be able to handle this project. The PC should be running Linux (though H.323 is also Windows-compatible). It can be the same PC that runs Asterisk, if you like.

Download and compile Open H.323

The best place to get Open H.323 is from its maintainer's web site, *http://www.openh323.org/code.html*. Compiling all of these elements is pretty straightforward on Linux. If you want to run Open H.323 on Windows, find the precompile executables. That said, these instructions are for Linux.

First, download and install PWLib. Save *pwlib_1.5.2.tar.gz* (or the filename appropriate for the version you download) to */root* as the root user. Then, unzip and untar it:

```
# tar xvzf pwlib_1.5.2.tar.gz
```

Now, you'll need to set some environment variables so Open H.323 software knows where to find the PWLib libraries:

```
# PWLIBDIR=$HOME/pwlib
# export PWLIBDIR
# OPENH323DIR=$HOME/openh323
```

```
# export OPENH323DIR
# LD_LIBRARY_PATH=$PWLIBDIR/lib:$OPENH323DIR/lib
# export LD_LIBRARY_PAT
```

If you plan on making this H.323 setup a permanent one, you should add the preceding environment variable commands to *.bash_profile* in */root*. Do this using Pico or your favorite text editor, without the pound signs (#) of course. Now, build the PWLib distribution using make:

```
# cd $PWLIBDIR
# ./configure
# make opt
# make install
```

Next, download the main Open H.323 file to */root*. Then, unzip and untar it, substituting the filename that's appropriate for the version you download:

```
# tar xvzf openh323_1.12.2.tar.gz
```

Now, build Open H.323:

```
# cd $OPENH323DIR
# ./configure
# make opt
# make install
```

The developers recommend a 128 MB swap partition to complete the build error-free. This need is minimized if you have enough physical RAM—256 MB of physical RAM should be plenty. This build could run for 30 minutes or more, so enjoy a delicious beverage.

Set up the Gnu gatekeeper (gnugk)

Once the Open H.323 build is finished, you'll need to download and compile the Open H.323 Gatekeeper (*gnugk*) software. Don't confuse this with the *opengk* that comes as a part of the Open H.323 distribution. *This* gatekeeper comes from a different source altogether, but is built using the same libraries as *opengk*. The big difference is that *gnugk* is a much more complete implementation of a gatekeeper, while *opengk* is a reference example and not very useful yet.

First, download and save the *gnugk* source code into */root* from *http://www.gnugk. org/h323download.html*. It will be named *gnugk-2.0.8.tgz* or something similar. After the download is finished, build the *gnugk* package:

```
# tar xvzf gnugk-2.0.8.tgz
# cd openh323gk
# make opt
```

Now, issuing the gnugk command will launch the *gnugk* gatekeeper. If you receive an error indicating shared libraries cannot be located, make sure you've got those environment path variables set in your login profile. If you run into compiler errors, try grabbing the x86 Linux executable from the *gnugk* site. Whether you compile it yourself or not, copy the distribution's bin directory's contents into */usr/sbin* and its etc directory's contents into */etc* as follows:

```
# cd openh323gk
# cp bin/* /usr/sbin
# cp etc/gnugk.ini /etc
```

To install a sample config file that allows any endpoint to register with the gatekeeper, copy *etc/proxy.ini* instead:

```
# cp etc/proxy.ini /etc
```

proxy.ini is far more permissive than the default configuration file and will allow you to register unauthenticated (i.e., passwordless) endpoints. Now, you can run the *gnugk* with the config file in */etc* by issuing:

```
# gnugk -c /etc/gnugk.ini
```

Register an H.323 softphone using OhPhoneX

If you're using a Windows PC, chances are you already have Microsoft NetMeeting. This is a very capable softphone, and it works well with Open H.323. In fact, the next section describes how to set it up.

But since the Open H.323 project produces a phone, too, we'll use it. It's called OhPhone, and it's distributed as an executable for Linux, Windows (*http://www. openh323.org*), and Macintosh (*http://xmeeting.sourceforge.net*).

These examples use screen grabs from the Mac OS X version. The Linux and Windows versions have only a text-based UI, but for those platforms, GnomeMeeting and MS NetMeeting make great alternatives.

The first thing you'll need to do with OhPhoneX is access its Preferences menu option. The Gatekeeper tab of the Preferences window will allow you to specify a gatekeeper, username, password, alias, and E.164 address (phone number).

In Figure 7-4, the address of the gatekeeper is 10.1.1.10—in prior projects, we've used this address for our experimental Linux VoIP server, so we'll continue in that vein. The ID is a superficial, free-form ID used like caller ID. The User/Alias ID is required only if *gnugk* is configured for authenticating registration attempts. The Password field is optional; its use is policy dependent, as *gnugk* accepts blank passwords. Finally, the E.164 Number is the phone number to which the endpoint is registering and, ultimately, the phone number that will be used to route calls to this softphone. Be sure to check the Use Gatekeeper checkbox, too.

When you close the Preferences window, click the Start Phone button, and then click the Console button: you'll see whether the softphone's registration attempt with the H.323 gatekeeper was successful. The console log of OhPhoneX, shown in Figure 7-5, contains the details of the registration attempt.

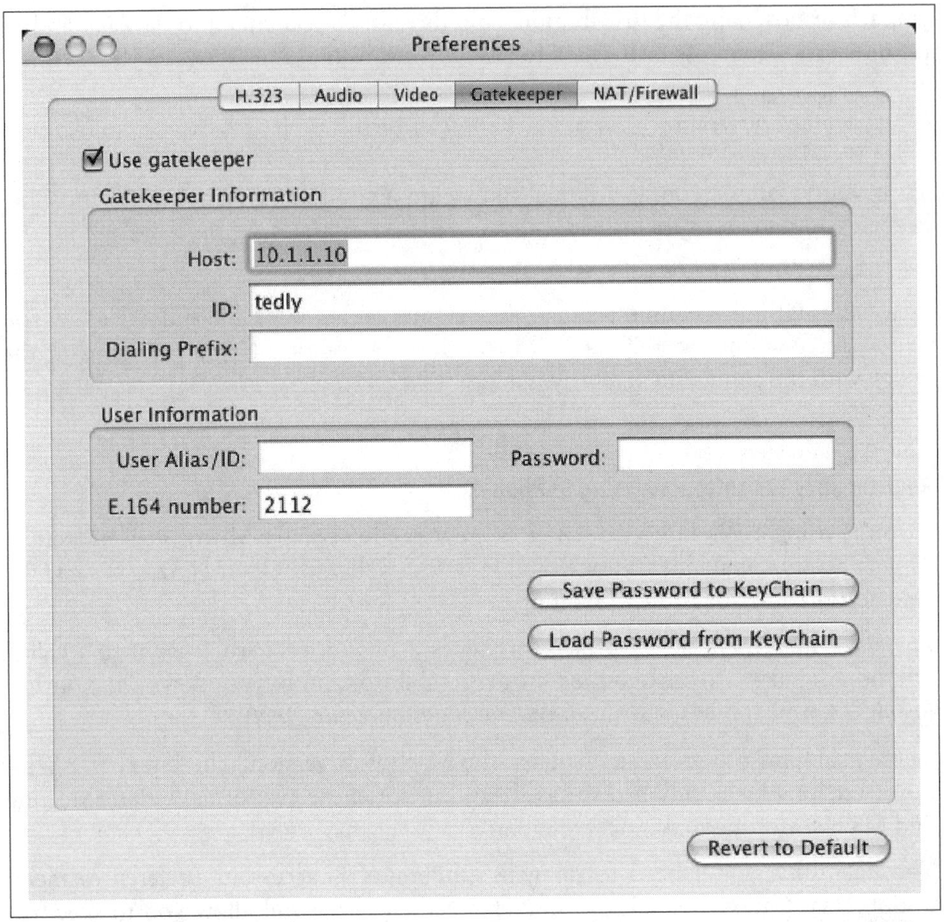

Figure 7-4. OhPhoneX's Preferences window has all the options an H.323 endpoint could possibly need to register with a gatekeeper

Now, if you register a second softphone from a second PC, you can call back and forth between them using the gatekeeper as the E.164 alias translator. This works the same way with H.323 hardphones. Callers dial the E.164 digits, and the gatekeeper provides the E.164 "resolution" that allows the software in the phone to do its H. 225, H.245, and RTP signaling to facilitate the call.

Once a call is in progress, the Connection Statistics window shows the status of the call in excellent detail, as in Figure 7-6.

Register an H.323 endpoint using NetMeeting

Microsoft NetMeeting is an H.323 softphone application that comes packaged with Windows Me, 2000, and XP. In order to run it on XP, however, you'll have to perform a slight hack to activate it. Click the Start menu, click Run, type **conf**, and click

---

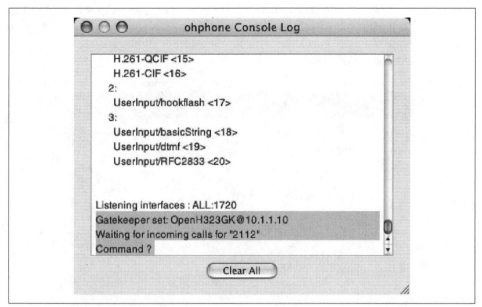

Figure 7-5. OhPhoneX's Console Log can help you troubleshoot the registration process

Type	Codec	Bitrate	Total Packets	Total Bytes
Audio In	G.711–uLaw–64k{sw}	63.3 kbit/s	825	193.4 KB
Audio Out	G.711–uLaw–64k{sw}	47.9 kbit/s	552	129.4 KB
Video In		0.0 kbit/s	0	0 Bytes
Video Out		0.0 kbit/s	0	0 Bytes

In Call With: Evil [10.1.1.10]   Call Duration: 0:25

Connection Quality

Packets Lost: 0 (0.0%%)   Round Trip Delay: 0ms

Packets Late: 0   Packets Out of Order: 0

Figure 7-6. OhPhoneX's Connection Statistics window tells you which codec your call has selected and how much bandwidth it's using

OK. Then, select "Put a shortcut to Netmeeting on my desktop" in the wizard that follows. Once this is done, NetMeeting is activated on Windows XP just as it would normally be on Windows 2000.

To configure NetMeeting to register with the gatekeeper, click NetMeeting's Tools menu, followed by the Options item. This will display the Options dialog, where you can click the Advanced Calling button. The Advanced Calling Options dialog will appear, as in Figure 7-7. Check the "Use a gatekeeper to place calls" and "Log on using my phone number" checkboxes. Enter the address of the gatekeeper, as well as the E.164 address you'd like to use into the Phone number field.

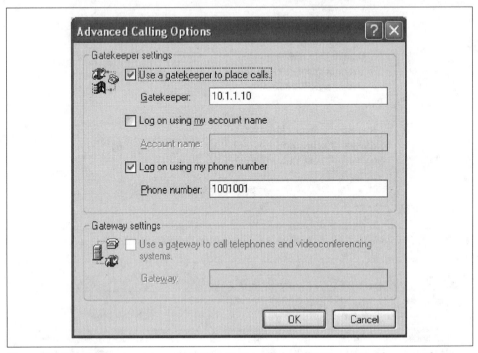

Figure 7-7. The NetMeeting Advanced Calling Options dialog allows you to configure gatekeeper registration

Microsoft NetMeeting is a very worthwhile H.323 softphone, and it's quite customizable. It allows video calling as well as audio calling and has a built-in T.120 whiteboard and instant messaging (text chat) applications. You can tweak the codec selection preferences by choosing Audio from the Options dialog and then clicking Advanced. The codec selection dialog is shown in Figure 7-8. If you're really looking to restrict codec selection, most compliant gatekeepers allow you to do it centrally.

Make the call

Once both phones are registered with the gatekeeper, you can call between them using their E.164 numbers since they're on the same zone. Now, if you like, download OpenAM from the Open H.323 project to set up an H.323-based personal message recorder.

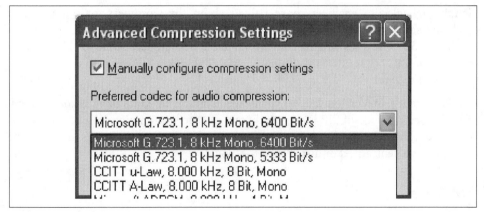

Figure 7-8. NetMeeting ships with a selection of five codecs, including G.711 (μLaw/Alaw) and G.726 (ADPCM)

SIP

The Session Initiation Protocol (SIP) was developed by the Internet Engineering Task Force as a way of signaling multiuser distributed telephony and messaging applications on an IP network. SIP has garnered much praise from IT professionals, while suffering some criticism from traditional telecommunications people. The main reason for its less-than-perfect repute with telecom pros is its origin outside the telecom world. But many telecom guys have had to forgive this, because they're learning that SIP has almost no shortcomings when compared to its ITU-inspired cousin.

The essential duties and formulaic pieces of SIP are the same as H.323. That is, there are VoIP endpoints of varying capabilities, and there are servers that participate in the signaling process and establish policy for the voice network. Unlike H.323, however, SIP is far more extensible. It is more than just a set of voice and video telephony protocols. Rather, it's a packaging framework for all types of message-based applications, from intercom calling to instant messaging and AV services.

Companies like Broadvox, Voicepulse, Broadvoice, Packet8, and others have emerged as frontier providers for dial-tone–style services delivered over the Internet, using SIP as the signaling system. Under these service offerings, consumers can purchase telephone calling capabilities that use the Internet, rather than a POTS line, as the transport for their phone service.

Avaya, Cisco, Siemens, Alcatel, and the major telephony hardware vendors have indicated a strong support attitude about SIP, while some have even backed away from H.323 investments. This bodes well for SIP's future, and there are already more SIP IP phones installed worldwide than there are H.323 ones.

This makes SIP both an easy decision and a challenging one. SIP's extensibility comes by way of a non-telephony mindset.

Traditional telecom engineers have balked about the wordiness, or bulkiness, of SIP's message structure. Instead of using compact, machine-friendly message packets like H.323, SIP uses lengthy, human-readable headers like SMTP or HTTP. Proponents of SIP counter that this human readability makes SIP easier to troubleshoot, and I tend to agree.

SIP is currently in Version 2.0. Its definition is found in RFCs 3261 through 3265. The defined purpose of SIP is to coordinate and facilitate monitoring of media sessions on the network. It supports a variety of addressing schemes and can be designed as a centralized or distributed topology.

SIP Nodes

SIP endpoints and servers are called *nodes*. A SIP phone is a node. SIP phones can communicate directly with each other in order to establish media sessions, just as H.323 terminals can establish direct channels. But more often than not, especially in an enterprise setting, SIP is used with a SIP server. SIP phones normally report to a dedicated SIP server node called a *registrar* upon boot-up.

SIP registrar

The SIP registrar is a database server that communicates with SIP nodes in order to collect, store, and disperse information about the whereabouts of SIP users. When a SIP node registers with a registrar, it tells that registrar how to get hold of the user, specifically what IP address and port to use for future SIP communication. You could think of the registrar as a router, because its main purpose is to give advice on how to reach SIP users, just as a TCP/IP router's purpose is to give advice on how to reach other networks.

URIs

SIP endpoints can be referenced using Uniform Resource Indicators, but so can SIP users. Consider this URI:

```
sip:lerxt@sip.bytor.com
```

This convention indicates both the user to be contacted and the server that is expected to know the address of that user's SIP endpoint. In this case, the user is lerxt and the server is sip.bytor.com. Secure SIP URIs, that is, those that indicate an encrypted signaling connection, use the *sips:* prefix instead of *sip:*. Encryption of SIP signals, if desired, occurs by way of Transport Layer Security, defined in RFC 2246.

A SIP URI doesn't always correspond to a single phone. If a user is available at one of several phones, then all of those phones can ring simultaneously, or in a specific sequence, based on the handling server's configuration. Most SIP registrars support simultaneous registration of the same user at multiple phones, and the most common way of handling this situation is to ring them all when a call is received for that user.

SIP methods and responses

SIP signals fall into 10 categories called *methods*. Each method accomplishes a different function for SIP:

INVITE
> This method is used to start sessions and advertise endpoint capabilities.

ACK
> This method is used to acknowledge to the called SIP peer that an INVITE has succeeded.

BYE
> This method occurs when the call is completed, that is, one user at a minimum wishes to end the call.

CANCEL
> This method is used during attempts to override a prior request that hasn't yet been completed.

OPTIONS
> This method is used to query a SIP peer for its capabilities information, without actually establishing a media channel.

REGISTER
> This method notifies the SIP server at which endpoints a particular user can be reached.

INFO
> This method is used to transmit telephony application signals through the SIP signaling path; these signals can include dialed digits.

PRACK
> This method (Provisional ACK) is used to notify an endpoint of intent to set up a complex call without actually providing an ACK. PRACK is the SIP equivalent of "all is well."

SUBSCRIBE
> This method provides a way of establishing event handlers within SIP telephony applications—i.e., "Tell me when Bob misses a call" or "Tell Bob when I am registered with the server."

NOTIFY
> This method delivers messages between endpoints as events occur—i.e., "Bob missed a call."

When a call must be started, ended, or altered, a SIP method is employed. The SIP methods in the preceding list are similar in concept to the HTTP methods GET and POST, and like HTTP, SIP expects response codes when it sends a method. SIP's numeric response codes are three digits long and break down into six categories:

1xx—Informational
2xx—Success

3xx—Redirection
4xx—Failure
5xx—Error
6xx—Availability

A complete list of SIP responses is found in Appendix A.

 Project 11.2 shows how to use a packet capture tool to observe SIP methods and responses.

Typically, a SIP caller initiates a method directed to a SIP callee, and that SIP callee initiates a response, according to the success or failure of the caller's method.

In Figure 7-9, you can see that a (highly simplified) call from an Internet host (A), in the form of a SIP INVITE, to *5150@oreilly.com*, would ordinarily result in a 200 OK response, clearing the way for the call to begin. Now, if the INVITE method specified a SIP peer whom the SIP server didn't know how to reach, a 404 Not Found response would be in order, as in Figure 7-10.

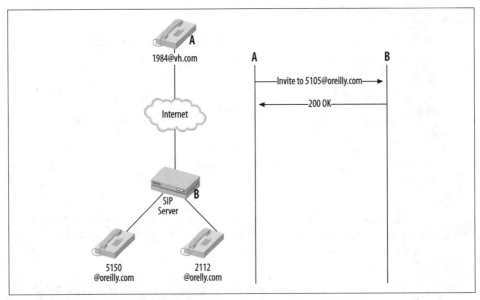

Figure 7-9. A call to 5150@oreilly.com would normally result in a 200 OK response, if 5150@oreilly.com were registered on the SIP registrar labeled B

A SIP INVITE header looks like this:

```
INVITE sip:5150@oreilly.com SIP/2.0
Via: SIP/2.0/UDP oreilly.com:5060;branch=9889gg1424
Max-forwards: 7
To: 5150 <5150@oreilly.com>
From: 1984 <1984@vh.com>
```

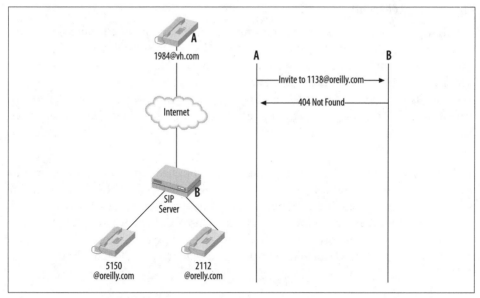

Figure 7-10. In the same setup as Figure 7-9, a call to 1138@oreilly.com, which is not registered in the registrar labeled B, will result in a 404 response

SIP proxies

Calls from one SIP endpoint to another can be considered local if they are both registered with the same SIP registrar, even if the physical endpoints are on different continents. The point is, they are on the same domain and are therefore peers on a local network of sorts. Nonlocal calls, however, are routed through specialized SIP server software known as a SIP proxy.

A *SIP proxy* is a server that routes or redirects SIP INVITE methods on behalf of one or more domains, just as a web server provides responses to HTTP methods for certain domains. So, when an incoming call from a foreign network is recognized, the SIP proxy's job is to connect it to the called user's endpoint, if possible.

Outbound SIP proxies serve the task of connecting calls, but on behalf of a local network of SIP users. Many users may share the same SIP proxy because they work in the same office or perhaps because they subscribe to the same SIP dial-tone service provider. Outbound SIP proxies are often used to overcome network communications problems posed by NAT firewalls.

Some local calls may benefit from being routed through a SIP proxy, too. Forcing even local SIP endpoints to use a SIP proxy allows for easy enforcement of a dial-plan, greater administrative control over the voice network, and the ability to do centralized telephony applications such as call recording.

SIP user agent elements

All SIP proxies and endpoints are comprised of two key software elements, the user agent client and user agent server (UAC and UAS). All SIP devices—be they softphones, hardphones, voice mail servers, or full-blown PBX servers—must be able to speak the SIP protocol, and the UAC and UAS elements are their mouthpieces. The UAC sends methods and receives responses, so its logical equivalent in HTTP is the web browser. The UAS receives SIP methods, processes them, and returns responses, so it's more like a web server. In varying degrees of completeness, all SIP endpoints and servers have both a UAC and a UAS.

In Figure 7-11, you can see the signaling process for a nonlocal call. This particular example uses an inbound call, which is fielded by a proxy server. By the time the SIP signaling begins, the calling endpoint already knows what host to contact for calls destined for the receiving endpoint's domain. This occurs by way of a DNS lookup for the hostname in the form *sip.domain.com*. The calling endpoint can then contact the proxy server for the domain in question, and send it an INVITE method.

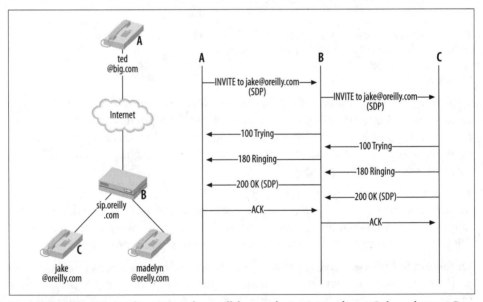

Figure 7-11. The SIP signaling process for a call from endpoint A to endpoint C through proxy B

The proxy server may immediately respond with an informational Trying response, or it may save the Trying response until after it has forwarded the proxied INVITE method to the appropriate endpoint on the local network. Incidentally, the way the proxy knows which local endpoint to contact is by performing a database lookup with the SIP registrar. If the user being called doesn't exist in the registrar, then the proxy will return a 404 Not Found response.

In this example, though, the user, *jake@oreilly.com*, does exist and is able to respond to the proxied INVITE method. His endpoint's responses are proxied back to the caller, and ultimately, the 200 OK response is sent, indicating the call is clear to proceed. One of the most important pieces of this startup signaling process occurs during the INVITE methods and 200 responses: SDP capabilities negotiation.

SIP redirect

When a SIP server responds to a calling endpoint's INVITE method with a 3xx response, that SIP server is redirecting the calling endpoint to a different SIP server. The calling endpoint should then contact that server with an INVITE method for further assistance in connecting the media stream. This feature is not implemented on all systems that support SIP. In fact, where complex, signaling-neutral dial-plan programming is available (like in Asterisk), SIP redirection isn't always necessary. The use of SIP redirects is more common in large, SIP-only networks with multiple servers, such as those that span the Internet.

Session Description Protocol

SDP is the de facto session capabilities protocol of SIP, similar to the H.245 protocol in H.323. It is defined in RFC 2327. When a call is placed from one SIP endpoint to another, an SDP capabilities construct is sent as text payload in a SIP INVITE, following the SIP packet header Content-type that indicates an SDP message is to follow. Here's a sample SDP payload:

```
v=0
o=HanSolo 7575 440 IN IP4 10.1.1.103
c=IN IP4 10.1.1.103:16385
m=audio 16385 RTP/AVP 1
a=rtpmap:1 G726/8000
```

This particular construct is requesting a G.726 media channel originating from user HanSolo at 10.1.1.203 using RTP for media packaging on UDP port 16385.

Using SDP, the calling endpoint can request certain codecs, sampling rates, or even a packaging protocol other than RTP (this is very rare, though). In the previous SDP block, the v token indicates the version of SDP being used, though SIP doesn't care, as SIP could theoretically use any version. The o token is a string of identifiers that uniquely name this SDP request, often including an NTP timestamp and the IP address and protocol designation of the sender. The c token tells which IP address to use for the media channel.

The m token describes the UDP port number and media framing protocol (RTP in this example), followed by a numeric identifier for this framing capability, called an RTP profile. More than one m token definition may be sent if the requesting SIP endpoint is advertising more than one set of RTP capabilities. The a token tells the RTP profile identifier, codec, and sample rate to use for the media channel. For an exploration of SDP messages captured by the Ethereal packet sniffer, see Chapter 11.

In Asterisk, the status of SIP peers can be displayed at the Asterisk command line, using sip show peers and sip show users. Calls in progress can be tracked using Asterisk Manager (*astman*).

Real-Time Streaming Protocol

Unlike H.323, the SIP protocol family provides a built-in recommendation for streaming prerecorded audio and video, in this case the RTSP protocol. This is the same protocol used by RealOne and similar media player applications. It's defined by RFC 2326.

SIP packet encoding

While H.323 uses the obscure ASN.1 (abstract syntax notation) encoding, SIP uses plain text. More specifically, SIP uses a text-based conversational approach for its signaling messages, whereas H.323 uses the Q.931 ISDN approach—one that is, by all accounts, meagerly understood outside telecommunications engineering circles.

Although SIP has many functional equivalents in H.323, it's fair to say that SIP's approach is more distributed, and certainly more Internet-like, while H.323's is more PSTN-like or mainframe-like. Table 7-2 illustrates this concept.

Table 7-2. Comparison of SIP, H.323, and IAX protocol families

Function/Characteristic	SIP	H.323	IAX
Endpoint discovery and admission	SIP REGISTER methods	RAS Protocol	IAX REG Control Frames
Call setup and teardown	SIP INVITE methods	H.225 Protocol	IAX NEW and HANGUP Control Frames
Capabilities negotiation, codec selection, and media session port selection	Session Definition Protocol	H.245 Protocol	IAX Capability Information Meta Frame
Packetization and sound sample transmission	RTP/RTCP Protocol	RTP/RTCP Protocol	IAX Voice/Data Full and Mini Frame
Streaming of recorded audio and video	RTSP	None recommended	None recommended
Frame encoding	Text (ASCII & similar)	ASN.1	Binary/proprietary
Messaging approach	HTTP-like	ISDN-like / Q.931	Proprietary
SoftPBX call path is called a...	Proxy	Gatekeeper-routed	SoftPBX
Call-routing reference device is called a...	Registrar	Gatekeeper	Server
Independent call path is called...	Redirect	Directed signaling	Direct signaling
PSTN interface approach	None recommended	H.323 gateway	None recommended
Encryption of signaling messages	TLS/TCP	None recommended	A work in progress

Function/Characteristic	SIP	H.323	IAX
Endpoint identification	SIP URI, email address, E.164 address, or alias	E.164 address	Email address, E.164 address, or alias
Connections through firewalls	Gatekeeper/softPBX call path	Proxy/softPBX call path	No proxy needed
Multiplexed trunks	Yes	Yes	Yes
UDP port number	5060/5061	1503,1720,1731	5036 (v1)

SIP versus H.323: the great debate

H.323 is older than SIP, and more considerate of legacy infrastructure, too. It defines standard procedures and best practices for interfacing with old-school telephony technology—practices like the use of gateways with built-in support of legacy protocols like ISDN and FXO/FXS.

SIP makes no such provisions, but is still the favorite among those schooled in Internet thought: it's extensible and reusable and does a whole lot more for telephony apps, ultimately, than H.323. SIP allows multiple endpoints to register the same alias, allows freedom from E.164, and enables other, non-voice applications like instant messaging and presence. For these reasons, some in the industry have pronounced SIP the winner of the battle for telephony signaling.

 SIP doesn't define gateways for interaction with the PSTN, so a legacy-aware signaling system like MGCP, MEGACO, or H.323 is often used alongside SIP in order to facilitate legacy gateways.

The fact is, both protocol suites are very much necessary. Parts of the H.323 recommendation—RTP and PSTN gateways in particular—are in use by SIP networks so the importance of H.323's features is obvious.

One day, SIP may contain some legacy interfacing of its own, but for now, H.323 fills that role quite aptly. All the big VoIP equipment vendors use its recommendations in order to interface with legacy systems like the PSTN. SIP, conversely, was built for the *new* network—the Internet network. Some mix of H.323 or MEGACO/H.248 (covered later) is in order if your SIP system plans on talking to a traditional telephony system.

IAX

The Inter-Asterisk Exchange Protocol, currently in its second revision, is a signaling protocol for VoIP networks, just like SIP and H.323. It provides endpoint and trunk signaling like those protocols, too. The chief difference between IAX and the other signaling families is that IAX doesn't implement RTP as the packetizing mechanism. Instead, IAX has its own way of packaging encoded voice.

IAX is also NAT-proof, so dozens or hundreds of simultaneous calls from behind a masquerading firewall will function correctly, just as HTTP does.

IAX is implemented in a far-simpler and less application-exhaustive manner than SIP and H.323. It is really intended *just* for telephony applications, while H.323 and especially SIP, include far more extensibility. IAX is therefore much more compact; complete implementations have been done with as little as 64 kb of object code.

While a complete cycle of registration, call signaling, voice transmission, and teardown can use several TCP and UDP ports and connections with SIP or H.323, IAX handles all of these functions using a single UDP port. When the IAX client (endpoint) registers with the IAX server or proxy, this UDP port is utilized. When a call is placed, this same port is utilized. When voice transmission occurs, this port is utilized once again. The way IAX distinguishes between registration, signaling, and voice packets is by including headers and metadata in each packet that defines what the packet's purpose is and whether it has a payload attached.

The IAX protocol documentation describes the order of these header and metadata elements as control frames, metaframes, and information elements, each with an IAX-specific syntax. IAX isn't encoded using ASCII or ASN.1, either. Instead, It uses a purely a proprietary performance-oriented binary-encoding scheme.

Unlike SIP and H.323, IAX is not a standards recommendation, but rather an independent protocol created by Mark Spencer, founder of Digium. Though proprietary, the specification for IAX is open and has been embraced by the VoIP community. As such, it is quite well-implemented in Digium's products. Asterisk, the open source softPBX, implements it fully, and Digium manufactures an ATA that is IAX based. A hardphone is also in the works. Refer to Table 7-2 for a list of the key characteristics of IAX.

MGCP

The Media Gateway Control Protocol is a first-generation signaling protocol for media gateway groups defined in RFC 2705. MGCP is used by phone companies and VoIP service providers in carrier-grade switching applications such as hosted PBX and IP Centrex (see Chapter 13 for more on this). Though supported in a number of telephony products from Cisco and others, including several early IP phones like the famously popular Cisco 7960, MGCP has been augmented by MEGACO/H.248.

Perhaps the most common place to encounter MGCP is an IP Centrex–hosted PBX service, wherein a group of IP phones at the customer premise are MGCP clients whose calls are routed by a gateway controller (GC). The GC is a high-density server that is connected to the customer premises by an IP transport like a T1. This essentially client-server setup delivers roughly the same kind of telephony functionality as H.323.

MEGACO/H.248

MEGACO is a protocol that describes a standard set of signaling functions for groups of media gateway devices that are *physically decomposed*—that's *distributed* for all us mere mortals—on a packet voice network. MEGACO was jointly developed with the IETF, which published RFC 3015, and the ITU-T, which published recommendation H.248: the documents are identical.

MEGACO provides a next-generation framework for building gateway devices, MCUs, and other types of telephony servers, but doesn't make specific provisions for any of them as H.323 does. In this regard, MEGACO is more extensible than H.323, but less so than SIP.

Depending on how they're implemented, MEGACO gateways can offer support for standard PSTN signaling technologies like DTMF, SS7, and ISDN. They can also provide modular or monolithic support of several packet-based network protocols, including TCP/IP, ATM, and frame-relay.

MEGACO fills in some of the telephony blanks in SIP. Since SIP doesn't make a specific recommendation for media conversion applications, MEGACO complements it nicely in this role. Likewise, an H.323 gateway may use MEGACO/H.248 for communication with the PSTN. Because of that, MEGACO and its predecessor MGCP aren't as commonly used for endpoint signaling as H.323 and SIP. In fact, MEGACO provides for dumb endpoints, the same way a CO switch or PBX does. This makes MEGACO more like the old-fashioned PBX (client/server) than SIP or H.323 (distributed).

MEGACO can be a transition piece that helps large networks of legacy telephone equipment migrate to VoIP, one gateway at a time. This gives it some enterprise appeal. But MEGACO was not designed with enterprise systems in mind. It was instead designed for switching in carrier networks, which have rigorous requirements for control of service, revenue generation, and billing for optional telephony features like caller ID.

Within the next 15 to 30 years, MEGACO could possibly displace the global network of last-mile CO switches and gateways that today use a mixture of in-band and SS7. In some areas, it already has. But there's a monumental cost associated with doing such a large-scale migration, and the phone companies must justify such upgrades with revenue. So don't expect to be using MEGACO signaling for trunks between your home phone and your local LEC any time soon.

The strategy that some phone companies have used thus far is to add MEGACO signaling to their CO switches at the same time they add DSL capability, which itself is in far higher demand. This could mean that, eventually, DSL will be the dominant last-mile vehicle for dial-tone services provided by CO switches that run MEGACO internally. Indeed, most phone companies that offer DSL services in a majority of their service areas have already rolled out some kind of last-mile VoIP service, too.

It *isn't* likely, however, that MEGACO interfacing will be installed in each home and office served by the telco. Rather, SIP, and to a lesser degree SCCP or MGCP, which can be provisioned as trunk media types on a MEGACO gateway, will eventually dominate last-mile signaling. Since SIP phones are already exploding in popularity, it will be far cheaper for the telephone carriers to bring SIP signaling, rather than MEGACO, to your doorstep, because SIP phones are already in great abundance and are cheap.

ICG, Qwest, AT&T, and many other large phone companies have started to offer VoIP dial-tone services in this very fashion—SIP or SCCP for the last mile and MEGACO connecting the CO switches.

Cisco SCCP

SCCP is a signaling protocol developed by Cisco. Though proprietary, open implementations of Cisco SCCP do exist. There is an SCCP module for Asterisk that is free and open source. The Skinny Client Control Protocol is so named because it is a far less ambitious suite than H.323. In this regard, SCCP is more akin to IAX.

Like IAX, SCCP is a leg-only signaling solution; it doesn't dictate how multiple legs of a single call path should be conducted as H.323 and SIP do. It doesn't mention anything about proxies either. In fact, SCCP is used in tandem with H.323 in most implementations, because many of the features SCCP does not implement, such as H.245, are required to provide a full PBX solution for a group of IP phones. So, in an SCCP environment, call setup and capabilities negotiation still occur by way of H.323, but these functions reside within the softPBX rather than being negotiated directly between endpoints.

The Cisco softPBX is called CallManager, and it provides the best reference implementation of SCCP. Vovida.org, the home of the open source VOCAL project sponsored by Cisco, has more detailed information and open reference projects on SCCP. There's also an open source SCCP signaling module for Asterisk.

Heterogeneous Signaling

Heterogeneous or mixed-standard signaling occurs when a single call path must be established using two or more devices of different signaling standards. A SIP endpoint, for example, could call an H.323 endpoint via a softPBX that understands both SIP and H.323. An H.323 endpoint could call an analog phone on the PSTN or connected to a media gateway. A SIP phone could call another office via the PSTN, only to end up ringing a second SIP phone at that office. All would be heterosignaled calls because they involved endpoints that support opposing standards: SIP, H.323, analog FXS/FXO, and so on.

The call path, that is, the logical path across the network that carries the sound stream for a call, must be softPBX-routed in hetero-signaled calls, because there is no capabilities negotiation between endpoints of opposing signaling standards. Figure 7-12 shows how a SIP server equipped with an FXO interface handles both a SIP signaling leg and a POTS signaling leg in order to set a call up from an IP phone to the PSTN.

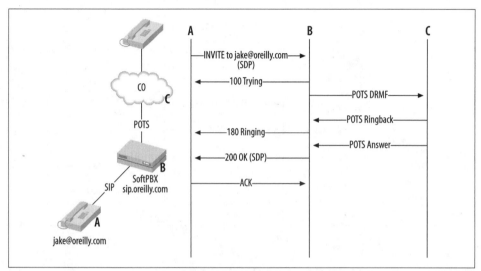

Figure 7-12. A SIP proxy server with a POTS interface (like an Asterisk server with an X100P) is able to handle a setup on both the IP leg and POTS leg of the call

Figure 7-13 shows how a single softPBX might handle a call from a SIP phone to an H.323 phone, if it supported both standards. Asterisk supports both standards, and so does a Cisco media gateway.

As in Figure 7-13, the call setup signaling for a call from 3001 to 3002 through a soft-PBX could work like this:

1. Endpoint 3001 sends a SIP INVITE method to the softPBX, which is its SIP proxy.
2. The softPBX responds with a 100 Trying SIP response while...
3. ...sending an H.225 Setup message to endpoint 3002, which has been resolved through the softPBX, which is also an H.323 gatekeeper.
4. 3002 sends a provisional H.225 Call Proceeding message to the softPBX.
5. 3002 sends an ARQ Verify request to its gatekeeper, which is the softPBX.
6. The softPBX responds with an ACF, indicating the call is legitimate.
7. 3002 sends an H.225 Alerting message to the softPBX; meanwhile, it rings.
8. The softPBX sends a 180 Ringing SIP response to endpoint 3001.

9. When the called party picks up the phone, 3002 sends an H.225 Connect message to the softPBX.

10. The softPBX sends a 200 OK SIP response to 3001.

11. 3001 sends a SIP ACK method to the softPBX.

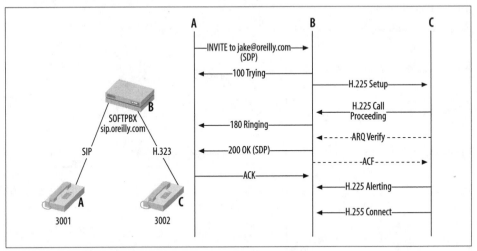

Figure 7-13. A single softPBX can facilitate calls between legs of opposing signaling standards

Heterogeneous Signaling and the SoftPBX

Different vendors implement heterogeneous signaling differently, and if you're really interested in tracing the signaling steps taken during a particular multiprotocol call, your best bet is to capture them using a packet sniffer or study the standards recommendations themselves. In the meantime, Figure 7-13 should give you a working idea of heterogeneous signaling.

Cisco AVVID/CallManager

Cisco's approach to signaling is distributed, meaning that specialized, separate devices are the preferred mechanism for signaling non-SCCP endpoint connections. A media gateway such as the Cisco 2650 is a modular chassis that contains a variety of interface hardware and IOS firmware that is specific to the type of signaling desired—H.323, SIP, and so on.

These media gateways refer to endpoints or remote gateways as *dial-peers* regardless of their individual signaling standards. A POTS link, an H.323 softphone, and a SIP hardphone are all called dial-peers.

Independently, Cisco media gateways can be used as very basic PBX systems, since they support a programmatic dial-plan, which is quite similar in concept to an IOS routing configuration and can use a variety of different phones. But the preferred, and more enterprise-savvy way to build a phone system is using CallManager, Cisco's full-fledged softPBX.

CallManager supports one endpoint-signaling protocol: SCCP. On the server, capabilities negotiation and other setup functions are handled by H.323, but it's SCCP that signals call progress events to the phones themselves. So, it's fair to say that Cisco's endpoint signaling is mainframe-like, because some big pieces of the signaling process occur on the server rather than "over the wire" between the server and the phone.

In order to use analog, SIP, or H.323 phones with a CallManager PBX, those phones must be connected to a media gateway, where their signals are trunked back to the CallManager, preferably using SCCP. In Cisco's environment, it's the media gateway that is multiprotocol aware—not the PBX.

Asterisk

A softPBX built with Asterisk is probably more like an Avaya or Nortel softPBX than like a Cisco one. Asterisk has all the signaling capabilities it needs on one PC server. Whereas a Cisco setup requires trunks to media gateways in order to support non-SCCP endpoints, Avaya's MultiVantage and Nortel's Meridian IP-enabled softPBX systems, like Asterisk, support all the necessary signaling standards in one machine.

Key Issues: Replacing Call Signaling with VoIP

- VoIP can replace phone-to-phone signaling that the PBX provides in traditional environments.
- Legacy protocol support is still required for analog phones and connections to the PSTN.
- The two most popular standards for phone-to-phone signaling are H.323 and SIP.
- H.323 was created by the ITU as a video conferencing standard but grew to become an ambitious PBX-replacement recommendation.
- Microsoft NetMeeting and OhPhone are H.323 softphones.
- A gatekeeper and gateway form the softPBX nucleus on an H.323 network. These two elements often run on a single server.
- H.245, H.225, and RAS are the three layers of a phone-to-phone signaling session on an H.323 network.
- SIP was created by the IETF as a media-session management protocol; it has proven a great match for telephony applications on the Internet.

- SIP defines less of the network than H.323 does, leaving to the application developer the details of application, session, and presentation layers.

- SIP doesn't address legacy interfacing at all.

- SIP is seen by many PBX vendors as a way of signaling trunk connections to other vendors' equipment.

- SDP is SIP's capabilities negotiation protocol.

- RTP is the packetization and framing mechanism used by SIP and H.323.

- IAX, MEGACO/H.248, and Cisco SCCP are other prevalent signaling protocols.

- IAX does not use RTP for packetization; it frames signaling and sound data in the same packet construct.

- Heterogeneous signaling is required when an endpoint of one signaling protocol wishes to communicate with an endpoint of another.

## CHAPTER 8

VoIP Readiness

Let's face it: VoIP isn't exactly new, but IP telephony's readiness for enterprise consumption is a fairly recent development. When it first appeared on the Internet scene, VoIP offered the ability for people to make free long-distance calls over the Internet. In fact, products like Internet Phone came with substantial buzz about how they let in-laws with Microsoft Windows have half-duplex speakerphone conversations through their PCs over the Net.

Lack of interoperability, poor quality of service, and a drop in traditional long-distance calling rates ultimately killed the first generation of consumer VoIP software. The short-lived voice-over-Internet craze of the late 1990s died. VoIP is still what historians might call a *disruptive* technology—it is changing the status quo—but as it becomes more standardized, quality-driven, and accepted, it also becomes a more sustaining technology, just as the PSTN has been for decades. In this regard, VoIP has proven much more valuable in the enterprise than in the home.

This chapter helps you define the business case for VoIP. It guides you through readying your network, and your business, for next-generation telephony. This chapter is aimed at people who are justifying VoIP adoption with the promise of lowering operating costs and making system users more productive. It helps project managers decide between VoIP standards and vendors at the outset of the project. If you're a technical type, one who doesn't care necessarily about the financial impact of the technology, you'll learn what IP telephony's implications are for your network infrastructure.

Network convergence can mean more than just voice applications: many are using packet-based solutions for video conferencing, instant messaging, and security/surveillance. For the balance of this chapter, as we talk about preparing for the converged network, we'll zoom in specifically on voice.

Assessing VoIP Readiness

Now that interoperable soft and hard IP phones and ATA devices exist and best practices for ensuring quality on the TCP/IP network have been established, the path to VoIP is paved with confidence. Thanks to voice-over-data pioneer Cisco Systems, the industry has gotten the proof it needed that Ethernet can indeed be used for enterprise voice traffic. The true appeals of VoIP aren't just a novelty, but rather real, solid propositions of dollars and cents. And the justification for any infrastructural change of VoIP's magnitude always boils down to dollars and cents.

Just as it took VoIP a few years to be ready for business, it may take a while for businesses to be ready for VoIP. Your business environment, network, and implementation plan must be up to par if VoIP is to succeed in your organization.

VoIP readiness can be heightened or lowered by these factors:

Business environment
> Is the business culturally ready for VoIP? Are its applications or cost models seeking benefits from VoIP? To what extent can VoIP really help the business—an IP-based trunk here and there or a full, end-to-end replacement network?

Network environment
> Is the network ready and able to deliver toll-quality voice services? What are the cost and timeline to get it ready? Is it desirable, or possible, to migrate the network to VoIP services a workgroup at a time, a LAN at a time, or a trunk at a time?

Implementation scenario
> Are the social, timing, and project-management ramifications of VoIP conducive to a successful rollout? Is there effective, mandated leadership and enough resources to complete the implementation?

Business Environment

Some business structures and philosophies are more compatible with VoIP ideology than others. When considering VoIP adoption, figure out where your business fits in the market adoption cycle for new technologies. If you're a persistent early adopter who loves skating along the leading edge, then VoIP is a no-brainer. If you're a very mitigation-minded, deliberate organization that is historically slow to invest, then your adoption of VoIP systems may take some time. Chances are, if you're reading this book, you don't fall into that latter category.

The Market Adoption Cycle

The *early minority* of organizations are those on the bleeding edge of technology: they always have their noses to the wind, are constantly seeking, and taking risks on, newer technologies (see Figure 8-1). The *early majority* of organizations are those that mitigate the risk of a business investment with a more conservative wait-and-see attitude. The *late majority* of adopters wait until the price of the technology has been brought down by the early crowd, and then they adopt. As far as VoIP goes, it will be a few years yet before the late majority begins to adopt the technology. The *late minority* of organizations tend to be very small, often one-person, shops that have no choice but to wait until a technology is so accepted, so standardized, that it can be implemented with a minimal investment and almost no risk. Is your organization historically a late majority or early minority adopter? Does your company's attitude toward technology match your desire as an administrator or system builder to embrace VoIP?

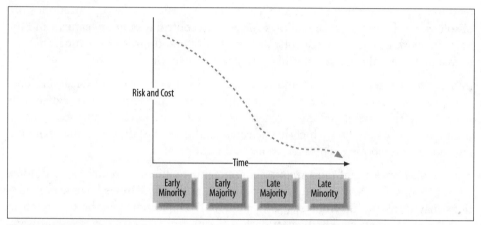

Figure 8-1. Early adopters of a newer enterprise technology like VoIP may enjoy the most benefit, but also take the biggest risk

Efficiency

When considering an investment in VoIP, you've got to justify the expenditure of implementing all that shiny new hardware and software against VoIP's benefits to the organization. Increased network efficiency is a possible benefit. Is your organization's network an efficient one? Would VoIP have any impact on existing levels of service to which your users are accustomed? Would voice traffic added to once data-only network pathways hurt other applications' performance?

Productivity

Another oft-cited appeal factor for next-generation telephony is productivity gains. Will VoIP enhance productivity? Will new telephony applications allow you to:

- Make more productive use of call center or reception-desk staff?
- Allow a greater volume of phone calls per user?

- Allow easier enforcement of end user agreements that govern monitoring and abuse?

- Result in more returned calls, thus, hopefully, more sales or prospects?

- Enable more centralized, company-wide contact management or messaging, such as a single voice mail server on the WAN rather than separate ones at each office?

Cost

The cost model of VoIP is substantially different than that of traditional telephony. The old cost model is economically fixed, meaning that network links and devices tend to have an unchanging cost and an unchanging value to the organization. VoIP obsoletes this model because it allows software to control telephony features and the voice network economy. Since software can evolve and improve over time, the cost to maintain your phone system is likely to decrease over time.

In a converged network, voice and data are carried by the same physical and data link layer components. This means that only one set of network support expertise is required, so the cost of staffing the network is said to be lower. Compare that to a traditional voice network, which has telecom staff, not IT staff, doing the maintaining and troubleshooting. This fact, too, can lower operating costs.

But the up-front capital (and training investment) required to replace a traditional voice network with VoIP is sometimes viewed as prohibitive. Indeed, some organizations may think the up-front cost of a total VoIP conversion may be too much to bear. In most situations, the biggest part of the up-front investment isn't hardware, but consulting. System design, training, and technical support can account for the lion's share of the budget—especially when the implementation is complex or your company lacks in-house expertise.

There are some ways to reduce that big up-front cost:

- Negotiate discounts that will be honored for the duration of the project, but don't require you to buy all the hardware at once. This way, you can purchase only the phone gear you need for each phase of your rollout.

- Lease telephony servers and phones to spread the cost out over time.

- Adopt VoIP incrementally. This may mean starting with a VoIP TSP to provide dial-tone for a legacy system, and then adding a softPBX and IP phones later on.

Cost models help sell IP telephony

A per-user cost model allocates the costs of an asset based on an even division among all its users. A $50 asset with 50 users is allocated at a cost of $1 per user. Allocating an even share of the cost of the initial investment to each user of the system allows you do several things:

- Establish how much of the cost each department or division within your organization is responsible for paying to finance the capital investment for a new voice system. (If 10 phone users work in Marketing, and only 5 work in Support, it can be argued that Marketing should bear a greater share of the cost.)

- Determine, based on revenue and overhead factors, whether VoIP conversion is really worth it for every workgroup in the organization. There may indeed be workgroups that cannot afford, or whose business models don't justify, a complete VoIP conversion at this time.

- Figure out the ROI-per-user time frame for a VoIP conversion. Some departments might get a return on their investment quickly, while others could take years.

You can apply the same method to the ongoing costs of the implementation: service and support, telephone company bills, internal training, and so on.

Actual consumption cost model

It's possible, in large VoIP plans, to miscalculate cost distribution because an even, per-user division isn't accurate enough. That is, the Marketing Department may have twice the number of phone users as the Support Department, but Support makes twice as many phone calls. Allocating per user in this instance would not be an accurate way to allocate costs. There is another way to divide up the pie—based on each person's appetite. Charge them "per bite" rather than "per piece."

This means looking at users' and departments' current *actual* utilization of telephony systems. This utilization can be expressed in number of calls, number of billed minutes, or just off-hook time. Utilization is a more precise method of distributing cost ownership. Most high-end PBX systems allow utilization reporting, and, if they don't, the phone company can perform a "traffic study" that may let you measure the same kind of statistics. That way, if Jake in the Marketing Department is using 75% of the system time, then the Marketing Department can be billed for 75% of the cost of the system, and so forth.

The actual consumption model can often uncover wasteful telephony spending, which makes it a great exercise in determining how VoIP can save your organization money.

Overhead costs

Overhead is that group of cost elements that aren't directly attributable to a particular user or department. A great example is a support agreement. These agreements are put in place, usually between the equipment vendor and the implementing organization, to replace faulty VoIP gear. These agreements tend to be paid for a year or two in advance.

Their cost cannot be allocated using user-based or actual consumption approaches. If an IP phone in the Marketing Department breaks some time during the year, but the Support Department has no such breaks at all during that year, you won't know how to charge the two departments for the maintenance cost until the end of the year when all break/fix episodes have occurred and the maintenance period is ended. The costs are overhead because, at the time they're expended, they can't be applied anywhere with a sense of actuality.

So figure on overhead costs being separate from your per-user and per-department cost. Instead, factor overhead costs into the big picture, the one that accounts for all users and all departments. But, most important, *don't forget them* in your budget.

The success delta

You may discover through these cost-modeling techniques that certain users or departments *cannot* benefit from a cost reduction through VoIP. But what you'll probably find is that a majority of users will experience a drastic drop in the cost of their telephony applications. Just how much of a cost reduction is the question that measures your success as a VoIP integrator—this success delta predicts your ROI, and it is often the key selling point for a VoIP solution.

Service provider cost savings

A great place to look for savings when switching to IP telephony is on your company's phone bill. There are a number of ways a transition to VoIP can decrease your monthly dollar commitments with the phone company:

- Switching from Centrex or POTS dial-tone trunks to less-expensive PRI trunk groups.
- Consolidation of many trunk locations into one or a few trunk locations with fewer total trunks. This way, you aren't paying for as many phone lines.
- Switching to IP-based trunks, which, due to compression, are more efficient than voice T1s.
- Eliminate trunks altogether by outsourcing PBX functions to a managed, off-site VoIP service provider. (This can also lower internal management costs.)
- Eliminate phone lines by replacing frequently used PSTN call paths with VoIP trunking—i.e., if your company calls a business partner 5,000 times per month, it might make sense to run a T1 directly to that business partner.
- Decrease long-distance calling charges by replacing frequently used long-distance call paths with least cost routing (LCR) on your WAN. This way, your remote offices can be used as connection points for long-distance traffic that is actually local for them.

Internal management cost savings

Before pitching VoIP in your organization, ask yourself if managing the voice network will ultimately be less expensive with VoIP than it has been without. Some who have integrated VoIP have discovered that, most notably in large environments, it takes less staff to manage the VoIP network than it did to manage the traditional telephony network. This may be because VoIP is "just another application" on the data network. It's also possible that it just takes a narrower skill set to support voice and data apps on the same network than it does to support them on separate ones.

How do you eat an elephant?

The old question "How do you eat an elephant?" is usually answered, "One bite at a time." And this may be the way you choose to roll out VoIP technologies: one segment, one workgroup, or even one building at a time. But don't forget the big picture. If your VoIP rollout is staged into manageable hunks, don't lose sight of the as-built plan, that comprehensive vision of the completed project. Your cost-benefit analysis is usually based on the endgame scenario, where the big investment payoff resides.

Table 8-1 shows a budget for a VoIP network. It considers all costs projected for a completely finished environment. It shows the monthly recurring and up-front costs and whether each cost is classified as capital (meaning it's an investment in durable goods like IP phones) and/or overhead. It's important to classify capital expenditures like IP phones and Ethernet switches because they have lasting material value for the company—a softPBX could be sold later on, and it's an asset to the bottom line because it is worth something. Consulting and service agreements, conversely, may enhance operations but aren't capital: once they're consumed, they are gone forever.

Table 8-1. Sample budget for a VoIP rollout

Cost element	Monthly recurring cost (MRC)	Up-front cost	Capital	Overhead
In-service IP phones (500)	0	$250,000	Yes	No
Hot-swap IP phones (10)	0	$5,000	Yes	Yes
SoftPBX (2)	0	$30,000	Yes	No
PRI media gateways (2)	0	$5,000	Yes	No
LAN cabling updates	0	$18,500	No	No
Telephone service assessment	0	$10,000	No	No
Service agreements	$3,500	$7,500	No	Yes
Project management/consulting	0	$65,000	No	Yes
Dial-tone trunks and LD	$4,500	$0	No	No

Table 8-1. Sample budget for a VoIP rollout (continued)

Cost element	Monthly recurring cost (MRC)	Up-front cost	Capital	Overhead
WAN links	0	$9,500	No	No
Total	$8,000	$400,500		

If you're rolling out IP telephony one department at a time, create a budget specifically for each department using per-user or consumption cost models. This will help you decide which pieces to bite off first and how best to spread out the costs on the more inclusive, projectwide budget.

 Your budget will reflect the contents of your desired VoIP topology design. Chapters 12 and 13 explore the details of topology issues.

There is no hard and fast rule for deciding which budgetary items are overhead and which items aren't. Generally, though, if it's difficult for you to associate the cost with a particular user or department, call it overhead. In this case, the hot-swap IP phones, which are there in case some of the 500 in-service phones break, are considered overhead because they aren't attributable to any one user or department at the time they are purchased. Through the magic of accounting, their costs can be tracked back to the right consumer at the time the phone breaks and the hot-swap phone takes its place. But, for now, they are overhead.

Recognizing revenue, productivity, and cost reduction

The success delta is the difference between what it costs to do business today and what it will cost to do business after the technology integration is complete. It describes, in dollars, what you hope to accomplish by adopting this new technology. It's also the difference between productivity today and productivity at completion. In some cases, it may be an expression of an increase in revenue—for example, a telephony service provider may convert to VoIP service in order to create a new revenue source. Table 8-2 is a sample worksheet that can help you recognize the sources of your revenue, productivity, and cost-reduction enhancements as a result of VoIP.

Table 8-2. Sample success delta worksheet

Cost or revenue element	Monthly recurring cost without VoIP	Monthly recurring cost with VoIP	Success delta	Explanation/notes
Private T1 voice links	$3,500	$2,000	$1,500	Using VoIP codecs to conserve bandwidth, fewer point-to-point links are necessary.
ISDN lines	$1,000	$0	$1,000	In offices where voice and data are trunked in on separate lines, it may be possible to eliminate voice-only lines.

Table 8-2. Sample success delta worksheet (continued)

Cost or revenue element	Monthly recurring cost without VoIP	Monthly recurring cost with VoIP	Success delta	Explanation/notes
Centrex lines	$1,200	$300	$900	Consolidation of soft switches to fewer demarc locations usually means fewer phone lines.
Long-distance fees	$850	$750	$100	Least-cost-routing using VoIP-over-WAN can yield a decrease in LD minutes.
Second phone lines for telecommuters	$2,400	$0	$2,400	Using broadband links to support remote IP phones can eliminate the need for secondary phone lines at home offices.
Cell phone voice mail expense	$500	$0	$500	Unified messaging and call bridging mean users have only one voice mail.
Missed calls impact on sales revenue (projection)	–$2,500	–$1,000	$1,500	VoIP-related features like presence and call bridging increase sales availability.
Voice system overhead costs	$2,600	$3,800	–$1,200	Maintenance fees will be higher than traditional PBX agreements. (This isn't always the case.)
Total			$6,700	

Like the budget, you can break the success delta worksheet into smaller ones that account for a specific department or for a specific phase of the rollout.

Calculate the ROI

Now that you've recognized the source of VoIP's cost and identified its beneficiaries within your organization, it's a simple matter to figure out your ROI. This is the rate at which the new benefits of VoIP balance with its cost. The ROI is your principal tool in expressing your organization's rate of return yielded by this technology. To calculate the time of your ROI, use this formula:

```
Up-front cost / monthly success delta = months to ROI
```

Since the up-front investment in Table 8-1 is about $400,000, and your monthly success delta is $6,700, it will take your organization about 60 months to complete the ROI and begin to reap the payback of the project.

```
$400,000 / $6,700 per month = 60 months
```

Convenience and Timing

Another question to ask when considering Voice over IP has to do with timing and logistics. The following are scenarios particularly favorable to a VoIP rollout.

- Is the company adding a new building or office location that does not yet have a PBX? This could be called a "green field" rollout—it's a great opportunity to introduce VoIP technology to your network.

- Have you been planning an upgrade to a new PBX because you're looking for new features? At this point, it would be rather convenient to make sure that new PBX is VoIP-enabled.

- Has your company been looking for ways to expand your existing PBX network? Adding VoIP trunks to a conventional PBX by way of ATAs and media gateways is a convenient way to begin the transition of the greater network to Voice over IP.

- Is your existing PBX lease up for renewal or buyout? Now would be a good time to step up to an IP-enabled softPBX.

- Are you looking at integrating CTI applications? Soft-based telephone systems, particularly Asterisk and other open source projects, can result in a less-expensive CTI development cycle than traditional equipment.

- Does your current telephone system pose a risk to your organization? Is it so out-of-date that, even though you're completely dependent upon it, it would be difficult to find replacement parts in the event of failure? Now's the time to upgrade to VoIP.

Network Environment

As a part of arguing the business case for a switch to VoIP, you've got to assess your network's readiness for it. Before you can deliver telephony, you need a fairly modern network. This can be done incrementally—a site at a time—or you can assess your entire global WAN all at once, if you have one.

LAN

Generally, LANs should be assessed as though they aren't even connected from the enterprise WAN, so that they are able, ultimately, to enable high-quality telephony in the event the WAN is disconnected.

In IP telephony, the LAN replaces the private voice cabling system used by the TDM bus on a traditional PBX system. The TCP/IP protocol running on an Ethernet LAN provides the endpoint leg of the voice transport. But not just any Ethernet LAN will work. Only modern, 100BaseT switched Ethernet (or better) is reliable enough for a production VoIP workgroup.

If you're using any LAN technology other than 10/100 mbps Ethernet, don't bother trying to run VoIP in your local area network. There are enough other reasons to dump Token Ring and 10Base2. If you haven't updated to 10/100 mbps Ethernet by now, you probably have bigger issues to deal with right now than VoIP. In a nutshell, it's time to forklift the Token Ring network and put in an Ethernet switch or two.

VoIPX? VoNetBEUI? Not a chance.

IPX/SPX just doesn't work with voice applications. Neither does NetBEUI. If possible, eliminate them completely from the network, as they don't play nice with all the available quality-of-service mechanisms. As you'll read in Chapter 9, using VLANs is one way to partition unwanted protocols away from the VoIP network.

Cabling

LAN cabling needs to be Cat5, Cat5e, or Cat6 twisted-pair. As with legacy protocols, legacy cabling needs to get the old heave-ho. In high-interference or long-distance segments where fiber must be used for desktop runs, you'll have to use an interface converter to support VoIP equipment. This is because IP phones and VoIP gateway modules are usually equipped with RJ45 connectors and because (presently) there's no way to deliver power using fiber-optic cables. Fiber is perfect for backbone connections, campus area links, and other high-speed LAN connections that don't have to carry inline power (see "Power over Ethernet," later in this chapter).

Switches versus repeaters

LAN segments where you plan on using VoIP must be switched. While hubs and other repeaters may work in low-traffic, less-than-critical situations, they aren't a best practice because they can't allocate or monitor bandwidth, and because they don't safeguard against Ethernet packet collisions, which cause latency and jitter.

While you're investigating Ethernet switches for your LAN, consider only managed switches and layer 3 switches, which give you more control over the traffic on your LAN than their predecessors. Your LAN switches (and LAN-connected routers) will also need to support class-of-service measures such as ToS, or IP precedence. You'll almost certainly need to use VLANs. Most managed switches provide all of the these. When reading the spec sheets of Ethernet switches, consider all of these standards. Chapter 9 offers more extensive coverage of them.

 VLAN means virtual LAN; it's a way of logically dividing an Ethernet switch in order to segregate LAN traffic at the data link layer. The standards that define VLAN were created by the IEEE.

Wireless Ethernet

802.11b/802.11g are great technologies for WLAN and LAN-to-LAN bridging, but they aren't very partial to VoIP. First off, they don't support QoS the way modern Ethernet switches do, and the packetization delay incurred by their framing and radio transmission introduces lag. Moreover, WLAN can't support nearly the number of endpoint devices as a wired Ethernet segment that's comparable in link speed. If you need wireless phones in the office, using analog cordless phones is still a very practical solution.

Power over Ethernet

In traditional telephony, the power to run each phone comes from the communications line that's attached to the telephone set. On the PSTN, 20 to 50 volts DC are required to power and signal analog endpoints. Voltage and frequency variances are what comprise analog signals for legacy protocols such as Ear and Mouth and FXS/FXO. Digital phone sets often have a similar voltage requirement, though their signals are carried over a digital bus. Legacy endpoints almost always get their power from the same line that handles their signaling, though one obvious exception is cordless phones, which need to be able to draw enough current to charge their handsets' batteries. Such draw isn't practical over DC communications wires.

With IP phones, the power situation is a bit more flexible. Most IP phones can be powered either through an AC/DC power transformer (like a legacy cordless phone), or through one of two common methods of receiving power through the Ethernet cable plant. These two methods are PoE, or power over Ethernet standards, and both do roughly the same task: provide a loop of DC electric power over an unused pair in the twisted-pair Ethernet cable that is connected from the switch to the IP phone.

Cisco power versus 802.3af

Cisco began manufacturing a PoE solution long before the IEEE's 802.3af recommendation was ratified in 2003. Consequently, most of Cisco's IP phones don't work with 802.3af. Instead, they use the Cisco PoE solution, which I call Cisco Power. The line voltage and pin arrangements of the two are what make them incompatible.

Meanwhile, 802.3af-compliant IP phones, which represent a majority, don't work correctly using Cisco Power. The two opposing standards just aren't compatible, though there are some wiring hacks that can allow a Cisco Power phone to be powered from an 802.3af power source and vice versa. Rather than make its own IP endpoints compatible with 802.3af, Cisco has provided another solution: make their switches offer 802.3af power.

Late-model Cisco switches can even autodetect the power requirements of each attached device and provide the right type of power on that particular port, so that you could mix and match 802.3af phones and Cisco Power phones on the same switch.

PoE is often just called inline power.

Consider the power requirements of your IP phone fleet. Will they work with 802. 3af, or do they require a proprietary power standard? Do the switches in place today provide the right kind of power, or any at all? Is connecting an AC/DC adapter at each phone location a reasonable idea? It may be in a 5-person office, but all those adapters could really add up in a 500-person office. How does the addition of switch-delivered power affect your VoIP rollout cost estimate?

Provisioning a central power source with PoE simplifies the task of supplying backup power. Three or four PoE switches in a single rack are much easier to connect to a UPS or backup generator than a building full of IP phones in separate offices.

Power injection

A number of vendors offer patch panels that inject inline power. Avaya and others offer 802.3af injectors that allow you to add PoE to your network without forklifting all of your non-PoE switches. Sometimes, PoE injectors are a less expensive proposition than PoE switches, especially when the switches themselves are young, can't be easily replaced, or are leased for a long term. That said, green field rollouts will likely want to avoid power injectors, because they take twice as much rack space as a PoE switch.

Properly implemented, PoE can save lots of overhead, and it lets you centralize your power backup strategy. (Imagine the alternatives—backing up the whole building or putting a UPS under everybody's desk.) Be sure each of your LAN segments can support PoE now or in the future, and plan appropriately for your chosen method of power provisioning.

LAN Readiness Checklist

Each location that has a LAN with which you'll support VoIP must adhere to the same minimum guidelines, according to a plan that you'll create:

- TCP/IP
- 100BaseT UTP (RJ45) Ethernet
- Cat5, Cat5e, or Cat6 cable plant
- Switched
- PoE if possible

It's important that you have a well-documented LAN or campus topology, including associated IP addressing plans, as well as addresses and physical locations of key servers such as firewalls, DNS servers, and DHCP servers. A description of your switches' VLAN setup is essential, if you're presently using VLAN. (VLAN is covered in greater detail in Chapter 9.) If layer 3 switches are used, well-documented routing and load-balancing plans should be produced, if these features are currently in use.

Where are your switches located? Are they distributed in phone closets like distribution frames, or are they all in a single room? Do people have hubs and switches under their desks next to their feet? All of these issues can be compiled into a readiness assessment, like the LAN readiness checklist in Table 8-3.

Table 8-3. Sample LAN readiness checklist

	Yes	No
Is the Ethernet segment comprised of Category 5 or better cable plant with UTP/RJ45 connections that terminate centrally where the switches reside?		
Are all of the Ethernet switches capable of VLAN? (See Chapter 9 for more on this.)		
Will the Ethernet switches provide PoE or is there ample space to add power injectors?		
Are there DNS servers accessible on this LAN segment?		
Does this segment have a DHCP server?		
Does the DHCP server have a large enough address pool to support the IP phone count?		
Is there a diagram of the LAN campus, including addresses and locations?		
Has the VLAN configuration been documented?		
Is the power source for this segment backed up using a UPS, a power generator, or both?		
Are switches all centrally located where backup power is available or do all intermediate distribution points have backup power?		
Do all switches and routers that touch this LAN segment support IP precedence and/or ToS?		
Has layer 3 switch routing policy been documented?		
Is the nominal packet loss on the Ethernet LAN currently below 1 percent?		

Once you've answered Yes to every item on the LAN plan checklist, you've got to identify which, if any, of the following required server devices will be located on the LAN:

- SoftPBX server
- Voice mail server, if separate from softPBX
- PSTN or TDM gateway, if separate from softPBX
- Analog endpoint gateway
- Trivial File Transfer Protocol (TFTP) server, if used
- Call accounting database server or syslog server
- Directory server (LDAP, Active Directory, etc.), if used
- DNS server

- DHCP server
- Firewall, NAT device, or Internet gateway router
- Default router

None of these need to be located on the same Ethernet segment as the IP phones (except the DHCP server), though disaster planning and bandwidth preservation often dictate that they be. (More on disaster planning is found in Chapter 13.) For example, if you're using a softPBX call path for all calls originating from a segment with 1,000 IP phones, it's probably a good idea to locate that softPBX on the same segment as the phones. This is because an IP-over-T1 link, for example, wouldn't have enough bandwidth to carry the media channels of even 50 G.729A/RTP calls. Such a link would be required if the softPBX were located on a remote LAN segment.

The TFTP, DNS, and DHCP servers are used to provide network configuration information and firmware updates to IP phones. Like a desktop PC, an IP phone needs an IP address, a subnet mask, a DNS address, and a default router address (if it will be used with a WAN or the Internet). These configurations can be centrally managed using a DHCP server and then automatically assigned to each phone during the DHCP host registration process. Some VoIP-specific configurations, such as SIP proxy addresses or H.323 gatekeeper addresses, can also be set on IP phones via DHCP or TFTP.

 TFTP servers are used to store firmware updates and configuration files for IP phones. This allows easy mass migration to new firmware and mass configuration.

So a TFTP server, DNS server, and DHCP are all essential for an enterprise VoIP LAN setup; account for them as you design your network. TFTP servers and DNS servers needn't be on the same Ethernet segment as the IP phones, but the DHCP server does. Count on one DHCP server per segment, just as you would in a plain TCP/IP LAN setting. Also—in many instances, TFTP, DNS, and DHCP services can all be run on the same physical server. DHCP is offered as a built-in feature on many routers.

WAN

Your WAN links, just like your LAN links, may have varying degrees of VoIP-friendliness. Some WAN technologies are particularly good for VoIP, like point-to-point T1, ATM switching, and optical links. Point-to-point T1 is good because its effective link speed doesn't fluctuate 1.5 mbps, because it's always dedicated to you and no other network users, and because it doesn't introduce much packetization delay. T1s can be sliced and diced as the administrator sees fit: half PCM DS0s, half data for TCP/IP, and so on. ATM switching works well for long-distance VoIP networks, especially through the use of VoATM, a way of removing some IP packet overhead before transport over the ATM network.

Other types of WAN links are very poor for VoIP, such as frame-relay, fractional T1 (56 k or 64 k), ISDN, and microwave radio. Frame-relay brings so much latency to the table that it can really frustrate callers, and bursts of packet loss and jitter are common on frame-relay networks. Fractional T1 (56 kbps) and ISDN just aren't fast enough to carry more than a single G.729A-encoded call at a time. Microwave radio bridges have some of the same problems that 802.11b/g WLAN networks have—especially packetization delay and jitter—and one that WLAN doesn't, namely sensitive antenna calibration.

Wide area network technologies don't always support the same QoS measures as Ethernet. Keep this in mind as you document your WAN. Does each point-to-point T1 have routers on either end of the circuit that can provide quality-of-service measures? Do those radio bridges support VLAN tagging? Do your ATM-WAN or frame-relay provider's other customers use the network for voice? Talk to the providers and some of their other customers to find out how well it works.

Table 8-4 shows a sample readiness checklist that you can use to evaluate your wide area network.

Table 8-4. Sample WAN readiness checklist

	Yes	No
Does the WAN backbone (if Ethernet) have less than 1% nominal packet loss?		
Is the WAN backbone capable of ToS and IP precedence?		
If this network's utilization will exceed 30% voice traffic, do all of the WAN routers support RSVP?		
If this network's utilization will not exceed 30% voice traffic, do all the WAN routers support DiffServ?		
Do all possible call paths across the WAN have a round-trip latency of 150 ms or less?		
Are all remote office links at least 128 kbps?		
Do all WAN connect points have backup power?		
If using frame-relay, is the typical maximum jitter on all PVCs less than 30 ms?		
If VPN is to be used to connect telecommuters to the VoIP network, do all telecommuters have broadband Internet access?		
Are the WAN connections fully documented, including service provider names, circuit IDs, router and DSU/CSU configs?		
If your WAN uses MPLS, are all shims/zones documented?		
If centralizing softPBX resources via WAN, does enough bandwidth exist between all client locations and the proposed location of the softPBX?		
Do WAN resources like routers and media gateways have access to a TFTP server for firmware updates and config changes?		
If using point-to-point T1s with legacy voice signaling, and planning to add VoIP to them on some of the DS0 channels, are the encoding and framing methods suitable for IP networking?		

VPN

Virtual private networks have proven to be both a substantial cost saver and, sometimes, a substantial headache. They're great because they allow the Internet to be leveraged in a wide area connectivity application and allow remote offices and road warriors to connect to the data center to check their email, update their database apps, and synchronize their Palm Pilots. Most of the time, VPN is great for these apps, because they are not sensitive to lag, jitter, and packet loss.

But VPN poses a much greater challenge to VoIP—for two reasons. First, VPN is especially laggy and ridden with the overhead of tunneling and encryption. Second, most VPN techniques don't provide for any quality-of-service monitoring or enforcement. So, like broadcast Ethernet or VoIP over dial-up, you may be able to use VPN to carry voice calls, but the quality and consistency of those calls will be unpredictable. Chapter 13 includes a configuration for a Cisco router that enables a VoIP trunk using a VPN tunnel.

Managed VPN

In response to the quality shortcomings of VPN, some carriers have begun to introduce a service called managed VPN. This service connects multiple sites within your organization to one another completely within a single carrier's network rather than the public Internet, using VPN technology to keep the traffic secure within that network. Having all of the VPN subscribers connected to the same network allows the service provider to maintain service-level agreements, something not possible with ordinary VPN. Managed VPN is therefore a much better carrier for VoIP traffic.

VoIP over dial-up

Save yourself the hassle and disappointment. No analog dial-up connection has enough bandwidth to support a full-duplex, enterprise-quality VoIP call. You may use Yahoo! Chat's VoIP over dial-up and think that, since it works, telephony-oriented VoIP can work over dial-up. But that's only because Yahoo! is using the Truespeech codec, which is not an enterprise-grade (or even toll-quality) codec. Modem speeds (56 kbps) just aren't fast enough to carry the voice stream, and besides, if your modem is plugged into a POTS line, just use *it* for voice calls.

Implementation Plan

Your project timeline, the labor resources required to complete the project, and the training plan comprise your implementation plan. Even with a truckload of top-of-the-line gear and all the technical knowledge you can absorb, your VoIP implementation can fail if you don't ensure that there's enough:

Time
    To assess the network, understand users' needs, and provision equipment

Manpower
> To program and deploy IP phones and servers and to provide project leadership

Training
> For the end users of the system

Standards basis
> To ensure proper interoperability with old and new voice systems

Knowledge Foundation

Your technical know-how will prepare you to project the size, scope, and commitments of time and money that will be required to complete your VoIP rollout. Learn the technical nuances of IP telephony, and you'll be much better prepared to qualify, delegate, and prioritize the tasks involved in the rollout. Technical depth also gives you a better foundation for dealing with your telephony system salespeople and contracted engineers.

Time

How long will it take to assess your network? Assess your voice applications? Think about the way your organization uses voice systems—do you have a receptionist who handles all calls? A small office or a large, multilingual call center? How will the size and scope of your user base affect the completion date you, as an integrator, are willing to commit to?

A good way to get a realistic grasp on *just how big* your voice project will ultimately be is to interview the end users who are most familiar with the usage of current voice systems in your office. The receptionist, if you have one, probably has a special wish list of features, while the salespeople or dispatcher has a totally different list of "nice-to-haves." In the end, it's these end user desires, and the desires of the organization, that will measure your success.

How much time will it take to program each phone? Each server? Each Ethernet switch and router? How long will it take to provision equipment, train administrators, convert or consolidate phone lines? How much of this can be performed by a third-party vendor versus internal staff? Get as good a grasp on these matters as you can.

Manpower, Vendors, and VARs

Your project-management style will vary depending on the size and scope of your VoIP project. Some VoIP adopters prefer to manage the project internally along with other projects; others may have a dedicated project manager lead the implementation full-time. That project manager may be an in-house employee or an outside contractor.

Who is going to provide leadership for the overall system conversion to VoIP? Will regional or departmental leadership be necessary? Will the implementation team be small enough for one project manager to lead, or will it be a large team?

Now, telephony system vendors may be able to assist you in making these man-power decisions. They may also be able to help you size your implementation team or provide supplemental support to your internal systems group.

Selecting a VoIP platform

Selecting a commercial VoIP equipment platform is probably easier than selecting a VAR or consultant to implement that platform. Equipment makers usually provide spec sheets that outline the technical capabilities of their solutions. Signaling standards, codecs, network compatibility, and capacity are usually worn on the sleeve, so deciding between Avaya, Cisco, Alcatel, Siemens, Shoretel, Nortel, Polycom, or SNOM is usually a deductive, feature-comparing affair.

A pretty painless way of comparing and contrasting all those equipment makers is to compare their support of the features needed by your organization, point-for-point. To do this, consult your user community and develop a list of all of the features and standards that are desired. Include any feature that is currently in use on the existing legacy voice components that are being replaced. Don't leave out anything just because it is taken for granted on your current system. Really dig in to your users' workflow in order to assemble this list. You'll probably find that it's at least a few pages long.

Next, assign a priority score, which we'll call a weight, to each feature on the list. Assign 3 to those items that are absolute must-haves. Assign 2 to the items that are very important but not showstoppers if they don't exist on the new system. Assign 1 to the items which are merely nice-to-haves.

Now, organize the features on the list in groups that make sense. Put call-handling features like conferencing and music-on-hold together. Put messaging features like email notification and all-points bulletin together.

Once all the features are grouped and weighted, examine each vendor's ability to provide each feature on the list. If the vendor provides the feature, add its weight to the weight of the other features that the vendor supports. As you examine each feature, you'll begin to develop a cumulative score of this vendor's ability to support your needs. Use this technique for each vendor that you're considering, and you can compare their scores head-to-head.

Using weighted scoring will result in an emphasis on the features you really need. Add up the scores by group and by vendor, and it should be pretty obvious which vendor you should select. Use Microsoft Excel or your favorite spreadsheet to automate the tally using formulas. Table 8-5 shows an illustration of this concept from an actual telephony feature comparison study.

Table 8-5. A partial feature comparison worksheet

	Cisco Call Manager Express	Avaya S8300 Media Servers	Asterisk in-house w/Cisco phones	Asterisk hosted w/Cisco phones
Message-waiting indicator on phone	Yes	Yes	Yes	Yes
Forwarding of messages to email	Yes	Yes	Yes	Yes
Flash firmware updates to other vendors' phones	No	Yes	Yes (TFTP)	Yes (TFTP)
Flash firmware updates without interrupting calls in progress	Yes	Yes	No	No
Automatic hardware failover on softswitch cluster	Yes	Yes	No	Yes
Ability to use SIP endpoints without closet gateways	No	Yes	Yes	Yes
Use MS Outlook to play back voice mail on endpoint speaker	Yes	No	No	No
Runs on Linux	No	Yes	Yes	N/A

When creating a feature comparison worksheet, use your own feature requirements, not those that appear on one vendor's spec sheet. Think about individual features you or your users have used. Do the phones need LCD displays? Do they need back-lit keypads? Do they need a speakerphone? What about inline power—does each vendor provide a solution that meets your needs? Does each vendor allow analog or T1 interfacing to your current PBX systems so that you can do a staged migration?

Choosing a VAR

Choosing a manufacturer's representative can be a daunting task. Corporate politics often play a role. If you're the sole decision-maker, then more power to you—but make sure you select (or recommend) a VAR for the right reasons. Chief among these reasons is a willingness to support published standards, and a standards-friendly attitude isn't always politically correct. If you're not the sole decision-maker, then the information that follows could be very valuable to you.

The people in the boardroom don't care about technical capabilities or geeky religious fascination with one brand or another. They care about productivity, efficiency, minimizing costs, and the potential for system growth. It's up to you as an integrator of VoIP to convince decision-makers who aren't savvy to the technology that a basis in standards is the best way to achieve these goals. What makes it tough is that VARs tend to claim just the opposite: that the standards "aren't good enough yet" and will cost the organization more than their own, possibly proprietary, solution.

You may be totally sold on building your VoIP network with a certain vendor's platform. But because of a political connection that's out of your control, you're forced to deal with a VAR who doesn't handle that vendor's equipment. Keep standards-compliance in mind from day one, and you'll have a better chance of avoiding this situation.

Creating an RFP

Unless you have a well-trained team of people who implement VoIP networks all the time (not likely unless you're a VoIP integration consultant), don't make the mistake of thinking you can implement the whole thing by yourself. Voices from outside your organization are very important—and accurately communicating your intentions and requirements to them is equally important. Hence, if your VoIP project is larger than a dozen phones or so, you should draft a Request for Proposals to define the scope of your implementation and qualify interested resellers and consultants.

Your RFP should clearly indicate:

- Your stated objective in seeking IP telephony solutions
- The general scope of your project (are you setting up an IP trunk to link two PBXs, or are you giving IP phones to 1,000 users?)
- What features and applications are required, and by which users
- The user count and departmental structure of your organization
- The structure of your current network
- Your standards preferences
- Which pieces of legacy equipment need to be integrated into the VoIP network
- A realistic time frame for completion based on your own technical knowledge and business requirements

Design your RFP so that standards—not brand names, makes, and models, are the prescription for success. Instead of selecting a particular manufacturer's platform ("We're standardizing on Siemens"), select a set of published standards upon which to build your platform ("We're standardizing on SIP, DiffServ, and 802.3af"), and then write your Request for Proposals so that standards-compliance is required.

Require each RFP responder to explain how she will support the standards you need. This way, if you prefer a different vendor than your senior-level management does, your argument doesn't need to be, "I just like Nortel better." Instead, it can be, "*ABC* system adheres to the standards that our company requires, while *XYZ* system does not." Your quarrel with the status quo will sound a lot better if you back it up with published standards.

Your RFP's basis in standards allows you to put dollars behind your preferences. If a particular vendor doesn't support a certain required standard, you can calculate the additional cost imposed on the project associated with that lack of compliance. For

> ## Long or Short RFP?
>
> Short. Writing a long, laboriously detailed RFP may be a good exercise to flex your mental VoIP muscles and refamiliarize yourself with voice terminology. Writing a 200-page RFP might be a minimum requirement at the Internal Revenue Service, but it isn't going to get *you* a better VoIP proposal. Keep in mind, even if you've become a VoIP expert and you've got a pretty detailed plan about how to proceed, it's still the RFP responders' job to fill in the blanks, offer their advice, and prove their knowledge.

example, if you can't use SIP-only endpoints, which are fairly cheap and interoperable, you could demonstrate *how much more* it will cost to implement the solution using SCCP endpoints instead.

Training

How will your end users be trained? How will your system administrators—yourself included—become familiar with the quirks of a VoIP network? Who will do the training—an outside vendor or an internal expert? How much time will it take, and how much will it cost?

Key Issues: VoIP Readiness

- The business environment, network environment, and implementation scenario must be adequately prepared for VoIP in the enterprise.
- The more technical VoIP knowledge you—as an integrator—possess, the better prepared *you'll* be for VoIP.
- TCP/IP is the only packet networking protocol used on a VoIP network, though VoIP networks are often connected by ATM.
- Enterprise VoIP guidelines vary by vendor, but the technology works best on switched, high-speed LANs and WANs.
- Wireless Ethernet and VPN will work with VoIP, but not with the reliability of switched Ethernet.
- VoIP can bring businesses gains of efficiency, productivity, and lower costs.
- Applying a cost-benefit assessment user-by-user or department-by-department is a good way to see which workgroups in your company will most benefit from these gains.
- For large VoIP projects, it's important to develop an RFP and allow several vendors to bid against it.
- The RFP should specify the standards you prefer, not the brands you prefer.

## CHAPTER 9

Quality of Service

Quality of Service is a subject of crucial importance to your success with VoIP. Not surprisingly, QoS technologies aren't often well-understood by traditional telephony people. But those with a data background have never had to use them, either. So this chapter will introduce you to QoS concepts and protocols, the problems they solve, and the complexities they introduce. Don't let that scare you, though. There are QoS approaches for networks that have a few dozen endpoints, and there are approaches for giant, high-capcacity networks, too.

QoS Past and Present

In traditional telephony, quality of service for each and every phone call is guaranteed by the constant availability of dedicated bandwidth. Whenever a channel or "loop" is established across the network, the bandwidth allocated to that channel is steadfast and unchanging. Most digitally encoded call paths on the PSTN use the same codec, G.711, so transcoding isn't necessary. Almost no processing bottlenecks will be found on the PSTN, and since the system isn't generally packet-based, there is almost never degradation in perceived call quality as a result of congestion.

> As a circuit-switched network, the PSTN provides quality of service by having almost no latency or congestion issues. The option of lowering call quality in order to increase call capacity never existed on the PSTN.

If the PSTN and SS7 can't establish a full-bandwidth path through the network, the call just doesn't get connected—and the caller hears a busy tone. The designers of the PSTN felt a breakdown of connectivity would be preferable to a breakdown of quality.

Of course, packet networks work the other way. When bandwidth availability drops, as more packets are sent on the network, throughput slows. Until a certain breaking point, bandwidth availability can be compromised while still allowing data through; the transmission just slows down. Some applications tolerate congestion and slow throughput better than others. The more tolerance an application has, the higher its *error budget* is said to be.

Slowness of transmission—latency—is the enemy of Voice over IP, and one of the key contributors to failure with the technology. Aside from careful network design and bandwidth provisioning, which are factors in building any IP network, there is an elegant solution to the latency problem, one that allows local and end-to-end guarantees of bandwidth and prioritization of real-time traffic over less sensitive traffic. This chapter covers that solution, which comes in the form of QoS protocols and standards: 802.1p, 802.1q VLAN, DiffServ, RSVP, and MPLS.

Call-quality scoring

Historically, the quality of phone calls' audio has been measured using the mean opinion score (MOS) from a group of listeners. These listeners hear sound samples from calls of varying quality, recorded during different sets of network conditions. A sound sample from each set of conditions is played for the opinion group, and each rates the sample's quality on a scale of 1 to 5, with 5 being the best quality. The conditions that can be used to alter the quality of a sound sample are choice of codecs, transcoding combinations, packet interval, and packet loss rate.

Using the MOS technique, researchers have determined that, with no packet loss, G.711's highest perceived quality score is 4.4. By comparison, G.729A's is only 3.6. But, when packet loss occurs, G.711's resilience stands out, as in Figure 9-1. There are other quality scales for gauging phone calls, but the MOS scale is still the most commonly used.

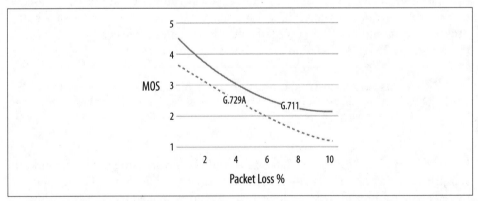

Figure 9-1. Packet loss has a direct negative effect on call quality (source: Nortel Networks)

It is unlikely that you'll need to use the MOS scale to grade call quality on your VoIP network. But it's important to note what the scale says. As you can see in Figure 9-1, you don't want to use G.729A across an ultrafast Ethernet link because the quality perceived by your system users will be lower than it ought to be. It also illustrates that you must arm your IP network against congestion.

How you can use the MOS scale

MOS can aid you in rating the perceived quality of your legacy system. Not all system builders will have the time or inclination to do this. But, if you support more than a few-dozen users or if call quality can vary depending on the call path, you should determine the MOS rating of all call paths on your current system. Ask a group of users to place calls across each call path and then record their MOS opinions of each call.

Do this before you replace your legacy links with VoIP. This way, you can use the scores as a guide when selecting standards and equipment for your new VoIP system. You can also tell whether you've succeeded in replicating the call quality you had before you started to replace legacy equipment.

After the implementation, particularly in large corporate or carrier-class networks, establish an SLA (service-level agreement) between you and your users that provides an MOS expectation for every call path that meets or exceeds that of your legacy system.

Noise

In traditional telephone networks, like the PSTN, the MOS scale was used to give engineers feedback that could be used to improve perceived quality. One of the biggest factors in perceived quality is noise. VoIP itself doesn't increase or decrease noise in the traditional sense, except where digital signal processing is applied specifically for noise reduction.

> Additive noise is the unwanted signals that accompany all transmissions of sound. Subtractive noise is an interruption or reduction of the sound transmission, such as that caused by packet loss.

Noise can be added to the phone circuit at many points. Consider a phone call placed from a loud factory floor. The person making the call will likely jam the phone receiver up to his ear as closely as possible, while plugging his other ear with a finger, all so he can hear the call more easily. What he's doing is *noise reduction*. There are electronic methods for noise reduction, too, but none are really specific to Voice over IP.

But VoIP does introduce new *kinds* of noise, broadening the traditional definition to include everything shown in Figure 9-2. Noise includes not only background noise and signal interference, but also a wide range of new elements that change the perception of the sound in the telephone earpiece. These noise elements can be momentary, like a short burst of packet loss, or permanent, like transcoding distortions that occur when the lossiness of a particular codec is compounded by the lossiness of a second codec. The resulting signal may sound robotic or machine-like.

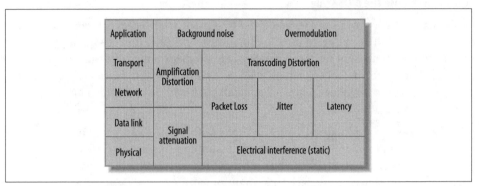

Figure 9-2. Sources of noise occur at almost every OSI layer

When noise alters a signal, the signal is said to be distorted. Distortion can be either additive or subtractive, meaning that the strength or amplitude of the sound is increased or decreased by the distortion. Additive distortion sources are things like background noise and amplification, while subtractive distortion is caused by signal loss and attenuation.

Thankfully, a PhD's knowledge of electromagnetics isn't essential. As an enterprise VoIP maintainer, it's probably enough to know that, while noise cannot be entirely avoided, it should be minimized. One of QoS's roles is to help you avoid situations in which poor service at the lower layers of the network results in additive or subtractive noise.

Some noise elements cannot necessarily be improved by VoIP QoS techniques, because they aren't exclusive to VoIP environments:

Background noise
> Sounds in the environment surrounding the caller or receiver that bleed into the call. Background noise can cause a phenomenon called over-modulation, which results in distortion because the sound being sampled is so loud that it is off the scale. Shouting into the telephone receiver is a good way to demonstrate over-modulation, too.

Signal level and circuit noise
> Elements that affect the way voice quality is perceived. If the caller's signal level is too low, her voice will sound faint; if it's too high, it will sound fuzzy or distorted. Circuit noise is caused by attenuation and interference at the physical layer.

Quantizing

Like the G.711 (PCM) codecs used in traditional telephony, all forms of quantization produce distortion. This is basically unavoidable, though this chapter describes how best to select codecs that minimize signal distortion.

Using an IP network as a transport for voice can indeed impair the sound signal, and this is where network QoS measures come into play. Since IP networks weren't originally designed for "reliable," real-time, and connection-oriented transmission the way the PSTN is, QoS measures can make up for IP's shortcomings and make for a network that is actually more resource-sparing than the PSTN could ever be.

QoS is Two Things

Quality of Service is both a network design concept and a set of standards for bandwidth reservation on the network. The QoS *concept* deals with fundamental detractors from quality—like packet loss and latency—and their cure: sound network design. This means providing enough bandwidth and proper physical and geographic organization of network traffic. Indeed, most network engineers, when faced with bottlenecks, instinctively seek to add more bandwidth. There's nothing wrong with that—it's just not the most elegant or cost-effective way of dealing with the problem. As a result, many network engineers overbuild from the get-go, and this can be a waste. After all, it's quality of service that's needed, not quantity of service.

The QoS *standards*, on the other hand, are specific network protocols that provide quality measures such as bandwidth reservation and packet prioritization.

Does overcapacity remove the need for QoS?

Some system builders approach the Quality-of-Service challenge by building a network that has so much capacity that quality problems are unlikely to occur. This may indeed work for web surfing and database apps, but it doesn't always work for voice. Besides, it's foolhardy to assume that, just because you're building your network big, you'll never need to manage bandwidth on it. QoS measures help you do the managing.

A second argument against overcapacity is in the cost of infrastructure. Installing a 10 gbps Ethernet switch is a less efficient method of ensuring on-time delivery of voice packets than a much cheaper gigabit switch that's coupled with a QoS measure or two.

Transistor technology doubles computer processing capacity once every few years, resulting in a constant upgrade cycle that yields ever more powerful computer hardware and software. When new, faster processors come to market, they are usually prohibitively expensive—and so are new, bigger-capacity networking devices. As is often the case with the latest Pentium iteration, it pays to wait a few months or a year until the cost of high-capacity network gear drops.

This way, your cost per megabit stays low. QoS can help you slow the costly cycle of infrastructure growth on your network. If money is tight, wouldn't you rather use your project budget for user-pleasing telephony features than for additional bandwidth that will be 30% cheaper a year from now?

QoS protocols approach this matter with an assumption of limited bandwidth resources. Their purpose is to increase the availability of the network for high-priority traffic. Sometimes, this just means handling voice packets before handling other kinds (precedence); other times, it means allocating a logical channel of dedicated bandwidth across the entire network from caller to receiver just as the PSTN does during call-setup for a long-distance call (bandwidth reservation).

 Multiple QoS standards can complement each other, but don't overdo it. Enforcing system policy for four of five of them, while possible, wouldn't be a very practical idea.

Several protocols—RSVP, MPLS, DiffServ, and 802.1p—provide different approaches to QoS. Some of them work on a coarse, or single-link basis. Others work on a fine, or end-to-end basis. These two basic types of quality measures are described in Table 9-1.

Table 9-1. Class of Service versus Quality of Service

Class of Service (CoS)	Quality of Service (QoS)
Coarse-grained.	Fine-grained.
Dumb.	Smart.
Simple traffic-shaping scheme.	Complex protocols.
Per-hop behaviors.	Traffic contracts.
Cannot guarantee bandwidth even across a single data link.	Can guarantee bandwidth across the entire network.
Runs on a single segment or on a single hop.	Runs on all hops between endpoints on every call.
Traffic precedence is used to increase network availability for all voice traffic.	Bandwidth is reserved exclusively for individual channels of voice traffic.

In Table 9-2, the QoS standards are compared and contrasted.

Table 9-2. Quality of Service standards compared

Protocol or standard	CoS or QoS	When to use	When not to use
802.1p	CoS	Single-segment Ethernet networks	Always use
802.1q VLAN	CoS	Private Ethernet networks	—
DiffServ	CoS	High-capacity routed networks	Networks with a higher percentage of voice traffic than data traffic

Table 9-2. Quality of Service standards compared (continued)

Protocol or standard	CoS or QoS	When to use	When not to use
RSVP	QoS	Bandwidth-limited routed networks	Networks with a higher percentage of data traffic than voice traffic
MPLS	QoS	Carrier-grade or ATM-switched networks	Anything less than carrier-grade networks

Latency, Packet Loss, and Jitter

Latency (also called lag), being the chief cause of poor perceived call quality, is caused primarily by slow network links. End-to-end latency, in the case of VoIP, is the time it takes from the instant the caller utters something until the time the receiver hears that utterance. Research has established that round-trip latency less than 150 ms is not immediately noticeable, but latency higher than 150 ms is discouraged, and latency higher than 300 ms is considered unacceptable. (Cisco Systems states that round-trip latency higher than just 150 ms is unacceptable.)

Latency has the following effects on telephony applications:

- Can slow down the human conversation
- Can result in caller and receiver unintentionally interrupting each other
- Can exacerbate another Quality-of-Service problem: echo
- Can cause synchronization delays in conference-calling applications

The best ways to beat latency are to use low-packet-interval codecs and maintain fast network links, because QoS protocols alone cannot directly improve latency's impact. That is, they can't speed up your network. One negative effect of latency in telephony systems is echo. It's discussed in more detail in the following section, "Echo."

 ITU-T recommendation G.114 defines acceptable latency budgets in various telephony applications.

Latency is the enemy of VoIP, but it's also the way some codec-based solutions to packet loss and jitter are made possible. This is because those solutions—namely, jitter buffers and packet loss concealment (PLC)—are both sources of latency. In fact, many things contribute to latency:

- Framing and packetization
- Software processing and PLC
- Jitter buffering
- Routing and firewall traversal

- Transcoding
- Media access and network interfacing

Minimizing latency is an important way to maximize the VoIP network's perceived quality of service.

The two biggest sources of latency are framing/packetization, which can add up to 30 ms of latency, and routing, which can add 5–50 ms per hop. Another big contributor is transcoding. The sample latency factors produced by transcoding in Asterisk 1.01 are outlined in Table 9-3. Source codecs are shown in rows, and destination codecs in columns.

Table 9-3. Asterisk transcoding delays in milliseconds, by codec[a]

Codec	GSM	μLaw	ALaw	G.726	LPC10	ILBC
GSM	N/A	4	4	12	14	59
μLaw	10	N/A	1	10	12	57
ALaw	10	1	N/A	10	12	57
G.726	17	2	2	N/A	19	64
LPC10	18	10	10	18	N/A	65
ILBC	19	11	11	19	21	N/A

[a] Sample latency factors are from a Linux 2.4.20-6 on a Pentium III 600 mHz.

Packet loss is very damaging to VoIP calls. Its chief cause is network congestion. As shown in Figure 9-1, different codecs have different packet loss tolerances or error budgets. PLC is a feature of some codecs that allows perceptions of a quality breakdown to be minimized through vectoring algorithms. These codecs work by replacing the sound that would presumably have been produced by a packet that was lost with sound that is predicted based on the sequence of packets received before it and (when extensive buffering is used) after it. But even with PLC in force, packet loss rates on a VoIP network should be kept below 1%. While QoS measures can improve the packet loss problem by providing reserved bandwidth or precedence for voice packets, it's still a plain-old good idea to conserve available network capacity and keep packet loss rates down. A drawback of PLC is that it can increase latency. Experimentation with PLC-equipped codecs should be done to determine how negative the latency-impact PLC is in your VoIP network.

Jitter is a more complex problem than latency and packet loss. It's the variation in latency time from one packet to the next. It causes packets to arrive out of order, leaving gaps in the framing sequence of the voice signal. Jitter is at its worst when voice traffic must travel through several routers on the network. The more hops, the worse jitter can get. Different routers, especially those at ISPs, may be configured to queue and forward different kinds of traffic in different ways. Others may be load-balancing, which can contribute to jitter. The chief goal of QoS protocols is to eliminate jitter. Devices called jitter buffers, in endpoints and VoIP servers, can minimize the effect of jitter, too. But, like PLC measures, they do so by increasing latency.

Echo

When you hear the words you've just spoken repeated back to you a split second later on the telephone, you're experiencing *echo*. Chances are, if the echo occurs less than 150 ms from the time you actually said the words, then you won't notice it. But when the echo occurs above this threshold, it can be particularly annoying. Echo is an unfortunate by-product of the gateway electronics that bridge soft-based PBX systems to analog or TDM links. It is caused by three conditions, and it's at its worst when they exist together:

- Interfacing between TDM and VoIP endpoints or analog and VoIP endpoints. The more interface points in the network, the bigger pain echo is likely to be.
- Long round-trip latency between caller and receiver. The higher the latency, the more annoying echo is likely to be.
- Interfacing of a call path between two-wire analog and TDM or four-wire analog devices. In this case, echo is caused by an inability of the TDM or four-wire circuit to cancel the local side-tone signal on the two-wire device (side-tone is covered in the sidebar).

Hybrids and echo

There's a lot of physics at the heart of the echo problem. When resulting from legacy interface conversion, the echo problem usually lies in the quality of the analog interface, called a *hybrid*. Hybrids may have built-in echo-canceling abilities. Some hybrids are more susceptible to echo than others. Many have noted that Digium's TDMxxx hybrids are less echo-ridden than their X100P hybrids.

While onboard echo reduction may exist on some hybrids, good network design is a more strategic way to deal with the problem. The keys to removing echo are minimizing the use of gateway devices that use hybrids (which, today, are quite hard to avoid) and reducing latency as much as possible. QoS and CoS measures will help you do the latter.

Zaptel's echo suppression

When you use a network that has little or no CoS/QoS support, like the Internet, you can still do a few things to minimize echo. Most IP phones offer varying degrees of echo cancellation, and Asterisk has a built-in echo canceller for VoIP and Zaptel channels. By modifying the echotraining setting in *zapata.conf*, you can raise or lower the echo canceller's sensitivity to echo. Experiment with each interface to see what value works best.

Side-Tone

One kind of echo is helpful. When you hear *your own* voice in the phone receiver, you have a tendency to assume the person on the other end of the call can hear you OK. When you don't hear yourself in the phone receiver, you tend to say, "Are you there? Can you hear me?" You're assuming that your partner can't hear you because you can't hear yourself. This "helpful echo" is caused by analog loopback in your own phone and is referred to as side-tone. Most late-model IP phones provide side-tone.

There's also an *aggressive suppressor* algorithm available—but you'll need to uncomment this header in */usr/src/zaptel/zconfig.h* and recompile the Zaptel driver module in order to use it:

```
#define AGGRESSIVE_SUPPRESSOR
```

CoS

CoS, or Class of Service, systems work to prioritize traffic on a single data link. While QoS refers to the greater network, CoS refers to only a single data link. So, an Ethernet switch might provide a CoS packet prioritization to/from a single host, but a group of routers might participate in a more elaborate, end-to-end QoS solution. The key difference is this: CoS is a single-link approach, while QoS is an end-to-end approach.

Class of Service systems define per-hop behavior, so they cannot guarantee a service level in terms of capacity or speed. Instead they give their "best effort" to deliver priority traffic according to those per-hop behaviors, which are established by you, the administrator. Class of Service solutions are great on data links where less than 30% of the traffic is voice, which is probably a majority of enterprise networks today. But they tend not to be as effective when priority traffic like voice represents the lion's share of the packets on the data link—as is often the case when the data link is a slow one.

Two key standards support CoS:

- 802.1p/ToS
- DiffServ

802.1p and ToS

802.1p uses a 3-bit portion of the Ethernet packet header to classify each packet into a particular level of precedence on the local data link. Each precedence level defines the per-hop behavior. 802.1p calls prioritization *traffic class expediting*. Type of Service (ToS) is the portion of the IP packet header that stores the same precedence information. Many vendors call ToS *IP precedence*. 802.1p and ToS tend to be used together when TCP/IP is the networking protocol.

If your VoIP network will be more than 70% data-to-voice and unlikely to reach capacity, packet prioritization techniques like LAN-oriented 802.1p and its WAN cousin DiffServ are adequate.

802.1p is a standard that benefits a single data link as opposed to a group of them using a single CoS policy. So it tends to be a feature of Ethernet switches, as opposed to IP routers, which instead support ToS. Since 3 bits are allocated for classification, 802.1p allows for eight classes of service. The names associated with these classes are arbitrary to the network they are used on, but Table 9-4 lists the suggested, generic names.

Table 9-4. Suggested 802.1p classes

Number	Name
0	Routine
1	Priority
2	Immediate
3	Flash
4	Flash-override
5	Critical
6	Internet
7	Network

The eight classes' individual per-hop behaviors are administrator-defined, so you could define class 5 for voice traffic (which is the prevailing wisdom). This would give voice traffic higher precedence on the local data link. Each hop along the path of the traffic would have to support the same per-hop precedence behavior in order to make the QoS policy consistent across a routed WAN.

Like many low-layer standards, 802.1p is a product of the IEEE.

Project 9.1, later in the chapter, describes how to check all the routers along a certain call path to see if they support 802.1p.

DiffServ

DiffServ, defined in RFC 2474, is a CoS standard that uses ToS tags in a more elaborate manner than 802.1p. While 802.1p tends to be used in switched Ethernet environments, DiffServ is used to support routed point-to-point WANs. DiffServ is an abbreviation for Differentiated Services.

When a packet reaches the edge of the network, either from an endpoint or from a remote network, DiffServ tags that packet's ToS header based on the priority established for that packet by policy. Once admitted into a DiffServ-equipped WAN, however, all subsequent router hops must enforce the priority set by the edge router that admitted the packet.

So, while DiffServ is considered a per-hop behavior CoS system like 802.1p, all Diff-Serv routers in the network core must uphold the prioritization decision that occurs at the edge. That is to say, core routers don't reorder or prioritize the packets, but they do forward them using a precedence policy that was established at the edge.

 Since a majority of bottlenecks occur on the edge of the network, on access links, rather than within the core network where there's usually tons of bandwidth, DiffServ may be the only Quality-of-Service standard necessary. Its policy decision points are always at the edge of the network.

In a nutshell, DiffServ furthers the concept of 802.1p so that priority policy can be established at the edge—once for the entire network—rather than on each individual hop. What neither of these protocols does, though, is ensure that the network is never overrun—that's a key difference between CoS (what 802.1p and DiffServ are) and QoS.

Policy servers

Common Open Policy Service, or COPS, is a way of storing and querying centralized policy information on the network. DiffServ can use COPS to obtain its marching orders for how to handle traffic coming into the network. In a COPS scheme, a centralized server called the policy server contains a policy record of traffic shaping and prioritization preferences that DiffServ or another CoS/QoS mechanism can retrieve. COPS itself doesn't enforce the prioritization—that's the job of DiffServ, which runs on routers; COPS just provides a way of maintaining a centralized record of the traffic policy. COPS is a product of the IETF Resource Allocation Protocol working group. Another IETF recommendation, LDAP (Lightweight Directory Access Protocol), can also be used as the basis of a policy server.

DSCP classes

DiffServ Code Points (DSCP) are IP packet headers DiffServ associates with different levels of importance. Since they're 6 bits in length, DSCPs can be used to define quite a wide scale of possible service levels. Most implementations support only 3 bits, replacing the 3 bits in IP's ToS header. The other 3 bits are earmarked for extensions to the DiffServ standard.

DSCP per-hop behaviors break down into three basic groups, interchangeably called PHB classes, traffic classes, or DSCP classes:

AF *Assured Forwarding*, a highly expedient DSCP class, sometimes used to tag signaling packets such as H.245/H.225 and SIP packets.

EF *Expedited Forwarding*, the most expedient DSCP class, used to tag packets carrying actual sound data.

BE *Best Effort*, a nonexpedient DSCP class, used to tag non-voice packets. Many DiffServ decision points don't use BE.

When packets hit the edge of the network, the DiffServ-equipped edge router decides the PHB for each packet based on a judgment of which DSCP class the packet should fall into. High-priority packets get AF or EF, while the others get BE or no tag at all.

> If you use DiffServ, configure your routers (or softPBX) to use Expedited Forwarding for all voice traffic. Project 9.1 describes how to set up a DiffServ edge router using Linux and iptables.

The DiffServ CoS process

DiffServ goes to work during call setup, when an RTP media session is being established. As the edge router handles the RTP session startup, while the calling party is transmitting its first packets of audio, a classification lookup is conducted against the COPS policy server in order to determine the priority of that RTP session.

> In networks with no centralized policy servers, the DiffServ router admitting the traffic into the edge can have policy rules programmed into its onboard memory.

The COPS server informs the edge router which of the three DSCP classes each packet should carry. Once classified, the edge router tags the packet with the DSCP determined by policy. Furthermore, the edge router remembers a "policy rule" that applies to all subsequent packets in this particular RTP session. This way, they all get marked with the same priority. DSCP classes are backwardly compatible with ToS classes, so DiffServ doesn't preclude the continued use of 802.1p on local area data links (see Figure 9-3).

As the RTP traffic flows through the core of the network, it has already been classified and marked, so that core routers are spared the task of having to look up the policy repeatedly. Specific per-hop behaviors will always be the same as each packet

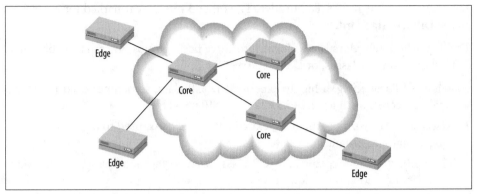

Figure 9-3. In DiffServ, edge routers determine the per-hop behavior as the packet enters the network

in the stream flows from one core router to the next. DiffServ has several advantageous qualities:

- As an IP-only solution, is usually managed with the same equipment used to manage an IP WAN: routers
- Can run transparently alongside other QoS solutions—including 802.1p and RSVP
- Is easy to set up on small- to mid-sized routed enterprise networks
- Can be optionally policy-based using COPS or LDAP
- Pushes QoS decisions to the edge of the network, resulting in less complexity
- Is compatible with IPv4 and IPv6

Project 9.1. Create a DiffServ Decision Point with Linux

What you need for this project:

- A Linux server capable of running the Netfilter firewall
- LAN

A Linux server that acts as a firewall or edge router can provide DiffServ policy enforcement just as a Cisco router can. It can also perform some packet-manipulation—say, to tag the priority of an 802.1p-tagged packet as it hits a locally connected Ethernet segment. Red Hat Linux 9 and Red Hat Enterprise Linux have DiffServ and 802.1p support precompiled. Other Linux kernels of Version 2.4 and above can have support for DiffServ and 802.1p compiled, as long as the kernel in question is compiled with these options enabled:

- Kernel/user netlink socket CONFIG_NETLINK
- Packet filtering CONFIG_NETFILTER
- QoS and/or fair queuing CONFIG_NET_SCHED

All options organized under that last QoS option must be enabled. If not using Red Hat 9 or Enterprise, you may also need to obtain, compile, and install the *iproute2* package. This package, and the kernel modifications introduced for QoS, together form the Linux Traffic Control system, a software library that you can control using a program called *tc*. Issuing the `tc` command is the quickest way to find our whether *iproute2* is installed. If you get a "command not found," then you don't have *iproute2*.

> Since it was designed to handle the QoS issues related to all kinds of network communications, not just VoIP, the Linux Traffic Control framework can be an overwhelming topic. While this exercise will show you how to mark RTP packets for a DiffServ domain, there are more practical (albeit more expensive) options for DiffServ than LTC, such as commercial routers.

Before you go any further, you should know that Linux Traffic Control supports a vast array of buffering, prioritization, tagging, and other traffic-shaping tactics. In fact, using LTC and Linux's kernel-based firewall, iptables, about a dozen different QoS standards can be applied. Some use a technique called *weighted* or *fair queuing*, which may hold and forward packets based on size or by transmission rate control—a technique sometimes called *leaky bucket*. LTC calls each technique for queuing or rate control a *discipline*. Other techniques use prioritization like 802.1p. Not all of LTC's capabilities are especially good for VoIP.

That said, configuring QoS enforcement points is more straightforward with brand-name routers because they offer well-documented, easy-to-administer QoS commands and integrate well with COPS. On Linux, a majority of QoS configurations are done using iptables, the command for altering the Linux kernel's built-in packet-filtering firewall, known as Netfilter.

> Using iptables to support DiffServ can create a policy decision point on the edge of the network. Core routers must also support DiffServ if the measure is to be successful from end to end.

Configure an iptables edge router for DiffServ

iptables can be configured to match packets based on origin/destination, protocol, or port numbers and then perform edge-style DiffServ classification before forwarding them. In this example, iptables matches all incoming or outgoing UDP traffic on a standard RTP destination port (5004) and assigns a DSCP class of EF using the iptables DSCP target:

```
iptables -A PREROUTING -p udp -D 0.0.0.0/0.0.0.0 --dport 5004 -j DSCP\
--set-dscp-class EF
```

The -A PREROUTING option tells Netfilter to apply this rule before the kernel routes the packet to its next hop. -p udp tells Netfilter to apply this rule only to UDP datagrams, ignoring TCP packets. The --dport 5004 option tells Netfilter that only packets destined for port 5004 (that is, RTP packets) should have this rule applied. Next, -j DSCP describes a target chain in which to modify the packet. Last, the --set-dscp-class EF option, which works only when the DSCP target chain is specified, changes the DSCP class of the matched packets to Expedited Forwarding.

If your Linux firewall is a point of connectivity for more than one network, it's possible to define DSCP classes on the basis of the destination network. If certain networks are associated with a different service level, you can tag the traffic as such. In this example, the 10.2.0.0 network gets Expedited Forwarding, while the 10.3.0.0 network gets Assured Forwarding:

```
iptables -A PREROUTING -p udp -D 10.2.0.0/255.255.0.0 --dport 5004 -j DSCP\
--set-dscp-class EF
iptables -A PREROUTING -p udp -D 10.3.0.0/255.255.0.0 --dport 5004 -j DSCP\
--set-dscp-class AF
```

Likewise, if protocols other than RTP are to make use of DiffServ, you can have your decision point tag them appropriately. This example tags IAX and RTP traffic as EF, while tagging all other UDP traffic as AF:

```
iptables -A PREROUTING -p udp -D 0.0.0.0/0.0.0.0 --dport 5004,5036,4569\
-j DSCP --set-dscp-class EF
iptables -A PREROUTING -p udp -j DSCP --set-dscp-class AF
```

The DSCP target of iptables will work only if the kernel has been compiled with the CONFIG_IP_NF_TARGET_DSCP option enabled, as described earlier.

 iptables provides DiffServ awareness for only IP Version 4.

Once a packet is marked with the appropriate DSCP class, it's up to the core routers to treat that mark with respect. If the DSCP class is EF, like the previous iptables example, then core routers need to expedite the traffic immediately.

LTC for DiffServ core routers

Indeed, if you intend to have your Linux firewall do something special with the packets depending on their DSCP class, then you can use iptables to tell Netfilter how to process them as a core router on a DiffServ domain would.

With a little bit of Linux Traffic Control diligence, iptables can be used to establish an even more extensive QoS enforcement point or a core router.

LTC on a softPBX

It can also be used, on a Linux-based softPBX, to tag traffic with the EF class. This way, the DiffServ decision point and the softPBX can reside on the same machine.

For more details, an excellent book on using Netfilter firewalls is *Linux iptables Pocket Reference* (O'Reilly). Another is *Building Secure Servers with Linux* (O'Reilly). LTC-specific resources are available at the Linux Advanced Routing and Traffic Control web site (*http://www.lartc.org*).

802.1q VLAN

Ordinarily, Ethernet provides one common broadcast domain per network segment. This means that when a packet comes across the segment destined for a local host whose hardware (MAC) address has not yet been resolved (ARPed) and associated with a certain switch port on the Ethernet segment, a broadcast to *all* ports is done in order to find a host with the right MAC address that's supposed to receive the packet. Once the port with the correct recipient is found, an ARP record is recorded in the switch so that all future traffic destined for that MAC address can go to that port rather than being broadcast.

One problem with this is that the broadcast traffic, while only a small percentage on most networks, can be a waste of bandwidth. Another problem is that, when broadcasts occur, every device on the network can receive them, which is a potential security hazard. But, thanks to IEEE 802.1q, both of these problems can be minimized.

802.1q VLAN (virtual LAN) isn't really just a CoS standard—it's a way to separate Ethernet traffic logically, secure Ethernet broadcast domains, organize the network by separating network protocols such as NetBEUI and IPX/SPX into their own VLANs, and all kinds of other cool local area wizardry. Each VLAN is a logically separate broadcast domain—even if it coexists with other VLANs on the same physical segment.

Layer 2 Switching

With most vendors' Ethernet equipment, to create VLANs, each switch port is assigned a VLAN tag—a numeric identifier that is unique within the network. This tag identifies the VLAN in which that port participates. Once the tag is assigned, the device connected to that port will receive traffic only from the assigned VLAN and will be able to send traffic only to the assigned VLAN. This concept is illustrated in Figure 9-4. Different groups of ports on each switch are assigned to VLANs 1 and 2.

VLANs are a layer 2 concept, because they operate at the data link layer, *below* the purview of network protocols like TCP/IP. This is a good thing, because it allows you to implement VoIP with a way of easily removing non-TCP/IP traffic from your

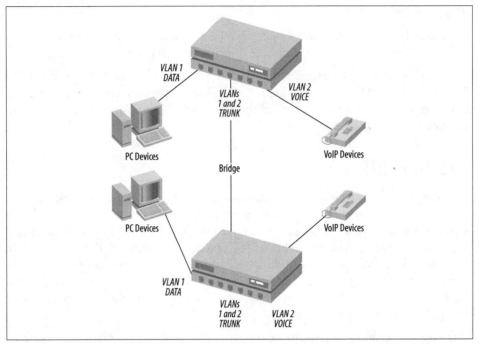

VLAN 1
DATA

VLANs
1 and 2
TRUNK

VLAN 2
VOICE

PC Devices

VoIP Devices

Bridge

PC Devices

VoIP Devices

VLAN 1
DATA

VLANs
1 and 2
TRUNK

VLAN 2
VOICE

Figure 9-4. VLAN trunk ports can be used to haul one or more VLANs from one switch to the next

VoIP network. The solution is to establish a VLAN strictly for VoIP devices and then enforce a policy of not using NetBEUI, IPX/SPX, or other non-TCP/IP hosts on that VLAN.

Layer 3 Switching

Sometimes Ethernet switches can be used to groom, inspect, or route traffic. This practice is called layer 3 switching, because, like a router, the switch must have some knowledge of the network-layer protocol. Layer 3 switching accomplishes some router-like activities: queuing, routing, and packet-inspection.

Queuing and packet inspection are of great interest to people concerned with QoS, because they can be used to shape the traffic on the data link based on each packet's characteristics. For example, it's possible to drop all non-voice traffic by filtering protocol types (UDP, TCP, etc.) and port numbers. This isn't the preferred way of giving VoIP traffic precedence on the Ethernet segment—just one way. 802.1p is probably a better way of handling prioritization, and just about all layer 3 switches support it.

But layer 3 switching can also be used to establish priorities for backbone traffic among different VLANs, and this is an important technique, especially when voice endpoints or trunk connections can't otherwise support a QoS standard. In one such situation, a data closet contains a hub or a non-QoS-aware switch, and that switch is connected to the backbone. Since that non-QoS switch can't do packet prioritization or DiffServ, at least it can glean the benefit of membership in an expedited VLAN.

Setting priorities for VLANs is a good way of bolstering QoS without a lot of overhead, but VLANs have one big caveat: excessive use can make your network hard to maintain. So don't use VLANs loosely. It's possible to build a network that's a VLAN monster. Complexity should be avoided when possible.

Many enterprise VoIP networks use two VLANs: one for traditional data devices like PCs and printers and another for VoIP servers and phones. The cabling, of course, is all the same. Other setups may add a third VLAN for administrator use or for QoS reporting.

VLAN trunking

Since VLAN itself is a logical way of separating Ethernet traffic, using it to combine, or trunk, Ethernet traffic must be a logical task, too. Just as you would plug a patch cable between two switches to bridge them, you must *logically* bridge two VLANs. So if you want to connect two Ethernet switches across the same VLAN, they must be bridged both physically—using a patch cable, wireless repeater, etc.—and logically—using a trunk port to connect the two VLANs to both switches.

When an Ethernet port is said to be a trunk port, this means that it can pass traffic on more than one VLAN. Therefore, it can send traffic only to Ethernet device that is able to receive traffic on more than one VLAN. In Figure 9-4, one port on each switch has been set up as a trunk port, able to carry traffic for both VLANs between the switches. So, the voice VLAN is a separate broadcast domain than the data VLAN, even in a bridged, multilocation segment like this one.

Ports that have a single VLAN assignment—i.e., ports that carry nontagged traffic for only a particular VLAN—are called access ports. In Figure 9-4, ports 1 through 6, and 8 through 12 are all access ports.

CoS over VLAN trunks

The Ethernet frame used to encapsulate trunked VLAN traffic contains 802.1q tagging, which identifies which VLAN each frame belongs to and which user priority the frame has. This user priority field (UPF) ranges from 0 to 7, with 7 being the highest priority. Based on the UPF, priority traffic is sent across the VLAN trunk first and, when received on the other end, is sent to the receiving switch's backplane first. Meanwhile, lower-priority traffic is queued and ultimately dropped if the switch runs out of buffer space.

To recap, VLAN accomplishes several things that can improve VoIP service:

- Divides and partitions Ethernet broadcast domains
- Provides a best-effort CoS prioritization scheme for traffic on each switch's backplane
- Creates a standard for dividing traffic logically across large routed and switched networks using VLAN trunking

For much more detailed information on VLAN, segmentation, and Ethernet in general, read O'Reilly's *Ethernet: The Definitive Guide*.

Quality of Service

Quality of Service is sometimes meant to refer to all measures that are geared toward improving the performance of voice on the converged network: 802.1p, VLAN, DiffServ, RSVP, and so forth. But within the context of VoIP, QoS is also a way of saying "more elaborate than CoS." That is, QoS takes additional steps beyond those CoS solutions that 802.1p and DiffServ take. The two key QoS standards are RSVP and MPLS.

Intserv and RSVP

Intserv (Integrated Services) is an IETF recommendation for provided dedicated bandwidth to individual flows, or media channels, on an IP network. The media channels are referred to by their sockets, just as they are in DiffServ. But, the difference is that all routers, whether at the edge or core of the network, play an active role in the policy decision process that results in dedicated bandwidth for each successful request.

RSVP (Resource Reservation Protocol) is the recommended signaling protocol for Intserv. This standard is a good choice for routed networks with limited core bandwidth. The Resource Reservation Protocol adds a layer of preflight signaling to IP networks that is similar to ATM (asynchronous transfer mode) switched networks. The purpose of RSVP is to ensure that the network has enough bandwidth to support each call, *before* any data is passed through the media channel. This is a decidedly different approach than DiffServ, which deals merely with packet prioritization.

And unlike DiffServ, RSVP adds decision-making points to the core network, increasing the processing overhead requirement on core routers. The increased overhead tends to discourage RSVP's use in bandwidth-glutted networks. But RSVP is the perfect solution for bandwidth allocation over slower links, because it guarantees availability for each RTP stream, rather than giving a "best effort." In this fashion, voice systems can tell routers, at the edge and the core, to reserve dedicated bandwidth for the duration of each call.

RSVP isn't needed in small setups. If you have a two-office VoIP network with 50 endpoints at each office, RSVP is overkill. Simplify your QoS approach by using 802.1p and, if you must, DiffServ on your routers and switches.

So RSVP is QoS, *not* CoS. It is an end-to-end solution that sets up a reservation across every hop that the call path crosses. Unlike DiffServ, RSVP runs after the call is set up. It uses the session ID of the RTP stream in order to identify a bandwidth reservation request. Each request forms a piece of a chain along the route between caller and receiver. This chain, or path, is put in place as soon as the RTP stream's session ID is determined during call signaling (H.245 or SDP).

In the case of Figure 9-5, the call path follows the logical path through the network cloud, crossing four routers along the way. Once the path is established, a traffic reservation must occur. Here's how RSVP works with an H.323 VoIP network, using Figure 9-5 and H.323 to illustrate the context.

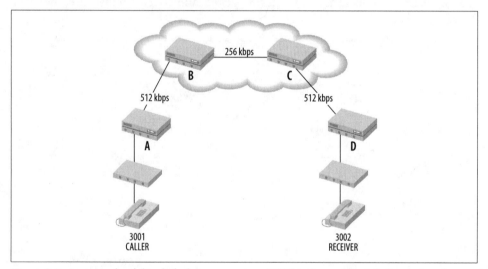

Figure 9-5. An example of slow links between routers. RSVP is the preferred QoS technique for slow networks.

1. H.245 negotiates the codec and establishes RTP sockets that will be used on either end of the media channel. These two sockets—the IP addresses and port numbers—together form the session ID that RSVP will use to refer to this RTP session. RSVP calls the session ID a *flow ID*.

2. The gateway router for the caller, B in Figure 9-5, sends a path message (PM) to the next hop, B, along the way to the remote gateway router, D. This PM will continue to be forwarded from one hop to the next in order to establish the QoS path.

3. B records the latency added as the PM reaches it, along with minimum latency, jitter ranges the router is willing to guarantee. Then, the PM is sent to the next router along the path, in this case, C.

4. C records the latency added as the PM reaches it, along with minimum latency, jitter ranges the router is willing to guarantee. Then, the PM is sent to the next router along the path, in this case, D.

5. When the PM reaches the remote gateway router, D, cumulative latency and jitter are calculated. The result is a profile call the ADSPEC, and the portion of the RSVP header used to accumulate QoS data during the PM is called the ADSPEC header.

So the ADSPEC portion of the path message for a call in Figure 9-6 would have a cumulative link delay (latency) of 75 ms and a cumulative maximum jitter of 22 ms. Because of jitter buffers and queuing, these two figures are added together in order to determine the effective latency that will be experienced during the call—not accounting for overhead on local data links or for packetization delay. In this case, there will be 97 ms of latency.

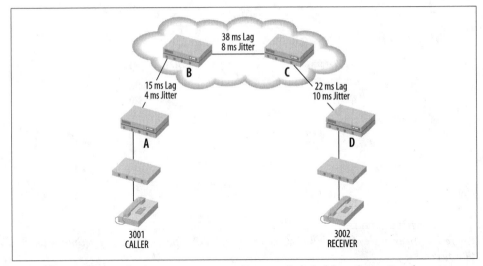

Figure 9-6. As the RSVP path message traverses the network, link delays (latency) and maximum jitter readings are recorded for each hop

When the remote gateway router reads the ADSPEC data and makes the determination, it can do one of two things:

- Give up, resulting in a busy tone for the caller
- Trigger the reserve message (RM) to set up the traffic contracts with each router in order to reserve bandwidth for the call

This decision can be driven by local programming or by policy established on a COPS/LDAP server.

In order to establish the traffic contract, the remote gateway (D) begins by sending an RM to the next hop (C) on the route back to the caller's gateway router (A). This path will always be the exact reverse of the path established during the earlier PM step. Now, that path is used to negotiate a traffic contract with each router:

1. The remote gateway router (D) sends the reserve message to the previous router in the path. The sender and receiver RTP sockets are confirmed, and a contract is established for the timeout value in seconds, sustained throughput, and peak throughput required by the RTP session.

2. The previous router in the path (C) sends a similar RM to its previous router in the path (B).

3. Router B sends router A another RM.

So the RSVP conversation has traversed the call path twice: once to establish the call path and once to backtrack the call path in order to send RMs. Now, the last stage of RSVP bandwidth allocation occurs as the call path is traversed a third and final time with RM responses:

1. Router A sends a reserve confirm message to router B if it agrees to guarantee the bandwidth and timeout values requested, or a rejection message if not.

2. Router B sends router C a similar response. If the first response, from router A, was a rejection, then all subsequent responses will be rejections as well.

3. Router C sends router D a similar response. If the first or second was a rejection, then this response will be a rejection as well.

If any rejection occurs, the session setup will fail, the RTP traffic will not pass, and the calling user should hear a busy tone. (*Should* hear, because different VoIP endpoints may handle this situation differently—even going as far as playing a recorded message like, "Your call cannot be handled at this time.")

Controlled loads versus guarantees

RSVP defines three service levels in RFC 2211:

Best Effort
> A class of service that has no QoS measures whatsoever, wherein RSVP ADSPECs and RMs can be used to *collect* data about network conditions but not actually enforce any bandwidth allocations. On Cisco routers, the fair-queuing feature is used to enable Best Effort service.

Controlled Load
> More like CoS, Controlled Load allows prioritization of traffic over multiple routers like DiffServ but includes core routers in the decision-making process.

Guaranteed

The ultimate application for RSVP, Guaranteed means that no packets will be lost, bandwidth will be constant, and delay will be within the prescribed ranges set up in the traffic contract.

The decision of which level of service to enforce with RSVP is up to you, the administrator. It makes the most sense, if you're going to go to the effort of setting up RSVP, to use the Guaranteed class, as the other classes are comparable to DiffServ—and DiffServ is simpler to implement. That is, if you aren't going to use Guaranteed, you may do just as well to use DiffServ.

 Project 9.2, later in this chapter, shows how to tell whether all the routers on a certain call path support RSVP.

RSVP has the following general characteristics:

- Is enforced with routers like DiffServ
- Can run transparently alongside other QoS solutions—especially 802.1p
- Is intended for networks with limited through bandwidth or networks with voice traffic typically greater than its non-voice traffic
- Can be optionally policy based
- Keeps QoS decisions to the core of the network and therefore has more complex signaling than DiffServ
- Is compatible with IPv4 and IPv6

All I Have Is a Single Office LAN! Do I Really Need All This QoS?

Merely enabling IP precedence/ToS on a stack of Ethernet switches will take care of almost all quality problems in a LAN-only setup. As long as the local area network isn't swamped with voice traffic, IP precedence and ToS are all you really need. That's good news—just about all modern Ethernet switches support them.

For point-to-point WAN links, IP precedence can work as long as the link isn't swamped with voice traffic. But, since these links offer far less bandwidth than a LAN, they have a tendency to get filled from time to time. If your routers are set to enforce IP precedence tags and you're still having audio problems, check your dial-plan to make sure the right codecs are being used so you're not wasting bandwidth over the link. And if all this doesn't solve your problem, a half-day visit from your local Cisco router engineer should get RSVP up and running.

MPLS

Multiprotocol label switching is the most advanced QoS measure available for enterprise VoIP networks. MPLS bears great similarity to ATM signaling but borrows heavily from RSVP. Unlike ATM, which incurs a 25% overhead on TCP/IP traffic (called the ATM "cell tax"), MPLS doesn't use its own framing format, just its own *labeling* format. This gives it the ability to work with Ethernet and other non-ATM switching technologies.

MPLS's role is in carrier-grade networks or extremely large enterprise networks with tens of thousands of nodes. Carrier-grade networks can use a mixture of MPLS, DiffServ, and RSVP all carried on a legacy ATM topology even as they seek to migrate away from ATM. MPLS can be supported by just about any modern topology.

The purpose of MPLS labels is to identify the paths and priorities associated with each packet. The paths correspond to the media channel of the VoIP call, while the priorities respond to the QoS level of service negotiated for those channels, just like RSVP.

But like DiffServ, MPLS can use a dumb network core. If a packet is carrying a label, all a router has to do is send it along the labeled path, rather than making a redundant assessment of the packet's payload. With MPLS, specialized, narrow-function circuitry can handle gobs more traffic than a traditional, CPU-based router can.

> For small- to mid-size VoIP networks with a few hundred endpoints, you can safely forego MPLS switching. MPLS is aimed at networks that operate with extremely high volumes of real-time traffic. Use MPLS switching in environments with carrier-grade requirements or thousands of endpoints.

MPLS inserts itself partially in layer 2 and partially in layer 3 on the OSI model. Its frame header sits between the IP header and the Ethernet header on an Ethernet network or between the label header and the payload on an ATM network. What's important to know is this: MPLS resides *outside* the reach of the network protocol, like 802.1p. But unlike 802.1p, MPLS also resides outside the reach of the data link's framing protocol (Ethernet framing, for example). This makes it invisible to the higher layers.

Like ATM, MPLS performs both switching (at a very high speed) and bandwidth reservation, so its ability to maintain fixed-bandwidth media channels for VoIP is superior. MPLS is as much a network routing and switching standard as it is a QoS mechanism. But, unlike ATM, MPLS can work with many kinds of data link/transport—Ethernet, ATM, frame-relay, and so forth.

If you're intent on using it, please read up on the subject at the MPLS and Frame Relay Forum's web site: *http://www.mplsforum.org/tech/library.shtml.*

MPLS has the following general characteristics:

- Is usually managed with switches
- Can run transparently within other labeled switching solutions such as ATM
- Is intended for carrier-grade networks
- Mimics ATM in function, but can be used across a variety of non-ATM networks, such as TCP/IP
- Is the most flexible, and the most sophisticated, QoS mechanism for VoIP systems

Residential QoS

Providing reliable voice quality using VoIP in the home or small office is a challenge for several reasons. First, residential broadband connections aren't always operated by the same people operating the VoIP service; this limits the ability of VoIP implementers to troubleshoot and support network-related problems. Second, most broadband Internet connections aren't supported by a backbone that has QoS measures. While it is common for many ISPs to prioritize in the style of 802.1p, almost none of them guarantee a service level. Finally, residential broadband routers haven't historically supported any QoS measures internally, though this is changing.

There aren't solutions to the shortcomings of the ISP's networks, but installing a QoS measure in the home is a good step to take. As with enterprise networks, the QoS enforcement points are routers. Some residential broadband routers now offer 802.1p packet prioritization, and a few even offer DiffServ support. Expect this feature to become more common.

In the meantime, experimenting with QoS measures doesn't require a pricey Cisco router. You could use a Linux computer configured as a gateway router to implement DiffServ as in Project 9.1.

Dial-Tone Providers That Offer VoIP Service

Some Internet service providers now offer telephone service via VoIP using broadband connections that they own. While this telephone service tends to be more expensive than that which comes from a third-party provider that doesn't own the "last-mile" data link, there's a greater likelihood that this type of service will have end-to-end QoS. Check with your ISP to see.

IP-Centrex is a way the telephone company can deliver dial-tone services to a PBX system or a group of IP phones with a VoIP gateway router. The dial-tone services are trunked from the phone company to the subscriber using IP-based packet streams. IP-Centrex offerings tend to be equipped with QoS. Check with your phone company.

Residential and Small-Office VoIP Routers

A new breed of residential gateway routers has begun to appear on the scene: routers with a built-in SIP or SCCP client and one or more analog RJ11 telephone connections. These devices enable routing and firewalling of Internet traffic, as well as ATA-type functionality so the connect analog phone(s) can make and receive VoIP calls over the Internet.

Inside the router's firmware, a SIP proxy and IP precedence measures may even exist, so groups of IP phones can be used to dial voice calls over the Internet. Some of these routers even offer to route calls over a local analog PSTN phone line if the Internet service fails. Check the specs to see what your VoIP router is capable of.

> A newer breed of broadband router has gained popularity recently. The IAD, or integrated access device, is a combination router and data link interface (i.e., DSL or cable modem).
>
> Most recently, IADs like the Zoom X5 series have begun to add VoIP capabilities and can even support QoS and analog phones. If you don't like clutter, consider using an IAD to connect your residence or small office. For more information about how these devices improve remote site survivability, read Chapter 13.

Voice QoS on Windows

Windows XP and 2003 can be used as routers to enforce Quality of Service on a VoIP network, though there are much better, and less expensive, ways to build a good QoS enforcement device. In one situation, though, Windows *should* be a QoS enforcement device—when it's used as an Internet gateway router with ISA services, a third-party firewall like CheckPoint Firewall-1, or a Microsoft Internet Connection Sharing feature. In this situation, you'll need to familiarize yourself with the Windows QoS Packet Scheduler.

QoS Packet Scheduler

The Internet Connection Sharing feature, which comes as a part of Windows XP, allows one computer to provide NAT/firewall services and gateway routing for a group of other computers on the same LAN. This way, these computers, which don't have a direct connection to the Internet, can still access Internet resources.

Using VoIP endpoints to place calls routed through the ICS gateway isn't really an enterprise-class solution, but it can be done. One might do it in a home office or hobbyist setting. The software element responsible for prioritization of packets passing through ICS, and through Windows in general, is called the QoS Packet Scheduler. Its job is to enforce the policy set forth in the computer's system policy profile, so that applications using Microsoft's QoS API can benefit from its QoS provisions.

The Windows kernel, which can perform routing and packet filtering like a standalone router, uses this API to support 802.1p, RSVP, and other, non-telephony-related QoS standards.

Windows RSVP Service

Windows NT Server, Windows 2000 Server and Advanced Server, and Windows Server 2003 can use a service-based RSVP software agent. The RSVP Service allows the Windows server to become a policy enforcement point, either as a router or as a softPBX.

Project 9.2. Audit a Network's Capabilities Using pathping and traceroute

What you need for this project:
- Microsoft Windows XP
- Linux
- LAN
- Access to the Internet

Though Linux is far better equipped to do VoIP network building and troubleshooting, Windows does have a nifty command-line tool that can be used to determine how well an IP route supports QoS measures. *pathping*, which ships with Windows 2000 and Windows XP, lets you see how well your Internet provider—or your corporate network—supports 802.1p and RSVP. It also lets you identify hops along the call path that are having excessive packet loss.

pathping is similar to *traceroute*. It first displays the IP route along all hops from the host where it's running to the host at which it's targeted. Then, it collects information from each hop along the way, like latency times, and displays it for you, the administrator.

This example returns the hostname and IP address from each hop along the route to the destination, if each hop provides an ICMP response:

```
C:\> pathping www.broadvoxdirect.com
```

The output shows the route to the destination, similar to *traceroute*:

```
Tracing route to www.bigvoxdirect.com [65.67.129.23]
over a maximum of 30 hops:
  0  kelly-6aizy9qd1.ce1.client2.attbi.com [10.1.1.202]
  1  10.1.1.1
  2  10.248.164.1
  3  bic01.elyehe1.oh.attbb.net [24.131.64.38]
  4  12.244.65.61
  5  12.125.176.121
```

```
 6  gbr2-p70.phlpa.ip.att.net [12.123.137.26]
 7  tbr1-p012601.phlpa.ip.att.net [12.122.12.101]
 8  tbr1-cl8.n54ny.ip.att.net [12.122.2.17]
 9  ggr2-p300.n54ny.ip.att.net [12.123.3.58]
10  so-1-0-0.gar4.NewYork1.Level3.net [4.68.127.5]
11  ge-2-1-0.bbr1.NewYork1.Level3.net [64.159.4.145]
12  so-0-0-0.mpls1.Cleveland1.Level3.net [209.247.11.134]
13  ge-6-0.hsa1.Cleveland1.Level3.net [209.244.22.98]
14  BIGVOX-DIS.hsa1.Level3.net [64.156.66.10]
15  penguin.bigvoxdirect.com [65.67.129.23]
```

This example checks to see whether each router along the path supports 802.1p precedence tags:

```
C:\> pathping www.bigvoxdirect.com -T
```

The output from *pathping* first shows the route, like the previous example, but also adds the 802.1p feedback as far along the route as is possible. Not all devices along every route support 802.1p. In this example, the sixth hop does not, because the router isn't configured for IP ToS and 802.1p. Since the 802.1p header can't be carried past the sixth hop, subsequent hops cannot be tested for 802.1p support:

```
Checking for connectivity with Layer-2 tags.

1  10.1.1.1        OK.
2  10.248.164.1    OK.
3  24.131.64.38    OK.
4  12.244.65.61    OK.
5  12.125.176.121  OK.
6  12.123.137.26   General failure.
```

The -R option will do a similar check for RSVP support in a similar fashion. You aren't nearly as likely to find RSVP-supporting hops on the public Internet as you are 802.1-aware hops. But, if RSVP is configured on your private network, you can use *pathping* to help you evaluate your network's hardware readiness for QoS. It will tell you which routers support RSVP and which routers either need to be re-programmed or upgraded to support it.

More information on *pathping* is available on Microsoft's TechNet web site (*http://www.microsoft.com/technet*).

Measure the latency time and jitter on a call path

The cumulative latency on a route is a good indicator of how latent it is and, therefore, how well it will work as a VoIP call path. An easy way to record latency between hops (routers) on a route is by using the traceroute (on Windows, tracert). The syntax is very simple:

```
# traceroute 15.54.21.112
```

This example will discover the route to the host at address 15.54.21.112, send several ICMP packets to each hop on the route, and then show you:

- The highest round-trip latency to each router, in milliseconds
- The lowest round-trip latency to each router
- The average round-trip latency to each router
- The IP address and/or hostname of each router
- Whether an ICMP ping response was received from each router

Whether on Linux, Windows, or Mac, *traceroute*'s output tends to be the same. The following sample output is from Windows, but all *traceroute*s show you the minimum, average, and maximum latency to each hop along the route:

```
Tracing route to www.macvoip.com [65.31.69.11]
over a maximum of 30 hops:

  1    1 ms    1 ms    1 ms  10.1.1.1
  2   14 ms   13 ms   18 ms  10.248.164.1
  3   18 ms   16 ms   12 ms  bic01.elyehe1.oh.attbb.net [24.131.64.38]
  4   19 ms   21 ms   34 ms  12.244.65.61
  5   31 ms   23 ms   24 ms  12.244.72.70
  6   25 ms   26 ms   28 ms  tbr1-p012401.phlpa.ip.att.net [12.123.137.45]
  7   32 ms   27 ms   27 ms  tbr1-cl8.n54ny.ip.att.net [12.122.2.17]
  8   28 ms   28 ms   34 ms  ggr2-p300.n54ny.ip.att.net [12.123.3.58]
  9   31 ms   30 ms   28 ms  att-gw.ny.aol.net [192.205.32.218]
 10   32 ms   28 ms   43 ms  bb2-nye-P1-0.atdn.net [66.185.151.66]
 11   29 ms   47 ms   34 ms  bb2-vie-P12-0.atdn.net [66.185.152.201]
 12   64 ms   48 ms   62 ms  bb2-chi-P6-0.atdn.net [66.185.152.214]
 13   60 ms   60 ms   62 ms  RR-DET.atdn.net [66.185.141.98]
 14   59 ms   54 ms   66 ms  os0-0.fmhlmi1-rtr1.twmi.rr.com [24.169.225.65]
 15   57 ms   53 ms   63 ms  ig0-1.fmhlmi1-ubr5.twmi.rr.com [24.169.225.22]
 16   64 ms   66 ms   68 ms  www.thelinuxfix.com [65.31.69.11]
```

Not all IP networks permit ICMP traffic, or *traceroute*s in particular, because some system operators prohibit them for security reasons. Most routes across the Internet should provide a valid response when using the traceroute command.

As you examine the output from the traceroute command, pay special attention to the variance in highest and lowest latency times (*not* the average latency time). This is a good, rough estimate of jitter between each hop.

Using *pathping* and *traceroute* during peak traffic periods, you'll be able to establish whether or not a particular IP route is a good place for voice traffic. You'll know the jitter and latency qualities of the network, you'll have identified problem routers and potential traffic bottlenecks, you'll know whether or not each router supports RSVP, and you'll know how well the network supports 802.1p precedence tagging.

 The variance between the minimum and maximum latency times in a traceroute command can be a good indication of jitter.

Best Practices for Quality of Service

When building or expanding a network for Voice over IP, there are many standards from which to choose, as you can see. Each has distinct characteristics that makes it good for certain situations. But, in the end, your success with QoS standards can be only as deep as your use of them is practical. Here's a list of best practices for ensuring Quality of Service on the network.

Choose the right standards

- Multiple QoS standards can complement one another, but don't overdo it. Using MPLS, RSVP, and DiffServ all on a 300-node network, while possible, wouldn't be a very practical idea.
- Don't expect the QoS policies you establish to be supported by legacy routers and switches.
- If you are coordinating several network managers, departments, or divisions, be sure you have their buy-in and a mandate to enforce the QoS policy you establish. Easier said than done, especially in large organizations.
- If your network traffic is more than 70% data to voice, use a packet prioritization technique like DiffServ or 802.1p.
- If your network is large (hundreds of voice nodes) and relatively busy, use DiffServ.
- If your network is especially congested or very large (thousands of voice nodes), use RSVP.

Use the standards correctly

- In DiffServ setups, classify IP voice traffic (RTP streams) as EF (Expedited Forwarding).
- In 802.1p setups, classify IP voice traffic using IP precedence tag 5.

Build the network to favor VoIP

- Use 802.1q VLAN, and establish one or more separate VLANs for IP phones and/or VoIP servers.
- To be on the safe side, assume every data link will need an additional 20 to 30 percent overhead bandwidth for call-signaling and -routing protocols (ARP, OSPF, etc.) aside from that normally used by the voice traffic itself.
- If you are able to manage queuing on your IP routers, use low-latency queuing (LLQ).
- Use network links with faster speeds. Adding capacity is an acceptable, albeit inelegant and sometimes noneconomical, way of solving QoS issues.
- Don't use slow (sub-128 kbps) links at all.
- Don't use wireless Ethernet for large workgroups of VoIP users.

- Avoid running VoIP sessions through a VPN unless absolutely necessary, as VPNs have a lot of overhead.
- Use digital PRI circuits or IP-based trunks rather than analog POTS or Centrex trunks to connect to the phone company. This can lessen the amount of digital/analog conversion and subsequent latency.

Use voice-coding techniques that enhance QoS

- Use the G.711 codec as often as possible; it is the most resilient waveform codec with a negligible packaging delay.
- Some codecs offer packet loss concealment (PLC). Use them to decrease perceived quality problems related to packet loss, but use them sparingly, as they can create latency.
- To minimize latency, decrease the packet interval. Intervals can go as short as 10 ms, but 20 is typical. Decrease the use of jitter buffers by avoiding features like PLC, if possible.
- Avoid transcoding if possible. If absolutely necessary, use IP endpoints and VoIP servers that support wireless codecs such as GSM.
- If echo is a problem (especially across low-speed links), use codecs that support echo cancellation.

Maintain utilization

- Keep total packet loss on each Ethernet segment below 1%.
- Keep packet loss on T1s and other point-to-point connections at 0%.
- Keep packet loss and jitter on frame-relay connections, VPNs, and other network clouds as low as possible. Negotiate an SLA with the service provider to enforce this.

Establish a service-level agreement

- Before you begin building your VoIP system, if you have time, determine the MOS (mean opinion score) rating of all call paths on your current system. Use these scores as a minimum effort when selecting standards and equipment for your new VoIP system.
- After the implementation, particularly in large corporate or carrier-class networks, establish an SLA between you and your users that provides an MOS expectation for every call path. Use this SLA and associated MOS scores as a metric to show your success.

Key Issues: Quality of Service

- Quality of Service is decreased when packet loss, jitter, and latency increase.
- There are two styles of Quality-of-Service mechanisms to deal with the problem: CoS and QoS.
- CoS solutions are coarsely grained, best-effort approaches the rely on packet prioritization.
- CoS solutions are best in situations in which voice traffic represents 30% or less of the total traffic, such as enterprise networks.
- 802.1p and DiffServ are the two most common CoS solutions for Voice over IP.
- QoS solutions are finely grained, guaranteed-delivery approaches that reserve bandwidth across the network.
- QoS solutions are best when there is limited bandwidth, or in carrier-grade networks.
- COPS is a system for centrally storing and maintaining QoS and CoS policies.
- RSVP is the most common QoS solution for enterprise-grade voice services over IP.
- Some ISPs offer QoS services to facilitate the use of the Internet and/or VPN for voice applications.
- Many small-office and residential broadband routers support 802.1p, and a few even support DiffServ.
- Linux's kernel-based firewall, Netfilter, can be used as a DiffServ edge router if the Linux Traffic Control components are compiled into the Linux kernel.
- In Windows NT, 2000, and XP, the QoS Packet Scheduler provides packet prioritization for API-compliant Windows applications such as a softPBX.
- The Windows RSVP service provides an RSVP policy enforcement point for voice and video software running on Windows servers.
- *pathping* is a Windows utility that can help you determine how well WAN call paths support QoS measures.

## CHAPTER 10

Security and Monitoring

Like the Web, email, and other Internet communications tools, IP telephony can be secured. This fact is one of its biggest appeals over old-school telephone equipment. Security means enforcing system policy, recording instances of abuse for forensic and litigation purposes, encrypting or otherwise hiding sensitive information in transit, bolstering call-management systems' resilience to exploitive attacks and computer viruses, and securing the access perimeter of the VoIP network.

Security tools and enforcement practices for VoIP applications are the same, essentially, as those for other IP-based apps, because they run on the same network. The security objective of VoIP systems is largely the same as those of other IP-based systems: in a nutshell, preserve the operational status of the system.

There are many threats to this objective and many countermeasures to the threats. Policy enforcements points, like firewalls, protect lower layers of the network, while authentication systems like RADIUS and application proxies provide higher-layer security. This chapter describes how to secure and harden a VoIP server, the basics of DMZs, how to enable logging of VoIP traffic with iptables, how to tweak the logging configuration of Asterisk, and how to log and monitor VoIP network traffic.

Security in Traditional Telephony

One of the big misconceptions about VoIP telephony applications is that they are inherently insecure. In truth, the VoIP technology family provides scores more security options than conventional telephones do. If anything is insecure, it's the *old* voice paradigm.

In the PSTN, there are several aspects of security: access control, call accounting/billing, and features. In these key aspects, the PSTN relies on the intrinsic characteristics of its own design as security controls.

Access Control

The PSTN permits network access via the physical loop component—the cable connection from the CO to the customer premises. This means that a person who has access to the customer's phone lines can place calls as though he is that customer. A friend comes over to your house, picks up your phone, and makes a call. The telephone company assumes he is authorized because he is *there*. While primitive, this is the basis of access security on the PSTN.

By comparison to a modern data network, this access control approach seems lax, but it's the way the PSTN has always done it. Indeed, even on PBXs and high-capacity voice circuits, physical logistics is still the most common method of controlling access to legacy telephony apps.

To overcome this weakness, some CO switches and PBXs can require users to dial a password of DTMF digits before a call can be placed or before certain telephone area codes can be dialed (for a quick review on phone numbers, refer to Chapter 4). Or the phone company can be made to force you to use a long-distance code before you can dial LD calls. Some telephone companies offer what's called a *receive-only* phone line, which controls outbound calling by not allowing it at all. Lots of PBXs let you limit outgoing calls on a phone-by-phone basis.

Snooping

With a lineman's set, a device used to test telephone circuits, and a pair of alligator clips, it is possible to clandestinely listen in on a PSTN subscriber's phone calls. This technique, while illegal, is quite easy to do, even from outside the subscriber's demarc. All that is needed is a point in the last-mile loop to tap in with the receiver, such as a cross-connect block or splice box. Since the signals transmitted from the CO to the D-frame are analog, snooping on endpoint legs of an analog CO switch (or PBX) is quite easy. All one needs is access to the right cabling. To prevent this kind of snooping, telephone cables tend to be buried or high up on poles where they are tough to access, and cross-connect points, if aboveground, are usually inside of sturdy, locked enclosures.

Phreaking

Of course, the ability to send DTMF digits is itself a bit of a security measure—after all, services on the PSTN are accessed by dialing them. And the only devices that can transmit DTMF digits are telephones, right? Well, not exactly. Tone generators are small handheld devices that allow the transmission of DTMF digits and other tones so that, for example, calls can be stolen from a public pay phone. So, in this case, access control is easily broken. This type of exploit, which carries the slang name *phreaking*, is considered the root of modern-day hacking.

The cell phone network has been abused by phreakers, too. Though cell phones have device-specific electronic serial numbers encoded into their firmware, it is possible, though difficult, to program an unauthorized phone with a different serial number so that it can make calls using a legitimate user's account. This practice is sometimes called *cell phone phreaking*. Now, there are better administrative measures to counteract phreaking than there were at the beginning of the cell phone era. Indeed, now that many of the cell carriers send voice signals digitally, they are able to interleave and encrypt them so that phreaking is more difficult.

Call Accounting and Billing

When you pick up the phone, dial the pizza place, state your order, and hang up the phone, a number of call accounting events are recorded. The PSTN, with help from SS7, records:

- Which number you dialed
- When you dialed the phone number
- When the call was connected
- The duration of the call
- When the call was disconnected

These bits of data are not crucial just to the billing process; they're also important because of what they indicate forensically. When a person's use of the phone system is used as evidence in court or during disputes between a phone company and its customers, call-accounting data is critically important.

Features

To address security concerns, telephone companies have implemented a number of calling features that improve privacy. Such features include *caller ID*, which allows the receiving party to know who is attempting to call her so she can decide whether or not to answer, and *privacy management*, which forces the caller to record his name so that the receiving party can decide how to handle the call without having to greet the caller.

Of course, while the phone company can increase security by providing privacy, it also provides security for anonymous callers, in the form of things like *caller ID blocking*. Security means different things to callers than it does to receivers. Yet, both are valued subscribers to the PSTN, so minimizing their aggravation of one another has become a regulatory headache. It's a losing battle, because the telephone company wants to protect privacy while also allowing anonymity—two concepts that are in conflict.

Security for IP Telephony

IP telephony security boils down to three risk factors—the application, the network operating systems, and the infrastructure. Secure these, and you secure the VoIP network. Here are the most common types of security risks to data networks today:

DoS (denial of service)
> Attacks that reduce or compromise the functionality of a software system via a buffer or bandwidth overrun or by exploiting a security bug. Generally, DoS attacks are accomplished using specific code to create certain conditions within a target host or network that trigger a denial of service. As a rule, DoS attempts are defeated with access control.

Man-in-the-middle
> Attacks that allow a third party to monitor, record, block, or even alter, a data transmission. Packet sniffing, or the capture of packets on a data link, usually accompanies this kind of attack. Man-in-the-middle attacks are defeated with encryption and authentication.

Trojan horses and malware
> Autonomous software processes designed to travel across the Internet and IP networks, infecting exploitable hosts in order to replicate themselves and, sometimes, to leave a footprint behind.

In the world of convergence, access control, call-accounting, and telephony features are aspects of the same extensible network. Depending on whether your IP telephony applications come from the phone company or from the local softPBX, your ability to control and customize them varies. If you've chosen a VoIP technology that is open and standards-based, like VOCAL or Asterisk, you can build exacting security policies, precise call-accounting and logging systems, and limitless authentication features.

Supporting the telephony application is a network operating system—usually Linux, FreeBSD, or Windows—that has its own security concerns. Anybody who's run an Internet server for even a short time knows the importance of hardening the network operating system against viruses, security exploits, and bug-ridden software agents.

Finally, the network infrastructure—the protocols and connectivity equipment—has a set of security issues that must be addressed. This can mean establishing policies for network access via firewalls or authenticating VoIP devices as they attempt to communicate. This also means auditing network traffic, discouraging network intrusion, and promoting privacy.

The application, the operating systems, and the network infrastructure each have many layers of security features and provisions. Like QoS systems, you may not use them all. As a VoIP network maintainer, your security duties will boil down into three categories: access control, software maintenance, and intrusion prevention.

Access Control

The biggest headache in building secure networks is often access control. It is by permitting access that rich applications are delivered in the network; it is by restricting access that those applications are made more secure. Unfortunately, just as in the outside, analog world, limiting access in the name of security can result in a poor experience or a lasting resentment by users.

Consider the security checkpoints at airports: very thorough, very time-consuming, and, possibly, a very big inconvenience. But the point of an airport security checkpoint is the same as an access control policy on your VoIP network: allow only the right people in—that is, people with a boarding pass and a clean X-ray scan. Likewise, if an access attempt that isn't authorized or is possibly damaging is made into the IP network, stop it immediately.

 Don't confuse access control with intrusion detection. Access control is a policy-based, preventive idea; intrusion detection is an intelligence-oriented, reactive idea.

There are many layers to the access control paradigm—credentials, origin and destination, timing, and physical presence controls. One or all of these may be taken into account when your network decides whether to let somebody have access to a registrar, a directory lookup, a media channel, or any data at all.

Credentials and Authentication

Access control credentials are special pieces of secret information that entitle a user, or a software process, access to the network. This may be a password, an encrypted string, or some other secret key that authenticates the user or software process, allowing it to use a resource.

On a SIP phone, this might be the registration password required in order to gain access to a proxy on the VoIP network, without which making phone calls would be far more difficult. It might even be a public key cryptographic relationship between two call-management servers. It could be the key that is used to encrypt a VPN tunnel between two remote VoIP networks. Credentials are used all over a secure VoIP network.

Credentials and authentication are the best way to prevent anonymous abuse of a properly functioning system, VoIP included. They also provide a digital signature that authenticates billing transactions associated with the use of a voice system. Regulatory agencies can require service providers and subscribers to recognize digital signatures during disputes.

Some SIP endpoints support MD5 authentication. So do Asterisk, VOCAL, and other softPBX systems. MD5 authentication allows you to hash (or scramble) SIP passwords so casual observers can't view them. Some SIP phones don't properly support this feature, so check with your vendor before building MD5 into your design. To try out MD5-equipped SIP clients with Asterisk, see Project 10.1.

It's also important to enforce access control policies on wireless Ethernet segments. This means using WEP keys on wireless devices that access the network by 802.11a/b/g means. Check your wireless IP phone's specs to see if it can be a WEP client.

Finally, credentials can be enforced by the endpoint. In a centralized authentication system, IP phones themselves can require a username and password or PIN from prospective callers. This way, users are authenticated before they are able to even get a dial-tone. Cisco, Avaya, and many other vendors' proprietary solutions support this kind of authentication, though usually through a built-in authentication scheme, as opposed to Active Directory or RADIUS.

Project 10.1. Use MD5 Hash to Secure SIP Passwords

What you need for this project:

- Asterisk running on a Linux PC
- The X-Lite softphone

Endpoint authentication helps you prevent somebody from simply plugging an IP phone into your Ethernet switch, directing it to your softPBX, and making phone calls on your network. All softPBX servers support password protection of endpoint access, but some go a step further, hashing the passwords so that they cannot be easily observed by the ill-intentioned.

Asterisk allows you use the MD5 algorithm to secure the SIP passwords that are stored in the Asterisk SIP configuration. This prevents clear-text SIP passwords from being the exclusive access control key for SIP endpoints. When the MD5 secret feature is used, Asterisk will require the MD5-hashed password in addition to the common SIP username and password.

First, configure a SIP phone to use an MD5 secret. This is accomplished differently on different SIP phone makes. Not all SIP phones have an implicit MD5 setting. On most, it is possible to enter the username, realm, and password and let the phone piece the MD5 secret together. On most IP phones, no additional configuring of the phone is necessary.

Next, edit Asterisk's *sip.conf* configuration file. For each peer that supports MD5 authentication, you can add the md5secret setting, which allows you to store the MD5 hash to compare when the SIP peer is attempting to register or place a call.

When entering the MD5 secret into the configuration, you must enter it in its hashed form.

The format for SIP MD5 secrets is *user:realm:password*. So, *3001:asterisk:MaBell* would have to be hashed before it was entered and stored in *sip.conf*. You can hash the string using the *md5sum* utility, a standard item on most Linux distributions:

```
# echo -n "3001:asterisk:MaBell" | md5sum
```

The output of this command is the hash, the part you'd enter for the value of the *md5secret* option in *sip.conf*:

```
bece6f2475b3239e877bbd2235332
```

Use a copy and paste for periodic updates to MD5 secrets. Otherwise, you're sure to be aggravated by data entry mistakes. Alternatively, you could write a shell or Perl script that automates putting MD5 hashes into your *sip.conf* file.

Origin, destination, and timing policies

Like a firewall, access to and from a VoIP network may be restricted due to a policy that takes into account the origin of the access attempt, the destination of a requested resource (such as which subscriber a node is trying to reach), and the timing of the request. There are two ways to restrict access based on origin, destination, and timing: rule-based firewalls and dial-plan configuration.

Rule-based firewalls generally have a clause for each rule that states the schedule in which it is enforced. It isn't really practical to use a firewall to restrict access for a particular endpoint, since that endpoint's calls would have to cross the firewall in order to enforce the rules. Instead, firewall devices are great for enforcing source/destination and timing-related rules on entire network segments. For instance, the firewall could permit calls from the Detroit office to the Chicago office only during certain hours of the day or only on weekends. Firewall rules could prohibit the use of remote VoIP users who use a VPN to place calls through the corporate phone system—during certain times of the day, perhaps.

Voice Spam?

Imagine an exploit that allows the attacker to make free telephone calls using your VoIP server. Just as email spammers exploit outdated and misconfigured mail servers to trick them into sending their messages, a new breed of voice spammers could take advantage of a poorly configured VoIP server to make fraudulent phone calls. Vigilance as a VoIP network manager is your only recourse: eliminate the most common risks by hardening your network, keeping your software up-to-date, and using detailed access control and call-accounting techniques.

Firewall rulesets are a form of access control that operate mainly at the network layer and below. Thus, they don't distinguish which codecs are being used, which phone number has been dialed to direct the call, or whether the caller and receiver are using ATAs, IP phones, or softphones. Instead, firewalls are concerned with:

- What IP address or IP network address packets are coming from (origin)
- What IP address or IP network address packets are destined for (destination)
- What time of day, date, or day of the week it is (timing)
- What IP protocol is being used (TCP, UDP, ICMP, etc.)
- What TCP or UDP port is being used

Higher-layer access control

It's the application layer—the dial-plan—that can distinguish dialed numbers, signaling protocols, codecs, and so on. These things can be used as access control factors, too—but on the PBX, not the firewall.

On the PBX, you might have a dial-plan that is programmed to behave a certain way depending on the origin, destination, or timing of the call. Let's say there's a 24-hour waiting area in a hospital with a courtesy phone that guests can use. But the reception desk that covers this waiting area is staffed only 16 hours a day. It's possible, using some combination of time-sensitive dial-plan rules, to secure this phone against abuse during times when the area isn't being monitored by a receptionist. Almost all PBX systems support this kind of programming.

Physical presence policies

One way to assure proper access security is to have the network verify the indicators that each device *present* on the VoIP network is *permitted* to access the network. This could mean accounting for a particular endpoint's MAC hardware address in an access control list that's centrally stored, so that when the endpoint is plugged into an Ethernet switch on the network, the switch knows to grant it access. This technique is quite common with 802.11b/g wireless Ethernet segments. On many wireless access points, the list of authorized MAC addresses is stored in the base station itself.

For the ultraparanoid system administrator, some Ethernet switches can be programmed to grant access only to a certain MAC hardware address on a *certain port*. This could be useful in a shared-facilities or bandwidth-lease environment where multiple tenants use the same Ethernet backbone and individual network ports are restricted to a certain VLAN.

Media Encryption

Encrypting the key piece of authentication info—the endpoint's access password—is one way to prevent unauthorized access to make and receive calls using an IP phone. But this measure alone doesn't protect calls in progress from being observed and possibly recorded.

Once an intruder has access to the softPBX or a critical piece of routing infrastructure, like a firewall, she may be able to store VoIP media streams packet by packet, and piece them together to listen to later. It isn't very hard to figure out which codec the call used and play the call back using a program that supports lots of codecs, like QuickTime or a decent sound file editor. This is a potentially devastating risk—a true cornucopia for the daring intruder. So, to remove the risk, the media streams to and from IP phones and the softPBX can be encrypted.

Encryption of media channels is something the bulk of VoIP vendors don't support, but the need is clear. The prevailing method of sending media streams in an encrypted fashion is via SRTP (Secure Real Time Transport Protocol), a mechanism that must be activated during capabilities negotiation (SDP or H.245). As with any other capability, all endpoints involved in the call must support the same standard of encryption.

Few IP phones on the market permit the use of SRTP, though Avaya and Cisco both offer support for media encryption through a proprietary means in some of their IP phones. The Zultys Zip 4×4 is a SIP phone that allows SRTP media encryption. Its documentation says that you can toggle back and forth between encrypted and unencrypted sessions during the same call, presumably using multiple SIP methods and setup exchanges. This would be useful if you had an encrypted call between the two 4×4 phones and you needed to transfer the call to a SIP phone that didn't support media encryption.

Software Maintenance and Hardening

A huge problem facing the software industry is unintentional vulnerabilities, or bugs, in software that can lead to security hazards. There are several kinds of faults, but the riskiest ones are those that permit a hacker or Trojan/worm agent to gain privileged access to a system. To deal with these faults, the makers of the software—either commercial programmers or the open source community—issue patches that fix them.

Of course, it may take some time to resolve a security bug once it's been identified. Linux and BSD-related bugs tend to be fixed rapidly, while commercial software bugs tend not to be fixed quite as quickly.

If a piece of software on your system is vulnerable because of a bug, it's better to disable that software until a patch can be installed than it is to keep running with the risk of attack!

Even if a patch is issued to resolve a security hole, it still takes the system administrator to apply that patch to the systems in question in order for the security risk to be removed. This can be a time-consuming process. Merely keeping aware of all the security bugs and patches can be a big job. The more complex the system—the more applications and users it supports—the more susceptible it is to exploitable bugs.

In order to come close to the elusive "five nines" reliability standard of old-hat voice systems, security bugs must not be allowed to hamper the VoIP network. There are several practical ways to avoid the threat of software security bugs:

- Avoid the use of VoIP servers without a properly configured firewall between them and the Internet. Firewalls stop most random Trojan and worm traffic. The firewall doesn't have to be a NAT device—remember that NAT can break SIP, MGCP, and H.323.

- Subscribe to a patch notification newsletter from a reputable source, such as the manufacturer of the software on your VoIP systems. If you use a Windows-based softPBX, get in the security loop at *http://www.microsoft.com/technet/security*.

- If possible, avoid using software that is frequent targets of Trojan, worm, and virus attacks. Instead, choose software that can be customized and highly hardened, such as Linux. If you're looking for a commercial alternative to a Windows-based PBX, try Avaya's MultiVantage softPBX, which runs on a proprietary build of Linux, or Cisco's CallManager Express, which runs on Cisco IOS embedded in a router. Of course, there's always Asterisk, which runs on Linux, FreeBSD, and Mac OS X.

- Create a very restrictive access control policy at the network layer to minimize the likelihood of a security compromise (see "Access Control," earlier in this chapter).

- Keep IP phone firmware up-to-date. In traditional telephony, the phone system was just a PBX that ran, and ran, and ran, without ever having a software patch or a security hiccup. Not so in the world of VoIP; do what you can to keep software well-maintained, up-to-date, and bug-free.

- If your system design calls for a TFTP server, use one that can control access by MAC address. Another approach would be to use a local firewall on the TFTP server to limit access. This way, the TFTP server's exploit risk is lowered. Plus, you don't want a casual user leafing through your files on the TFTP server.

- Don't connect your IP phones to the Internet if all your phone users are in one building and there's no reason to connect them to the Internet. This is a last resort, however, as the goal of every noble VoIP integrator ought to be leveraging the Internet as much as possible.

Hardening VoIP Servers

An important aspect of software security is hardening: the act of proactively making your operating system and application software more secure. On a softPBX server, hardening means removing unnecessary services and software agents, so that you have less of a garden of possible security vulnerabilities to worry about.

On Windows servers, this often means removing unneeded services. If Microsoft Internet Information Services isn't needed, shut it down or remove it. That's one less risk. The same is true of Linux systems. Notoriously exploitable software like Apache and BIND should be carefully patched or just removed from the system.

But hardening isn't just disabling or removing software from telephony servers. It also means optimizing the system configuration of every relevant device on the network. On an Asterisk server on Linux, for example, this would mean establishing a very restrictive local firewall policy. That way, only authorized traffic can get into, and out of, the softPBX. This is sort of access control at the host level—the last line of defense, if you will.

Even if a would-be intruder is sitting three feet away from your softPBX server, he won't be able to break in if it's hardened correctly. Here are some hardening tips that work on Windows and Linux:

- Remove unnecessary software from each critical system in the VoIP network.
- If possible, move software that is more vulnerable to security compromise onto servers that are separate from high-availability applications like VoIP. This might mean running DNS and softswitch services on different machines.
- Local user accounts and passwords should become disabled after three or four incorrect login attempts.
- Establish local firewall policy on each machine that runs a critical piece of the VoIP network to reduce risky traffic flows.

Project 10.2. Harden a SoftPBX

What you need for this project:

- Asterisk running on Red Hat Linux

Though this project is geared toward an Asterisk server, it is principally accurate for any softPBX.

Two basic aspects of the softPBX need examination during hardening: the software that's installed and the software that's running. As far as hardening the software that's installed but not running, or not *needed*, the course of action is quite simple: remove it.

Production servers shouldn't have Asterisk running as root due to security concerns. Refer to Chapter 3 to set Asterisk up so that it can run as a nonroot user.

Remove unnecessary software

That means getting rid of Bind if you're not using it, removing Apache if it isn't needed, and unloading MySQL if you've no need for it. Just because the software isn't running doesn't mean it can't be used to facilitate a sophisticated security exploit, so remove it altogether if you don't absolutely need it. Use a package-management tool if you're not certain which software is required on your system. Red Hat provides RPM for this purpose. Chances are, you don't need the Gimp, a graphics tool, on a softPBX, so use RPM to remove it.

Since we're dealing with Asterisk, you can disable a number of modules in order to reduce the risk of security exploits. Asterisk provides modules for all kinds of signaling protocols and telephony applications, and you may not need them all. Use the noload directive in *modules.conf* in order to specify those that you'd like to disable:

```
noload => pbx_kdeconsole.so
noload => chan_modem.so
```

In this case, the two modules being disabled are the KDE log console module, which provides a graphical console for the KDE desktop environment, and the modem module, which is used for ISDN connectivity with Asterisk.

Clean up xinetd

xinetd is Red Hat Linux's catchall daemon for *telnet*, *finger*, and a number of other Unix network applications. (It's the successor to *inetd*.) Its configuration files, in */etc/xinetd.d*, are used to enable or disable support for a long list of network access services. Use this configuration directory to disable all but those that you absolutely need. Here's the contents of a file in this directory, */etc/xinetd.d/imap*. It disables the IMAP server from running:

```
service imap
{
        disable                 = yes
        socket_type             = stream
        wait                    = no
        user                    = root
        server                  = /usr/sbin/imapd
        log_on_success  += HOST DURATION
        log_on_failure  += HOST
}
```

Check all the files in this folder for the disable=yes line, or, if you prefer, you can altogether remove the config files for the services you don't need.

The idea is to eliminate unnecessary TCP/IP listeners, reducing the likelihood of an attacker discovering a vulnerability. So, if you don't need *telnet*, TFTP, *talk*, and *finger*, then, for goodness sake, disable them. The fewer services that have listening ports, the more secure your server will be.

Kill shell access for daemon users

If you set Asterisk up to run as a nonroot user (see Chapter 3), then you'll need to make sure that user has no shell access. This setting is found in the */etc/passwd* file along with the list of users on the system. Change the shell setting for each non-root user that's meant only for running software processes to */bin/nologin* or something that doesn't exist. */bin/bash* and */bin/sh* are common shells that do exist, and regular user accounts, like root, still need some form of shell access, so don't change those. This way, inadvertently logging on as that user will not result in access to Unix. Here's a snippet from */etc/passwd* that contains a number of such users with disabled shells:

```
root:x:0:0:root:/root:/bin/bash
bin:x:1:1:bin:/bin:/sbin/nologin
daemon:x:2:2:daemon:/sbin:/sbin/nologin
adm:x:3:4:adm:/var/adm:/sbin/nologin
asterisk:x:8:0:asterisk:/usr/bin:/sbin/nologin
```

Optimize the local firewall on the softPBX

In order to build a local firewall policy on the softPBX server, you'll need to identify which VoIP protocols you're using, and plan a policy based on the kind of TCP and UDP port access needed by each one:

SIP
: 5060 and 5061 TCP and UDP

H.323
: 2099 TCP and UDP, 2517 TCP and UDP

H.323 Video (H.263)
: 2979 TCP and UDP

MEGACO/H.248
: 2944 and 2945 TCP and UDP

TFTP
: 69 TCP and UDP

ASTMAN
: 5038 TCP

RTP
: Depends on configuration of capabilities negotiation preferences of the endpoint RTP implementation (most RTP agents use 5000/5001, 5004/5005, 8000/8001, or high-numbered ports).

IAX

   5036 UDP

RSVP

   3455 TCP and UDP

RTSP

   1756, 1757, 4056, 4057 TCP and UDP (RTSP can vary by session like RTP)

So, if you are using SIP, you need to permit inbound SIP signaling on UDP ports 5060 and 5061.

Consider the following iptables policy commands:

```
iptables -P INPUT -j DROP
iptables -A INPUT -p UDP --dport 5060-5061 -j ACCEPT
iptables -A INPUT -p UDP --dport 5036 -j ACCEPT
iptables -A INPUT -p UDP --dport 5004 -j ACCEPT
iptables -P OUTPUT -j ACCEPT
```

This set of iptables commands manipulates the kernel's firewall so that only RTP, IAX, and SIP traffic can be accepted by the server, while all outbound traffic (OUTPUT chain) is permitted. This policy is based only on UDP port numbers. If incoming traffic isn't on ports 5060, 5061, 5036, or 5004, it is dropped. A truly hardened server would restrict outbound traffic, too.

Check for security risks in the dial-plan

It's possible to create a dial-plan that unintentionally allows incoming calls to use the Asterisk server to dial out, possibly placing expensive and resource-consuming outbound calls. Examine the sample configuration for Asterisk to see how its designers recommend separating outbound contexts from inbound ones. Then, make sure your dial-plan separates them, too. Here's part of a dial-plan snippet from */etc/ extensions.conf* that allows any incoming call to dial out from the server. It's an example of how *not* to set up your dial-plan:

```
[default]
exten => s,1,Answer( )
exten => s,2,GoTo(outbound)

[outbound]
exten => _XXXXXXX,1,Dial(Zap/1/${$EXTEN})
```

Familiarity with dial-plan programming will breed better security, so brush up on your Asterisk commands (or your Cisco IPT commands). Asterisk dial-plan configuration is covered in more detail in Chapter 17.

DMZs and Firewalls

If an Internet-facing server, or *bastion host*, as they're called, is locally firewalled and properly hardened in its own right, then it should be quite difficult to compromise.

But local firewall policy is the last line of defense. The first line of defense is a separate, standalone firewall appliance or server that sits between the bastion host and the Internet, monitoring and filtering traffic. This appliance could even be a router with firewall firmware installed, though in high-availability designs, the firewall and router should be kept separate.

By design, a firewall should sit between the attacker and the server she's attacking, which allows the attacker's actions to be observed and logged. The idea is to make it difficult to go unnoticed when carrying out a malicious access attempt from the outside world. The administrator just has to be paying attention to his firewall's logs.

But threats can come from the inside, too—the private side of the firewall, where all the servers are. Targeted, punitive threats often originate from within one's own organization. Many enterprise networks are set up so that a large group of workstation PCs has no firewall between it and a segment of servers, possibly including VoIP servers; see Figure 10-1. The naïve system administrator believes that an attack attempt is less likely from a workstation PC on this segment because she has greater physical access control, building security, and the like—so attacking from within the enterprise is thought to be harder than hacking from the outside—i.e., "Who would dare attack me from my own network?" But the random outside threat rarely has emotional baggage as motivation, making the inside threat a worrisome issue.

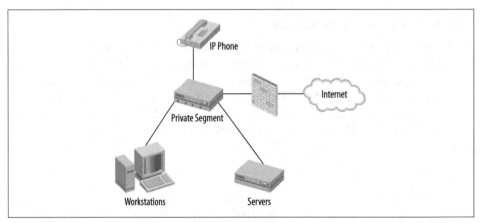

Figure 10-1. A typical small network firewall setup

The truth is, a VoIP network must be protected from both origins of attack. This is where DMZ segments (demilitarized zones) come into play. A DMZ is a network segment, usually Ethernet, that resides between two firewalled routers: one that provides protection from the Internet and another that provides protection from the private network, as in Figure 10-2. Normally, the DMZ is addressed with a publicly routable IP subnet, and is therefore a true part of the Internet. But the firewall between the DMZ and Internet filters blocks certain ports while logging access attempts.

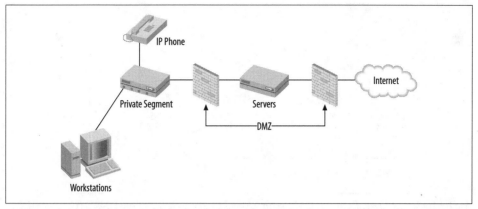

Figure 10-2. A DMZ setup provides protection for the servers from both private and Internet-based threats

Building a DMZ

There are many possible approaches to setting up a DMZ segment that are well beyond the scope of VoIP. Suffice it to say that entire books have been written on the subject of firewalls and security, such as *Building Secure Servers with Linux* (O'Reilly). But there are some common threads among all DMZ setups:

- One or more firewalled routers restrict traffic into, and often out of, bastion hosts.
- A DMZ often has limited access to the private network, while the public network (Internet) has no access to the private network.
- The private network often has limited access to both the public network (Internet) and the DMZ (see Figure 10-2).

When critical servers reside in the DMZ, the firewalled interfaces on the routers can enforce a policy of rules that protect them from risky traffic that originates from the Internet or from the private network. Keep in mind, though, that every hop in the VoIP call path adds some lag—and with two firewalled routers, you've got two extra hops. As a result, network performance on these two machines is important. Just like any other device in the call path, firewalls and routers must be able to handle the traffic load demanded by your VoIP network (see Figure 10-3).

In Figure 10-3, you can see that a SIP server and DNS server reside in the DMZ. They are firewalled on both the Internet side and the private side of the DMZ. Let's assume that the SIP proxy server in the DMZ is the softPBX for all IP phones on the private segment. This means that all IP phones and SIP clients must register with it in order to become aware of one another's existence.

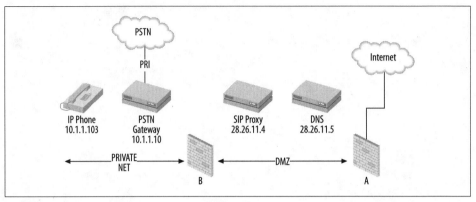

Figure 10-3. The firewalls in this DMZ setup (A and B) allow traffic to the DNS server and SIP proxy to be firewalled and monitored, whether it comes from a private network or the Internet

---

Port Scans

A *port scan* is a systematic process that attempts to identify all of the open ports and available services on a TCP/IP host. Most hackers use port scans to help them identify potential vulnerabilities on a target host. The hacker either wants to exploit them to gain control or deny access, or he wants to identify and fix them. Fortunately, there's a pretty good way of preventing VoIP services from being located through port scans.

Since TCP is connection-oriented, scanning for TCP ports is easy: if a connection is established, the port is considered open and available; if not, the port is closed and unavailable. Secure TCP-based services as well as you can, or disable them so they can't be located with a port scan, especially on mission-critical servers.

But VoIP mainly uses UDP, which is connectionless. In an insecure network, it's possible to identify UDP ports that aren't available by sending them data and then waiting for an ICMP unreachable message. Ports that don't produce such a response within a reasonable time frame are considered open and available. The moral of the story is this: disable ICMP traffic on your firewall, and nobody will be able to tell which UDP ports your VoIP services are using.

Now, let's assume IP phone 10.1.1.103 needs to be able to make and receive calls to and from the PSTN. The firewall rules for both sides of the DMZ must permit all the signaling and call paths required to allow these:

- In order for the IP phone to communicate with the SIP server, UDP port 5060 and 5061 must be permitted from 10.1.1.103 to 28.26.11.4 and vice versa on firewall B.

- In order for the SIP server to communicate with the PSTN gateway (via SIP), UDP port 5060 and 5061 must be permitted from 10.1.1.10 to 28.26.11.4 and vice versa.

- In order for an RTP call path to exist between the PSTN gateway and the IP phone, no rules need to be set up on the firewall, since both are on the same segment.
- As a matter of course, all hosts accessing services on the DMZ should have the ability to resolve DNS hostnames, and so TCP and UDP port 53, for DNS, should be permitted from 10.1.1.10 and 10.1.1.103 to the DNS server, 28.26.11.5.

At this point, assuming the VoIP network itself is configured to handle the calls via dial-plans and channels, and so on, PSTN subscribers can call the IP phone, and the IP phone can call the PSTN, even though the SIP proxy is on the DMZ. But what was really accomplished through all this was security. Use a similar exercise to identify the additional TCP/IP traffic flows required to support an Internet-based SIP calling application through firewall A:

- In order to communicate with Internet-based SIP hosts, the SIP proxy at 28.26.11.4 must be able to send and receive SIP traffic (UDP ports 5060 and 5061).
- In order to provide DNS name resolution for SIP services, the DNS server at 28.26.11.5 must be able to send and receive DNS traffic (TCP and UDP port 53) to and from the Internet.
- The SIP proxy can already communicate with the IP phone and the PSTN gateway (see earlier step).

The way a DMZ protects its hosts is by putting them behind firewalls, which limits the amount of public Internet traffic they're exposed to. However, if somehow a DMZ host is compromised even with firewalls in place, it can be easily used to launch attacks against neighboring servers on the DMZ. So, computers that sit on the DMZ, called bastion hosts, must be individually hardened. Rather than trust all traffic originating from within the DMZ, the bastion host should have a very restrictive local firewall policy: permit only the traffic that is needed for that host's application, and never assume that traffic is friendly because it's coming from a firewalled segment. Project 10.1 deals with local firewall policy.

Intrusion Prevention and Monitoring

The VoIP administrator's best intrusion detection tactic is log reading. Just about every device that handles network traffic can account for it in logfiles or through centralized logging services like syslog. By reading logfiles, where information about the traffic is saved, the administrator can pick up on misconfigured software, potential security vulnerabilities, and patterns of illicit or unauthorized behavior on the network.

Watching logs is critically important in a VoIP environment. Consider that, even if you are using DiffServ, a precedence-based QoS measure, it is possible that your network could be swamped in a DoS attack, robbing the available bandwidth for telephony apps. That would not be a good situation anywhere: voice is expected to work 100 percent of the time.

But rather than respond to threats after you've already become a victim, you can use a few techniques to proactively monitor for problems. These techniques are applied at places where network traffic is concentrated: routers and softPBX servers.

PSTN-to-IP Attack?

Some sysadmins and VoIP skeptics are concerned that a perpetrator might try to gain access to a private IP network through the PSTN. Even if it were possible for an attacker to fatally exploit a bug in the VoIP infrastructure—say, a codec—her only means of transmitting data into the compromised host would be through the analog or TDM connection to the PSTN.

Once compromised, it is possible this connection wouldn't be running any longer, thus cutting off the attacker's pathway into the network. The attacker's available bandwidth would be less than 64 kbps, and he would have no means of sending IP traffic, because his pathway into the system wouldn't even be TCP/IP-enabled. Even if he could crash the host, he couldn't transmit any data to it through the PSTN. So, aside from a denial of service due to an exploited bug somewhere in the VoIP network, the threat here is understandably low.

Project 10.3. Logging and Controlling VoIP Packets with iptables

What you need for this project:

- A Linux PC capable of running the NetFilter firewall (iptables)
- LAN

When a Linux NetFilter firewall is used to protect a group of VoIP bastion hosts or just as a gateway router for a segment where VoIP is used, a lot of VoIP-related events can be monitored and logged. Logging from the firewall is useful for the security-minded, but it's important for other reasons, too. It lets you get a feel for which remote networks and hosts are communicating with your VoIP services and how often they are. This can improve your understanding of bandwidth consumption and traffic patterns on your network, besides giving you a keener awareness of security.

NetFilter's default configuration provides for no logging. If you want a particular type of packet logged, say, from a specific network or on a specific port, you must tell NetFilter to log it. When a packet is logged, its pertinent information is sent to syslog to be stored. Syslog is the system-wide logging daemon that is a staple in most Unix-variant operating systems.

Logging packets using NetFilter doesn't save the contents of the packets—just information from the packets' headers! If you want to *capture* packets, you'll need other software, like Network Associates' Sniffer or the open source tool Snort.

To enable logging, you must set up a rule that specifies which packets you want to log. This rule says to log all packets sent *to* the machine running the NetFilter firewall (keep in mind, this will eat up tons of storage space *fast!*):

```
# iptables -A INPUT -j LOG --log-prefix "Log it all baby."
```

The log prefix option allows you to specify what will appear at the beginning of the log entry for each packet. That way, when you comb through lengthy databases of these entries, you can find specifically what you're looking for. The following rule is very broad—it captures any and all SIP traffic going *through* the firewall (FORWARD chain) and logs it:

```
# iptables -A FORWARD -P udp --dest-port 5060,5061 -j LOG --log-prefix "SIP"
```

Let's say that you are operating a SIP proxy that facilitates VoIP calling via SIP directly to two other proxies. Let's say all three SIP proxies are in the same organization, and that site-to-site VPN is used to connect them all. The three proxies support three VoIP LANs at separate offices. The LANs they support have network addresses 10.1.0.0/16, 10.2.0.0/16, and 10.3.0.0/16, as in Figure 10-4. The configuration examples given in this section are assumed to be running on the firewall in the 10.1.0.0 network.

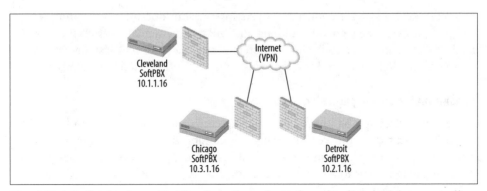

Figure 10-4. Packet logging on NetFilter firewalls can be configured to distinguish between traffic based on its destination or origin

Assuming a VoIP WAN like the one in Figure 10-4, it's possible to do some interesting logging and prefixing. Say you want to log SIP traffic by remote network:

```
# iptables -A FORWARD -P udp -s 10.2.0.0/16 --dest-port 5060 -j LOG --log-prefix \
"FromDetroit"
# iptables -A FORWARD -P udp -d 10.2.0.0/16 --dest-port 5060 -j LOG --log-prefix \
"ToDetroit"
```

```
# iptables -A FORWARD -P udp -s 10.3.0.0/16 --dest-port 5060 -j LOG --log-prefix \
"FromChicago"
# iptables -A FORWARD -P udp -d 10.3.0.0/16 --dest-port 5060 -j LOG --log-prefix \
"ToChicago"
```

The preceding example tags all the Detroit traffic separately from the Chicago traffic, making it easier to discern later on when you're viewing the packet log.

A simple modification to the previous example would allow you to log RTP traffic (port 8000 or whatever your endpoints use). On a strategically placed Linux firewall, this could provide valuable information about bandwidth consumption at the Detroit and Chicago sites in the example.

Another technique is to differentiate VoIP traffic that is to/from the private network from that which is to/from the Internet or another foreign VoIP network:

```
# iptables -A FORWARD -P udp -s 10.0.0.0/8 --dest-port 5060,8000 -j LOG --log-prefix\
"Private VoIP"
# iptables -A FORWARD -P udp -d 10.0.0.0/8 --dest-port 5060,8000 -j LOG --log-prefix\
"Private VoIP"
# iptables -A FORWARD -P udp -s 0.0.0.0/0.0.0.0 --dest-port 5060,8000 -j LOG \
--log-prefix "Internet VoIP"
# iptables -A FORWARD -P udp -d 0.0.0.0/0.0.0.0 --dest-port 5060,8000 -j LOG \
--log-prefix "Internet VoIP"
```

In the preceding example, private traffic—that is, traffic to or from a 10.x.x.x host, is tagged separately from Internet traffic, which is indicated by the catchall network address 0.0.0.0.

In a VoIP network that's connected to the Internet but doesn't use the Internet as a call path, it would be a good idea to log all VoIP traffic originating from the Internet. Such traffic could be an indicator of system abuse, just as email systems are abused by spammers. Refer back to Project 10.2 for a list of VoIP-related port numbers.

Reading and analyzing packet logs

Once you've used iptables to tell NetFilter to catch some VoIP traffic and log it, the log data is stored in the kernel facility of a syslog database file, where it can be retrieved using the dmesg command on Red Hat Linux. Other operating systems may provide different tools for viewing system logs.

When packets are logged, several bits of information about each packet are stored:

- The protocol of the packet (UDP, TCP, ICMP, etc.)
- The date and time the packet traversed the NetFilter chain where it was logged
- The size of the packet
- The source and destination addresses and ports (sockets)
- The originating MAC hardware address (when the packet comes from an Ethernet interface)
- Of course, the prefix that you specify in the --log-prefix option, if any

---

iptables Quick 'n Dirty

iptables is the program used to manipulate the Linux kernel's built-in firewall, NetFilter. iptables uses tables, which are stages of processing for each packet on the host. These tables are called *filter*, *nat*, and *mangle*. The default table, and the one with the most relevance for traffic monitoring, is *filter*. Each table has chains that represent the general characteristics of each type of traffic. In the *filter* table, the chains are:

INPUT
> The chain that processes packets received by the host from the network. This chain is frequently used to define local firewall policy on Linux-based softPBXs.

FORWARD
> The chain that processes packets that the host is routing. This chain is often used to define network policy for a group of hosts being protected by the NetFilter host.

OUTPUT
> The chain that processes packets created by the local host that are destined for the network. This chain can be altered to protect other hosts from attacks originating from the NetFilter host or just to minimize network chatter.

While iptables is an exhaustive subject deserving its own volume, here's a synopsis of its syntax.

Syntax:

```
iptables [-t table] -[ADCP] chain rule-specification [options]
```

Options:

-t Specifies which table to alter or examine.

-A Appends a rule to a chain.

-P Sets the policy for a chain. If no other rule matches the packet, then this policy will be enforced.

-D Delete the rule from the chain that matches the rule specification.

-L List the rules. Optionally, supply a chain name, and only that chain's rules will be shown.

-F Remove all the rules. Optionally, supply a chain name, and only its rules will be removed.

The -A and -P actions always have a filter to match packets on their protocol, destination, port number, and so on. The filters are set up using these options:

-d Destination network address

-s Source network address

-p Protocol, i.e., UDP, ICMP, etc.

-i The name of the local interface that first processed the packet

-o The name of the local interface that will process the packet if and when it is routed

(continued)

iptables rules often have a target specified by the `-j TARGET` option, which tells NetFilter what to do with the packets matched by the rule. Some useful targets are:

LOG
> Log the packet according to your system's kernel-logging configuration.

REJECT
> Don't accept the matched packet, and send an error packet in response.

DROP
> Don't accept the matched packet, and don't send any response.

DSCP
> A target used with the *mangle* table to alter DSCP code points and classes in packets.

TOS
> A target used with the *mangle* table to alter IP Type of Service headers in packets.

There are dozens more options that can help you leverage iptables in Voice over IP. Refer to the iptables manpage for more details.

---

dmesg's output is flat text that you can redirect to a file. Suppose you wanted to isolate the traffic prefixed with *Chicago* into a file by itself:

```
# dmesg | grep Chicago >chicagoVoIP.txt
```

Or better yet, email that log to somebody, perhaps so they can import it into a spreadsheet for further analysis. In the following example, pressing Ctrl-C will stop the *dmesg* application and an email will be sent containing the *Chicago* entries:

```
# dmesg | grep Chicago | mail chicagoVoIPadmin@oreilly.com
```

SNMP

Simple Network Management Protocol is a lightweight method of collecting traffic and performance data from network devices such as servers and switches. Different kinds of data use different parameter schemas, called management information bases, or MIBs. MIBs define how SNMP refers to metafields specific to a certain kind of data, such as Ethernet traffic or DNS-lookup statistics.

MIBs exist for SIP (*http://www.iana.org*), VOCAL (*http://www.vovida.org*), MEGACO (*http://www.ietf.org*), and other VoIP technologies. There are some useful VoIP-related SNMP monitoring tools (OpenNMS, Multirouter Traffic Grapher, etc.) that can be customized to make use of these MIBs, too. There isn't yet a fully integrated SNMP MIB for Asterisk, though, which leaves Asterisk administrators only one performance data-collection option: log reading. Fortunately, Asterisk is very flexible in the logging department.

Project 10.4. Tune Up Asterisk's Logging Configuration

What you need for this project:

- Asterisk

Log analysis should be the core of your daily system monitoring and security activities. Like other softPBX servers, Asterisk supports flexible logging, providing several levels of logging detail in several different files. It also supports using syslog.

 By default, Asterisk stores its logs in */var/log/asterisk*.

Configuration of Asterisk logging is done in the */etc/asterisk/logger.conf* file, which Asterisk reads at boot time or whenever it is started. The first section of the file is [general], where you can assign a value to the dateformat option to specify what date format to use in Asterisk's logs. To figure out the syntax of the date formats, read the manpage for strftime( ) by running *man strftime*.

The next section, [logfiles], describes which files should be used for logging output, and how detailed each should be. The syntax for this section is:

```
filename => level,level,level...
```

Consider the following logging configuration:

```
[general]
[logfiles]
messages.log => notice,warning,error
debug.log => notice,warning,error,debug,verbose
```

In this example, *messages.log* will contain a digest version of Asterisk's logging output, while *debug.log* will get everything in minute detail. Be careful with logs, though—Asterisk won't start once the logfiles reach 2 GB in size. On a busy system, a file like *debug.log*, resulting from the previous code, would hit that size pretty quickly, so make sure your logfile rotation includes Asterisk.

The console keyword

If you use console as a log filename, Asterisk will assume you mean the console device, not an actual logfile. So, if you added this to the [logfiles] section, the desired level of logging would be output to the console session where Asterisk is launched:

```
[logfiles]
console => warning,error
```

Use a non-default log directory

Some attackers cover their tracks by removing commonly used logfiles that could contain evidence of their tampering with the system. So it's generally a good idea to keep logfiles in a non-default place. This way, if the attacker uses an automated program to remove logfiles, it will be less likely to find and destroy Asterisk's. To change Asterisk's default log location, edit */etc/asterisk/asterisk.conf* and change the *astlogdir* directive to a path of your choosing. (Then make sure that path has appropriate permissions to allow Asterisk to write files in whichever path you choose.) A sample *asterisk.conf* follows:

```
[directories]
astetcdir => /etc/asterisk
astmoddir => /usr/lib/asterisk/modules
astvarlibdir => /var/lib/asterisk
astlogdir => /var/log/asterisk
astagidir => /var/lib/asterisk/agi-bin
astspooldir => /var/spool/asterisk
astrundir => /var/run/asterisk
```

Enable syslog

syslog can be a target for Asterisk logging output, too. To enable it, use a syslog keyword in the [logfiles] section, similar to the console keyword:

```
syslog.local0 => warning,error
```

Snort and Nagios

Snort is an open source intrusion-detection system (IDS) and packet-logging apparatus. Unlike NetFilter and syslog, Snort allows more customizable logging of traffic. Instead of merely logging the headers of rule-matched packets, Snort lets you record and store the entire rule-matched packets. It also lets you configure alarms that can be triggered by signature characteristics of known exploit attacks and suspicious conditions.

Like many of the other things we've discussed in this book, Snort is a deep subject in its own right. Indeed, to fully integrate Snort and other security software into your VoIP arsenal, you'll want to read O'Reilly's *Managing Security with Snort and IDS Tools*.

Nagios (formerly called Netsaint) is an open source network-monitoring suite. Its web-based interface allows network managers to watch activity on a single host or network-wide basis. It uses a combination of an SNMP collection and open Nagios-specific add-ins that make it aware of new vulnerabilities and intrusion signatures. Nagios even offers environmental monitoring interfaces for your server room's HVAC system, so you can be alerted when the temperature gets too high or low. You can find out more about Nagios at *http://www.nagios.org*.

Key Issues: Security and Monitoring

- Security in traditional telephony is quite limited and much less effective than that of modern data systems. Security in IP telephony applications is infinitely controllable and expandable due to its basis in software.

- Snooping and phreaking are fairly easy in traditional telephony environments.

- Call detail records and call accounting can provide forensic data when investigating an instance of system abuse.

- A properly configured and well-tested dial-plan is a good way to prevent anonymous abuse.

- Credentials and authentication are often required in order to use IP telephony applications, just as they are required to use other kinds of applications.

- MD5 hashing is a good way to hide passwords from casual observation, and most SIP implementations support authenticating endpoints using MD5-encrypted credentials.

- Media encryption can stop would-be eavesdroppers from listening in on your VoIP conversations.

- Software maintenance is crucial with VoIP servers in order to avoid virus and worm attacks that exploit known vulnerabilities.

- Hardening VoIP servers is important, just as it is in any other mission-critical application.

- Removing unnecessary software and optimizing local firewall policy are two elements of hardening a server.

- DMZs can provide limited, controlled access to a softPBX from the Internet and from a private network. A server on a DMZ is a bastion host.

- Intrusion prevention is accomplished through regular testing for known vulnerabilities and through vigilant monitoring of logs.

- iptables can be used to configure VoIP-specific logging with the help of syslog. To read VoIP logs from syslog in Red Hat Linux, use the dmesg command.

- SNMP can be used to assist you in monitoring a VoIP network.

- Packet-log reading can be used to "dig deeper" when you suspect illicit access on your network.

CHAPTER 11
Troubleshooting Tools

The word *troubleshooting* may trigger memories of desperation when a critical system was down and you couldn't figure out why. Troubleshooting often justifies high-priced consultants. The mere mention of it can make fainthearted IT executives squirm... because if you're troubleshooting, that means something's wrong. Regardless, troubleshooting tools are an important ingredient in the VoIP success recipe.

If you're building your standards-based VoIP network from the ground up today, then you're probably using SIP and not H.323. SIP is clearly the prevailing standard for VoIP call signaling, as it provides more interoperability and easier troubleshooting. The tools used to troubleshoot SIP and H.323 are largely the same, though: packet sniffers, log analysis, and softphones.

Since SIP is a framework for real-time media applications, the stability of one SIP-based system to the next can vary greatly. Problems are most likely at the application layer. Troubleshooting them may require a specific knowledge of the application, or even access to its source code. This isn't always practical or available. Many system engineers aren't hard-core C programmers. Of course, for those who want to probe the mechanics of IP telephony with C, a great book to read is O'Reilly's *Practical VoIP with VOCAL*.

This chapter therefore focuses on generic, more common, troubleshooting scenarios: inspecting SIP registration and call-setup exchanges using a packet analyzer and interpreting log output to find the root cause of interoperability problems.

VoIP Troubleshooting Tools

Here's what you'll need to troubleshoot a VoIP network:

- A softphone, because softphones tend to be easier to configure than hardphones. They also provide more useful logging output. Xten Network's SIP softphone with logging output is available free from their web site (*http://www.xten.com*). A free H.323 softphone with logging output is OhPhone.

- A packet sniffer such as Ethereal, which runs on several platforms. It's available at *http://www.ethereal.com*. Snort can be useful for troubleshooting security issues, too.

- If you're troubleshooting from behind a firewall, you'll need access to an unfirewalled public IP address, too. SIP can be stymied by NAT firewalls (more on this later).

- A SIP server like SIP Express Router (*http://www.iptel.org/ser*) or a SIP-compatible registrar like Asterisk or VOCAL. In this book, we've used Asterisk. For H.323 setups, a free gatekeeper and gateway like those included in Open H.323 are recommended.

The Three Things You'll Troubleshoot

Latency

On the Internet, the round-trip latency of a phone call can be relatively stable from one moment to the next but vary heavily from one day to the next. You'll use troubleshooting tools to find the sources of latency that can make VoIP call quality fluctuate.

Jitter

On the Internet, moment-to-moment changes in latency are called jitter. Jitter is a huge problem. You'll use troubleshooting tools in order to identify the root causes of jitter.

Signaling and interoperability problems

Perhaps the most difficult to isolate, interoperability problems require a working knowledge of the signaling protocols you use on your VoIP network. Avoid interop problems by sticking with a single standard, or even a single vendor, if you can. But when interoperability is required, you'll use troubleshooting tools to isolate problems.

SIP Packet Inspection

If you enjoy sifting through megs and megs of captured packets, you're either a nefarious cracker searching for an elusive password, you're a network QA analyst who gets paid by the byte, or you have a sick sense of pleasure. Sifting through mounds of packets that have been captured from the network isn't entertainment, after all.

Even among network administrators, the use of packet-sniffing tools is often a last resort. When logfiles, packet loggers, and other analysis tools fail, the ol' packet sniffer gets the dust blown off, and the administrator has to refamiliarize himself with it in order to figure out exactly what's wrong on his network.

But, since you're serious about VoIP, you should cozy up to that sniffer. In packet telephony applications, capture tools aren't just the last resort—they're a primary troubleshooting tool. Using a sniffer (or *protocol analyzer*), like the ubiquitous Ethereal or Network General's aptly named Sniffer, you can capture and dissect call setups, teardowns, and other signals. You can zero in on capabilities negotiation—say, to figure out why two particular phones can't call each other while the rest of the network functions fine.

Project 11.1. Inspect SIP Traffic with Ethereal

What you need for this project:

- SIP registrar such as Asterisk
- Xten Network's X-Lite SIP softphone
- Ethereal software running on Linux, Windows, or Macintosh

Ethereal is a packet analysis tool that you can use to inspect network traffic, from the Ethernet layer all the way up to the application layer. Packets are captured in a buffer and displayed on the screen. Filters can be applied to restrict the capture to packets matching a certain source, destination, size, protocol, or service.

To obtain Ethereal, download it from *http://www.ethereal.com* and install it on your Windows, Mac, or Linux box. The screenshots and examples here assume the Windows version. You'll also need a SIP softphone such as X-Ten Network's X-Lite or X-Pro. This project assumes X-Lite. Both X-Lite and Ethereal will need to be running on the same machine, because switched networks permit you to sniff only packets that are being sent to or from your own host. (If you're using a non-switched network, like a hub, Ethereal can observe packets not bound for or originated from your PC.)

To demonstrate SIP packet observation with Ethereal, we'll set up a filter that allows us to capture SIP registration signals in two scenarios—one for a successful SIP registration, another for a failed SIP registration. As in the other projects, the SIP server with registrar is 10.1.1.10. In this instance, Asterisk is used as the SIP server.

 X-Lite offers excellent diagnostic logging, too. Some of the packets you observe with Ethereal in this project will correlate with entries in the X-Lite diagnostic log, which you can view by selecting Diagnostics from the right-click menu in X-Lite's UI.

Configure the SIP softphone

If you're setting up X-Lite for the first time, you'll need to click the configuration button, right of the center, next to the Clear button (see Figure 11-1). Once you click this button, you'll see the configuration menu appear.

In the configuration menu, double-click Menu, then System Settings, then SIP Proxy, then [Default]. This will take you to the SIP client configuration, as in Figure 11-2.

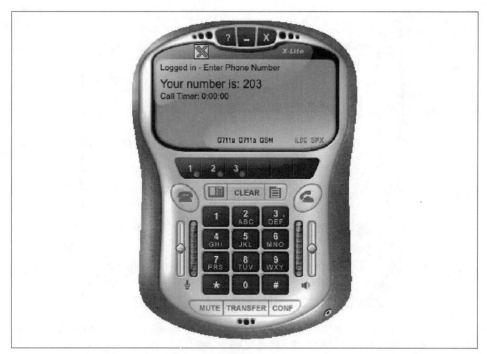

Figure 11-1. The X-Lite softphone's user interface

Here, you can configure the softphone to register using a number and/or password to match what you've established in the dial-plan configuration on the SIP server.

Once configured as in Figure 11-2, the SIP softphone will automatically register with the SIP registrar as soon as you close the configuration menu. If registration was successful, you'll see Logged In in the UI display, as in Figure 11-1. If not, make sure the SIP proxy profile called [Default] is enabled and configured to match a SIP peer on the server.

Configure Ethereal

Once Ethereal is installed, launch it. Next, begin a capture by clicking the Capture menu and the Start menu item. This will show you the Ethereal: Capture Options dialog, as in Figure 11-3. In order to limit the kind of traffic that Ethereal will capture, you'll need to use a filter string. Ethereal has a rather sophisticated syntax for this string, which instructs Ethereal what to capture and what to ignore.

In this case, our SIP server is 10.1.1.10, and the standard port for SIP traffic is UDP 5060. We want to capture traffic in both directions—that is, to the SIP server and to the softphone running on the same host as Ethereal. The string that achieves this is:

```
host 10.1.1.10 and udp port 5060
```

Check the "Update packets in real time" and "Automatic scrolling in live capture" options in order to see the packet capture log occur immediately rather than waiting

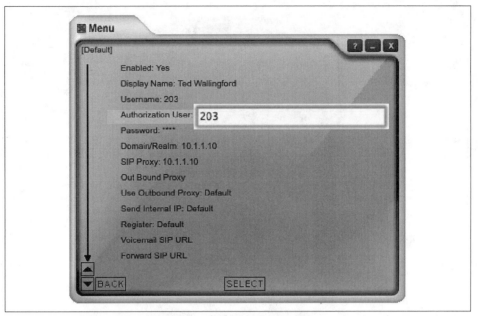

Figure 11-2. X-Lite's SIP client configuration allows you to set up many proxy profiles that you can use to test different combinations of settings or to test many SIP servers

until the capture session is complete. Then, click OK and the main capture window will appear.

Observe SIP registration

Now, restart X-Lite. It will attempt to register automatically with the SIP server upon startup. By the time X-Lite says you're Logged In, you can stop the packet capture by clicking Ethereal's Capture menu and Stop menu item. The main capture window should be filled up with a number of packets, as in Figure 11-4.

In this instance, Ethereal shows the first packet, packet number 1, as a SIP REGISTER method. Newer versions of Ethereal, such as the 0.10.7 used here, can parse SIP packets and tell you which methods and responses they contain. Packet 4 is a second registration request (the first one failed because X-Lite tries anonymous registration first).

Packet 5 is the 100 Trying SIP response sent from the SIP server back to the softphone. Packet 6 is the 200 OK SIP response sent from the SIP server back to the softphone, indicating the registration was successful. Packet 7 is a SIP NOTIFY method asking for username 204 at the SIP registrar. Packet 8 is the 200 OK response. At this point, registration is complete. The additional packets (9 and 10) are keep-alive packets that X-Lite sends to its SIP registrar. Not all SIP phones do this.

The bottom pane of the main capture window shows the actual hex-encoded content of the packet and the ASCII-encoded content of the packet that corresponds to

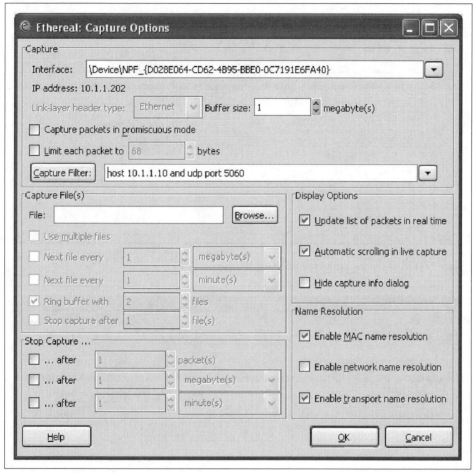

Figure 11-3. Ethereal's Capture Options dialog

it. The hex is on the left, and the ASCII on the right. This is where you can usually pick out problems: an incorrect password or a botched username would be easy to spot this way.

Observe registration failure

Outside the test lab, you probably won't have occasion to capture SIP packets unless something is working incorrectly. But then, why else would you need a packet sniffer? Simulating a registration failure is very easy. Just alter the registration username of the SIP proxy profile in X-Lite to one that doesn't match a SIP peer on the SIP server. (Be sure that your softPBX requires SIP clients to use a username and password; Asterisk can be configured to allow anonymous registration!) Then, start the Ethereal capture with the filter string used earlier, restart X-Lite, and watch the registration crash and burn (see Figure 11-5).

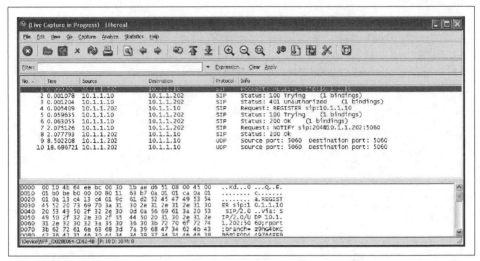

Figure 11-4. An Ethereal capture of a SIP registration

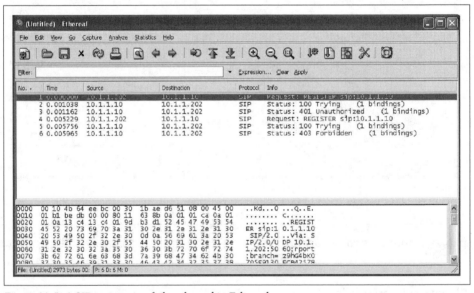

Figure 11-5. A SIP registration failure logged in Ethereal

Packets 1, 2, and 3 in Figure 11-5 are a result of the de facto anonymous registration attempt and can be ignored. Packet 4 is the SIP REGISTER method using the incorrect username. Packet 5 is the customary 100 Trying response. Then, packet 6 contains the 403 Forbidden response, which indicates service is unavailable for this username. (An incorrect password should result in a 401 Unauthorized.)

To spot the incorrect username, highlight packet 4 and scroll down the bottom pane until you see the SIP User header in the ASCII-encoded column of the packet contents on the right side.

Similar capture and analysis techniques can be used to troubleshoot access control and dial-plan configuration problems on the SIP server, or misconfigurations on SIP phones. If you're daring enough to decode ASN.1, you can even use this technique to troubleshoot H.323 signaling.

Project 11.2. Inspect SDP Capabilities Negotiation

What you need for this project:

- A SIP-compatible softPBX such as Asterisk
- A SIP phone such as X-Lite
- Ethereal software
- LAN

The Session Description Protocol is an essential part of SIP call signaling. Its elements are text tokens sent in SIP packets with the SDP content-type header. These tokens advertise the capabilities and requirements of each endpoint according to the parameters of the application, be it a telephone call, instant message, or something else.

During call setup, specifically during the SIP INVITE method, the SDP payload is sent from one endpoint to the other. A SIP 200 OK response indicates agreement with the SDP parameters, while a 4xx response indicates disagreement or incapability. If you need a refresher on SIP signaling, refer back to Chapter 7.

Inspect successful capabilities negotiation

Using Ethereal configured with the same filter string from Project 11.1, you can capture a successful capabilities negotiation. In its default configuration, Asterisk supports G.711 so that just about any IP phone, X-Lite included, can place calls to it. In this case, X-Lite will be used to call Asterisk extension 201, and the SDP exchange for this call will be captured.

If you don't have such an extension on your dial-plan, then you can call Asterisk's default autoattendant demo at extension 500 instead. (If you've removed this extension in your hacking of Asterisk, just run a make sample from your Asterisk source directory in order to get the default config back again.)

When you place the call on X-Lite, use Ethereal to capture the SIP packets and zero in on the SDP content carried in the INVITE methods and 200 OK responses. In Figure 11-6, you can see that the call setup was successful.

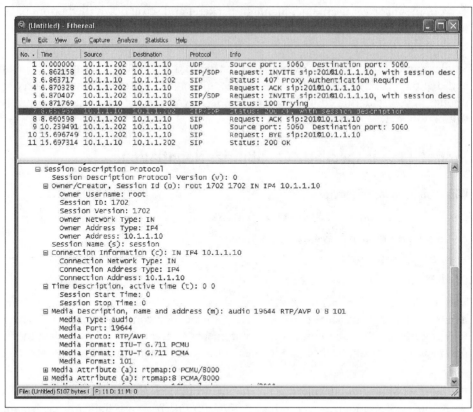

Figure 11-6. Ethereal can parse SDP content so it's easier for you to troubleshoot call-setup problems

Packet 5 is the authenticated INVITE method. The user in this example is calling SIP user 201. Included in packet 5 is an SDP payload. Ethereal indicates this in its Protocol column in the top packets pane of the main capture window, shown in Figure 11-6. Packet 6 is the 100 Trying response. Packet 7 is the 200 OK response, which also includes an SDP payload. If there's a codec match in the media attributes list of the SIP INVITE and the 200 OK response (shown in the bottom pane), then all that's needed is a SIP ACK method sent by the caller to confirm agreement on the first matching attribute. That's what packet 8 is.

Inspect failed capabilities negotiation

If there were no capabilities match, call setup would fail. This scenario can be produced easily by temporarily crippling the capabilities of the Asterisk server. To make it impossible for the X-Lite softphone to negotiate an audio stream with the Asterisk server, you can disallow all codecs supported by X-Lite and permit only GSM codec, which X-Lite doesn't support. This snippet of */etc/asterisk/sip.conf* does just that:

```
[general]

port = 5060            ; Port to bind to (SIP is 5060)
bindaddr = 0.0.0.0     ; Address to bind to (all addresses on machine)
disallow=all
#allow=ulaw
#allow=alaw
#allow=gsm
```

The G.711 and GSM codecs have been commented out. This will simulate a codec capabilities mismatch, so the SIP client won't be able to pass the SDP negotiation, and call setups will fail. (Don't forget to issue a reload command at the Asterisk command line.) Now, we can run the capture again while X-Lite tries to call extension 500, as before. Only this time, the call will fail because there's no suitable codec common to both the caller (X-Lite) and the receiver (Asterisk).

Packet 7 is shown in Figure 11-7. It's a SIP INVITE carrying SDP content that includes a list of *a* tokens.

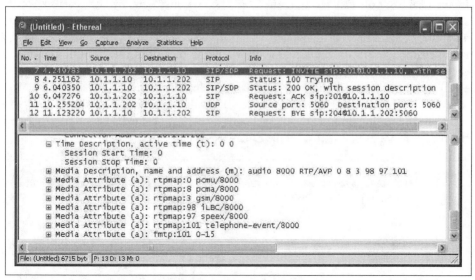

Figure 11-7. The bottom pane of the capture window shows the media attribute list: the SDP text payload that advertises the capabilities of the calling endpoint

These represent media attributes, or capabilities. Ethereal presents the SDP content in a parsed, hierarchical fashion. But the raw ASCII SDP payload of this SIP packet, which can be seen in X-Lite's diagnostic log, actually looks like this:

```
...
Content-Type: application/sdp
User-Agent: X-Lite release 1103m
Content-Length: 290

v=0
```

```
o=203 146336832 146337009 IN IP4 10.1.1.201
s=X-Lite
c=IN IP4 10.1.1.201
t=0 0
m=audio 8000 RTP/AVP 0 8 3 98 97 101
a=rtpmap:0 pcmu/8000
a=rtpmap:8 pcma/8000
a=rtpmap:3 gsm/8000
a=rtpmap:98 iLBC/8000
a=rtpmap:97 speex/8000
a=rtpmap:101 telephone-event/8000
a=fmtp:101 0-15
```

The capabilities are listed with a reference number following the rtpmap keyword. 0 pcmu/8000 indicates that 0 is the reference number that RTCP will later use to refer to this G.711 μlaw at 8000 Hz capability. The other capabilities are advertised with other numbers. (These numbers are reserved like commonly used port numbers in TCP/IP, and they can be overridden.)

In Figure 11-8, you can see that the 200 OK response sent by the receiver to the sender has an SDP payload that presents no audio codecs at all in its media attributes. This is because they have been purposefully disabled, of course. Packet 10 is the customary SIP ACK method acknowledging receipt of the 200 OK and giving the go-ahead for RTP to begin. But without any matching SDP media attributes to establish the RTP media channel, the receiver selects attribute reference number 101 using SDP's M token. 101 means there's no valid capabilities match. RTCP will report to the calling endpoint a few seconds later that no media channel exists, and the receiver "hangs up" with a SIP BYE method in packet 12.

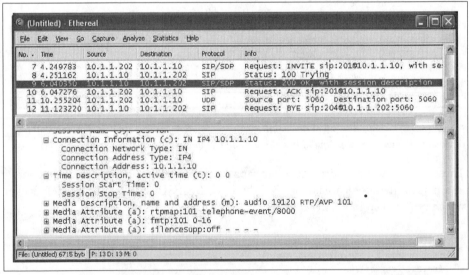

Figure 11-8. The bottom pane of the capture window shows the media attribute list: SDP's listing that advertises the capabilities of the receiving endpoint, in this case, Asterisk

Don't forget to reenable the codecs after doing this experiment, or you'll have a *real* problem to troubleshoot!

Call-signaling issues can be frustrating, especially when using a mixed bag of SIP products from different vendors and vintages. Just as you can reveal SDP failures as in Project 11.2 and authentication problems as in Project 11.1, packet capture is the best tool for exposing any and all signaling problems.

Interoperability

Interoperability is a sore spot among some VoIP admins because of the multitude of signaling standards, RFC revisions, and competing commercial standards. Interop problems are the main reason lots of enterprise decision-makers used to talk down VoIP technology and perhaps the key reason VoIP wasn't a rising star of enterprise telecom sooner.

Of course, talking about interoperability problems is easier than pinpointing them in the field. During a signaling exchange, two participating endpoints may not be "on the same page," so to speak. Caller and receiver may be assuming a different approach because, though they're both using SIP, there's enough leeway in the spec to allow for app-specific "dialects."

Mismatched capabilities is one example. A video phone might have a *preference* for video calls, but it could also be configured to permit *only* video calls. Audio-only calls could be crippled due to a configuration mistake. Or one manufacturer's preferred authentication scheme might not work very well on another vendor's PBX. It's also possible that a particular user agent may not fully implement the protocol—a softPBX that doesn't support SIP redirects, for example.

There are a lot of interoperability pitfalls here. The test lab is the best place to discover and isolate problems, before that shiny new equipment's in the field. Not all problems can be solved in the lab, so interop troubleshooting is something you'll soon be doing, if you aren't already. To troubleshoot VoIP issues, put yourself in the endpoint agent's situation. Find out what the endpoint is seeing and how it is reacting. Do the same for the server. The key to troubleshooting interoperability between two VoIP agents is asking, "What does this device think it's doing?"

Troubleshooting VoIP problems is like debugging a misbehaving computer program to find the root cause. Programmers can diagnose issues in their software by watching its output for clues that lead them to the bug. In the same way, watching the logs and packets generated by your VoIP components will lead you to a faulty config or bug.

Looking at captured packets will often tell you *what* signals a device is sending, but not *why* it's sending them. For that, you'll need to go to the device's logfiles and diagnostic data. Comparing the caller's account of the interop event with the receiver's account will help you reveal the root causes of interop problems.

Project 11.3. Trace Both Ends of a Call Setup with Log Comparison

What you need for this project:

- Asterisk
- Xten Network's X-Lite SIP softphone
- LAN

To set up an interop problem in the test lab, we'll configure Asterisk to require an MD5 secret for a SIP peer, without informing the SIP client, X-Lite, what the MD5 secret is. This will simulate lack of support for MD5 SIP authentication by the calling endpoint, a common incompatibility. You can require an MD5 secret for any SIP peer by finding the section for the peer in *etc/asterisk/sip.conf* and adding:

```
md5secret=whatever
```

whatever signifies any string you want, since our purpose is merely to require MD5 support from the X-Lite SIP client. Once you've added this token to the *sip.conf* file, issue a reload at the Asterisk CLI and then launch X-Lite on a nearby PC. Attempt to place a call using X-Lite configured for the extension number whose SIP peer you just gave the md5secret token.

 To get the level of log output needed for this project, your logging configuration will have to be extremely verbose. Refer to Project 10.3 for a refresher if needed.

Using X-Lite's Diagnostic Log window, you can see the REGISTER method that the SIP client is sending to the Asterisk server:

```
SEND TIME: 158566682
SEND >> 10.1.1.10:5060
REGISTER sip:10.1.1.10 SIP/2.0
Via: SIP/2.0/UDP    10.1.1.201:
5060;rport;branch=z9hG4bKA98898C43B1E11D99DF4000A958240F6
From: Ted Wallingford <sip:203@10.1.1.10>;tag=1055640066
To: Ted Wallingford <sip:203@10.1.1.10>
Contact: "Ted Wallingford" <sip:203@10.1.1.201:5060>
Call-ID: 383F74D93AA611D99DF4000A958240F6@10.1.1.10
CSeq: 62767 REGISTER
Expires: 1800
Max-Forwards: 70
User-Agent: X-Lite release 1103m
Content-Length: 0
```

The call ID, shown in bold, is the SIP user agent's unique identifier for this particular signaling exchange. It will be used for the duration of the exchange, both in requests and responses, regardless of the outcome. In this example, the outcome is a 403 Forbidden response from the registrar because of failed registration. Of course, in a real troubleshooting scenario, we still wouldn't know *why* the registration is failing. We would know only that, for one reason or another, we aren't authorized to register. Here's the Forbidden response as recorded in X-Lite's diagnostic log:

```
RECEIVE TIME: 158747423
RECEIVE << 10.1.1.10:5060
SIP/2.0 403 Forbidden
Via: SIP/2.0/UDP 10.1.1.201:5060;branch=z9hG4bK15429F973B1F11D99DF4000A958240F6
From: Ted Wallingford <sip:203@10.1.1.10>;tag=1055640066
To: Ted Wallingford <sip:203@10.1.1.10>;tag=as0e35e19d
Call-ID: 383F74D93AA611D99DF4000A958240F6@10.1.1.10
CSeq: 62774 REGISTER
User-Agent: Asterisk PBX
Allow: INVITE, ACK, CANCEL, OPTIONS, BYE, REFER
Contact: <sip:203@10.1.1.10>
Content-Length: 0
```

The call ID matches the request with the response. Up until this point, the reason for the 403 Forbidden result is still unknown. The next troubleshooting step is to compare X-Lite's version of the failure, which reveals little, with Asterisk's version, which we hope will reveal more.

Depending on Asterisk's *logger.conf* configuration, the level of granularity and location of the logs may vary (Project 10.3 describes log tweaking). In this instance, let's say we've got a highly detailed Asterisk message log at */var/log/asterisk/full*. By issuing this command, we can quickly find all the entries that have the call ID that matches X-Lite's diagnostic log:

```
# cd /var/log/asterisk
# cat full | grep 383F74D93AA611D99DF4000A958240F6@10.1.1.10 | more
```

You can use copy and paste to bring the call ID from X-Lite into a terminal window so you don't have to type that length call ID string. The output of the cat command will look something like this:

```
Nov 20 12:23:17 DEBUG[1090935488]: Auto destroying call
'383F74D93AA611D99DF4000A958240F6@10.1.1.10'
Nov 20 12:23:17 VERBOSE[1090935488]: Destroying call
'383F74D93AA611D99DF4000A958240F6@10.1.1.10'
```

By itself, this output doesn't tell us much, only that Asterisk decided to terminate the call associated with that call ID at some point during the exchange. But this isn't a dead end in the troubleshooting process, because this output reveals Asterisk's identifier for the Linux thread handling this call, 1090935488, shown in bold. (Non-Linux Asterisk uses the PID instead of the thread ID.)

Asterisk doesn't use the SIP call ID in log entries nearly as often as it uses the thread identifier, so we'll need to comb the log again for entries that contain the thread identifier. Entries that occur with the same thread identifier in rapid succession within 1 to 2 seconds following the instance of the SIP call ID are likely to be the entries we're looking for. If you have the time frame of the event you're troubleshooting narrowed down to a 2 or 3 minute window, so you can further grep the output with a reference to the time window in question:

```
# cat full | grep 1090935488 | grep "Nov 20 12:2" | more
                                       ^^^^^^^^^^^
```

Now, you'll be able to examine all the output related to thread 1090935488:

```
Nov 20 12:26:01 VERBOSE[1090935488]: 12 headers, 13 lines
Nov 20 12:26:01 VERBOSE[1090935488]: Using latest request as basis request
Nov 20 12:26:01 VERBOSE[1090935488]: Sending to 10.1.1.201 : 5060 (non-NAT)
Nov 20 12:26:01 VERBOSE[1090935488]: Found RTP audio format 0
Nov 20 12:26:01 VERBOSE[1090935488]: Found RTP audio format 8
Nov 20 12:26:01 VERBOSE[1090935488]: Found RTP audio format 3
Nov 20 12:26:01 VERBOSE[1090935488]: Peer audio RTP is at port 10.1.1.201:8000
Nov 20 12:26:01 DEBUG[1090935488]: Peer audio RTP is at port 10.1.1.201:8000
Nov 20 12:26:01 VERBOSE[1090935488]: Found description format pcmu
Nov 20 12:26:01 VERBOSE[1090935488]: Found description format pcma
Nov 20 12:26:01 VERBOSE[1090935488]: Found description format gsm
Nov 20 12:26:01 VERBOSE[1090935488]: Found description format telephone-event
Nov 20 12:26:01 VERBOSE[1090935488]: Capabilities:
- us 0x60e(GSM|ULAW|ALAW|SPEEX|ILBC),
- peer audio=0x60e(GSM|ULAW|ALAW|SPEEX|ILBC)/video=0x0(EMPTY),
- combined 0x60e(GSM|ULAW|ALAW|SPEEX|ILBC)
Nov 20 12:26:01 VERBOSE[1090935488]: Non-codec capabilities: us - 0x1(G723), peer -
0x1(G723),
combined - 0x1(G723)
Nov 20 12:26:01 DEBUG[1090935488]: Setting NAT on RTP to 0
Nov 20 12:26:01 VERBOSE[1090935488]: Found user '203'
Nov 20 12:26:01 NOTICE[1090935488]: Failed to authenticate user Ted Wallingford
<sip:203@10.1.1.10>;tag=435868921
```

The first dozen lines of output are Asterisk's logging of the SDP attributes sent when X-Lite tried to place a call with a SIP INVITE method. The giveaway that our 403 Forbidden is caused by an authentication problem is in the last two lines of the output.

Here, we have it narrowed down to an invalid password, a mismatched MD5 secret, or nonsupport of MD5. We've eliminated an invalid username as the root cause, because the username is reportedly valid. To determine whether the problem is related to MD5 or just a plain-old password mismatch, we need to find out whether the SIP peer, in this case extension 203, is required to use an MD5 secret. Since we set up this failure on purpose, we know that it is. (Without MD5, the same steps could have been used to troubleshoot a simple password mismatch.) In a less concocted troubleshooting situation, you would go to /etc/asterisk/sip.conf and check for an MD5 requirement for the peer labeled [203].

To resolve the simulated incompatibility, remove the md5secret token from *sip.conf* and attempt to place a call with the X-Lite softphone again.

Troubleshooting Quality-of-Service Issues

Not all troubleshooting situations involve signaling. As you may have read in prior chapters, noise problems and dropouts stemming from jitter and packet loss are cause for troubleshooting. Fortunately, the standards for sound transmission—RTP and codecs—are simple compared to call signaling, so troubleshooting sound issues is generally easier.

That's not to say all sound-related issues are a snap to dismiss. Sound issues are affected by many disciplines of networking: switching, routing, QoS, and packetization. The way congestion on a routed link shows up is in dropouts, or in busy signals if you're using a QoS solution. On a switched link, like Ethernet, busy signals may occur long before dropouts are perceptible. So symptoms are different on different links.

As with signaling, good diagnostic logging is valuable for troubleshooting. Packet analysis is less useful when examining sound quality problems. One can certainly identify jitter using packet capture, but it's just easier to identify jitter across IP links using tools like *traceroute* and *pathping,* which succinctly show the variances in round-trip latency on every hop of a route. If you need to dig deeper, capture tools like Ethereal and its commercial equivalents can assist you in isolating jitter at the RTP level.

Commercial packet analysis tools

If firewalls or other filters block ICMP or TCP traffic along the call path you're troubleshooting, then you may use a packet capture tool with advanced VoIP capabilities, like Network General's ubiquitous Sniffer or WildPacket's EtherPeek VX. Both permit analysis of signaling and media packets, readily understand the structure of RTP, can identify jitter, and can reveal sources of congestion on your network.

When the going gets tough, call in the end users

The human ear is a great instrument for troubleshooting. You can carry an IP phone from one server closet to the next, listening to calls to the same destination from different points along the call path. This process of elimination will help you narrow down a group of potential congestion points until you've identified the one that's to blame. On large, expansive networks, where this can't be done by a single person, you could enlist help from other people within the organization. Even end users are pretty reliable when it comes to assisting with this type of troubleshooting. Ask them to make a call that you expect to traverse the call path you're looking at, and then ask them how it sounds or if they get a busy signal. Both can be indicators of congestion, though a busy signal will happen only in networks with QoS.

When, Not if, You Have Problems...

Each manufacturer of Voice over IP products, be they open or proprietary, has a certain amount of software and firmware code to maintain. Supporting Asterisk's development is a developer community with a mountain of C source code undergoing constant revisions. Cisco has a core development team for Call Manager, a firmware team that works on the IP phones, a developer staff for Call Manager Express, and an open source development coordinator. Because your VoIP network is made of software, it's only as good as the software that runs it. At some point, through diligent troubleshooting, you may pick up what you suspect is a bug.

For open source projects like Asterisk and VOCAL, the odds are good that another user has also noted the bug. The odds are also good that the bug has already been fixed. Then, it's just a matter of downloading the correct source code revision, compiling, and installing the revised distribution. Visit these sites to sign up for mailing lists where you can monitor the status of bug fixes and report bug-related issues and get general troubleshooting problems:

> *http://lists.digium.com* (Asterisk)
> *http://www.vovida.org* (VOCAL)
> *http://www.openh323.org* (Open H.323)
> *http://www.iptel.org/ser* (SIP Express Router)

Commercial systems, like CallManager, Avaya's Media Servers, Nortel Meridian, Shortel's IP-enabled PBXs, and others, have the traditional commercial approach to bug catching. When enough customers complain, the development team ups the priority of a certain fix and then the company releases a patch. You, the user, download the patch to your PBX, reset your system during off-hours, and (usually) the problem is resolved.

Though a tech-head might prefer the freewheeling open source community's quickness in bug fixing, most IP telephone adopters go the commercial route for software maintenance. Either way, the system administrator should report any bugs encountered to her system's development community. After all, as the old saying goes, it's the squeaky wheel that gets the grease.

Simulating Media Loads

The bandwidth requirements of a VoIP call can be calculated by adding overhead factors such as RTP and IP headers and Ethernet framing to the size of the actual audio frame. That gives you the packet size, which you can multiply by packet rate in order to figure out the bandwidth required for the call (this was expanded upon significantly in Chapter 6). But that's all academic.

A great tool for adding load to a network is IPerf, which you can download from *http://dast.nlanr.net/Projects/Iperf*. This package can create streams of UDP traffic that use up the amount of bandwidth you specify—perfect for simulating VoIP load. IPerf has a server and client that each set up a socket for sending/receiving a large stream. You can control the port number, protocol, and size of the stream.

In a real-world scenario, you may need to fill, or nearly fill, a call path with traffic in order to test its behavior under heavy loads. By tweaking the amount of load you place on the call path, you can figure out the failure thresholds for it. Use IPerf to simulate a fixed load, and continually increase it, recording the call quality and completion rate each time you do.

Once you hit the performance ceiling, you should be able to state that a particular link will carry no more than x G.711 calls or that a certain call path through the network will carry no more than y G.729A calls. You would almost certainly have to do this kind of simulation during off-hours.

But how can you account for the bandwidth consumption of non-VoIP applications on the network? One way is to measure it during regular business hours, and then add it into your IPerf simulations.

Use your managed switches or MRTG (multi-router traffic grapher) to record traffic levels during regular hours. Then, later, use an IPerf client/server pair to simulate that traffic during off-hours. Use a second IPerf client/server pair to simulate VoIP traffic, and continually raise the traffic level, until your test call breaks down or fails.

Key Issues: Troubleshooting Tools

- Basic troubleshooting tools for VoIP networks include packet sniffers like Ethereal, log analysis, a softphone like X-Lite, and a VoIP server like Asterisk that allows detailed debugging output in its logfiles.
- Ethereal has built-in parsers that allow you to easily view SIP methods and responses, as well as SDP attributes and capabilities negotiation signals.
- Ethereal's packet capture filters are configured using a string syntax. This syntax is described in Ethereal's documentation and in O'Reilly's *Managing Security with Snort and IDS Tools*.
- Packet capture often reveals *what* is going wrong, but you may have to investigate diagnostic output on both ends of the signaling conversation in order to find out *why*.
- X-Lite's diagnostic log records every SIP packet it sends or receives.

- Asterisk uses an internal thread identifier that doesn't necessarily correspond to the SIP call ID, but can aid you in tracing an individual call through Asterisk's logs.
- Troubleshooting quality-of-service issues is aided more by process of elimination than it is by packet analysis.
- Network General Sniffer and WildPackets EtherPeek VX are packet capture tools that offer inspection of RTP traffic if needed.
- IPerf can let you simulate traffic loads on a call path in order to determine its true VoIP performance capability.

## CHAPTER 12

PSTN Trunks

While private trunks connect voice switches on your private network, PSTN trunks serve another purpose: connecting your PBX or your VoIP network to the outside world. They can be analog phone lines, digital phone lines like T1s, ATM connections, or VoIP based, depending on what kinds of service are available from your PSTN carrier.

Legacy telephony purists will balk at the use of the word *trunk* to describe a T1 or an ATM connection, arguing that a trunk is nothing more than a phone line connecting two switches. In fact, the definition has grown to mean any connection between two voice networks. A 5-mile-long T1 between two old-school PBXs is a trunk, and so is a UDP pathway between two VoIP servers. Even in non-voice scenarios, the word *trunk* is used to describe a pathway between two switches—take VLAN trunks as an example.

The way you think about trunk connections is different when they're PSTN trunks. While privately owned trunks are relatively cheap or free, PSTN trunks incur service fees. Careful design, utilization, and monitoring of PSTN trunks is important to your bottom line. PSTN trunks can also offer calling features that let you do things that may be less easy to do with private trunks: features like distinctive ring and three-way calling can be integrated into your voice network to simplify your PBX design or to enable functions that you otherwise couldn't provide.

In this chapter, we'll cover the fine art of connecting to the phone company: choosing a dial-tone connectivity technology, locating dial-tone trunks, and providing ample voice bandwidth for next-generation telephony apps. We'll also cover a variety of techniques for integrating PSTN trunks into your voice network, including distinctive ring, automatic call distribution, and find-me-follow-me.

Dial-Tone Trunks

Choosing from the plethora of legacy voice connectivity technologies isn't a no-brainer. When you choose a dial-tone trunk solution to supply your voice switch with a path to the outside world, you should consider the capacity of the solution, the implications for quality of service, geographic availability, and, of course, the cost.

POTS and Centrex Trunks

Plain old telephone service, introduced in Chapter 1, is an analog phone line from the CO that connects to your PBX using one copper pair. This is the kind of phone service you most likely have in your home. The POTS line is cost effective when fewer than 10 lines are concentrated in one location. Most POTS lines are paid for monthly, with overage charges added when a certain level of utilization (minutes) has been surpassed. POTS lines are available just about anywhere there's electricity and running water. Each POTS lines can support one phone call at a time.

Centrex phone lines are POTS lines with special, business-related calling features like four-digit private endpoint dialing. These are the lines on which you have to "dial a 9 to get out." Centrex has the same economics as POTS—you generally pay a monthly fee that covers a certain amount of utilization on the line; after that, you pay by the minute. Centrex is less widely available than POTS. It's often absent from residential and rural areas. Like POTS, each Centrex line can support one phone call at a time.

T1/PRI trunks

While T1/PRI trunks vary greatly from one carrier to the next, a safe rule to use when comparing their cost to that of traditional POTS lines is this: a T1/PRI is a cost-effective choice for locations needing 10 or more PSTN trunks connected to the same voice network. In fact, using 10 voice channels on a T1 is often cheaper than using POTS or Centrex lines because of most telephone companies' price structure. (Check with your phone company to see if this is the case for you. In some areas, PRI still isn't cost competitive with POTS.)

T1s in the United States use PRI signaling to support up to 23 simultaneous calls. With DID (direct inward dial), hundreds of E.164 phone numbers can be used with PRI. DID, combined with T1's high capacity and cost appeal, makes this solution very appealing. Since T1s dedicate a fixed slice of the bandwidth pie to each voice channel, quality of service is never an issue.

The interface at the subscriber's demarc where a T1 ends is called a smart jack.

If your voice network originates enough traffic, you may be using separate T1s for long-distance and local traffic. One or more T1s connect to your LDC, while another T1 or two connect to your LEC. This arrangement can enable you to avoid regulatory fees imposed for cost recovery purposes by LECs (called PICC charges) and enjoy a lower LD billing rate.

ISDN BRI trunks

Supporting up to two PSTN calls simultaneously, the BRI signaling technology provides an essentially obsolete option for PSTN trunking. BRI circuits tend to be less cost-effective for voice calls than POTS or Centrex and are always less cost-effective than PRI. Yet, because of BRI's vintage, you may still encounter it with some legacy systems. Consolidating BRI connections with PRI connections is just about always a good idea, unless you've only a half-dozen or so trunks to support. Even then, POTS is often cheaper than BRI.

VoIP trunks

Using a T1 (PRI) trunk to bring in 23 channels of digital dial-tone is more cost-effective than bringing in 23 POTS lines to do the same thing. But using that T1 as an IP point-to-point to link the CO to your PBX can bring even greater efficiency, especially if your telephone company supports bandwidth-conservation goodies like highly efficient codecs and silence suppression.

If you use a residential VoIP dial-tone provider, the SIP or IAX connection from your ATA or softPBX to the TSP is, by definition, a VoIP-based trunk. But these types of VoIP trunks have no quality-of-service measures, and their proprietors cannot guarantee a level of service. This is where a traditional phone company (SBC, XO, Verizon, etc.) has a big advantage over the upstart TSPs. Some TSPs do not yet offer 911 emergency dispatch service yet, either.

In North America, phone lines serviced by state-regulated phone companies (ILECs) are required to allow quick dialing of a public safety dispatcher. The phone number used in case of emergencies is 911.

In early 2005, the FCC mandated that CLECs provide access to the 911 call-routing system on behalf of VoIP-only TSPs such as Vonage. As a result, competitive carriers like CLECs and TSPs can be required to pay the CLEC a fee for routing of 911 calls to the local dispatch center. Competitors pay this fee because they fall under state regulatory agencies that require them to provide 911 service on a minimum of one line at each subscriber's premises. If you use a competitor, you may see this fee, which is usually only a buck or two, added on to your telephone statement as a tax.

An ILEC has dedicated trunks connecting to PSAPs (public safety answering points), which it must allow the CLECs and TSPs to use. Of course, not all TSPs have integrated 911 call routing as well as the ILEC's we've all relied on for decades. As access to 911 dispatch services are enabled, TSP support of 911 will become more consistent. Check with your TSP to see if they support it.

The big difference between TSPs like Vonage and phone companies like Verizon is QoS. TSPs usually can't offer quality-of-service measures because they don't own the infrastructure that they use to deliver their VoIP trunks to your network. Phone companies, on the other hand, own the "last mile," even for IP links. They have complete control over each phone call, from its point of origin on your phone network, to its PSTN handoff point on the phone company's network. This total ownership of the last mile means the phone company can guarantee quality for you.

If you're intent on using a TSP, you won't get QoS unless you're willing to pay for a direct network connection to that TSP. More often than not, this connection would be a T1. Of course, that would likely end up being more costly than just using the phone company for dial-tone service, because if you have a T1, it's touching the phone company's network anyway. Remember, the phone company, not the TSP, owns the last mile of connectivity, and this gives the phone company the quality advantage.

Hosted PBX

A step beyond VoIP trunks from the CO switch to your PBX is outsourcing PBX services to an application service provider. Hosted VoIP PBX services do this, allowing you to support only IP phones (and not a softPBX) at your premises. These phones communicate directly to the hosted PBX server at the provider's data center using a direct IP link or the Internet. The use of VoIP trunks for your on-premises PBX is a relatively new idea. There are lots of variations on the VoIP trunk theme, though a SIP-based service called IP Centrex is getting the most marketing attention by the big local phone companies.

ATM trunks

A number of long-distance carriers now offer local dial-tone service using T1 circuits connected to an ATM network. In high-density situations, this kind of local service can save money, especially in long-distance fees. If your system has a lot of long-distance traffic, you may consider a VoATM solution for your PSTN trunks.

Television cable and fiber trunks

Cable television operators like Adelphia and Comcast offer telephone service via VoIP, though it remains to be seen whether this service will evolve into a dial-tone trunk offering for PBX operators. Obtaining dial-tone service from a cable operator is likely to get you quality of service that is on par with a phone company. That's because cable operators own last-mile infrastructure as ILECs do. Because of cable's high bandwidth yield, it has the potential to carry many times more voice traffic than a VoIP trunk over a T1.

Some cable operators are introducing fiber-optic cabling to the customer's demarc. This will result in even higher capacity. Aside from voice calling, videoconferencing and eventually television programming will be delivered using a packet-based, probably IP-based approach.

If that's not enough bandwidth...

If some combination of the preceding technologies doesn't solve your PSTN connection needs, then there are even faster kinds of connections. Optical STS connections, also called OC (optical carrier), offer call capacities in the tens of thousands. Chances are pretty good though, if you're seriously considering OC, you're a telephone carrier, a very large ISP, or AT&T. Very few corporate call centers need that caliber of voice bandwidth.

Table 12-1 compares trunk technologies and their characteristics.

Table 12-1. Characteristics of dial-tone trunk technologies

	POTS	Centrex	T1 PRI/E1	VoIP via T1/E1	VoIP w/ QoS via T1/E1	VoIP via cable
Simultaneous call capacity	1	1	23 / 29	G.711: ~18/24	G.711: ~18/24	Depends on carrier
Who provides calling service	ILECs, CLECs	ILECs, CLECs	ILECs, CLECs	TSP	ILECs, CLECs	TSP or cable company

Table 12-1. Characteristics of dial-tone trunk technologies (continued)

	POTS	Centrex	T1 PRI/E1	VoIP via T1/E1	VoIP w/ QoS via T1/E1	VoIP via cable
Who provides connectivity	ILECs, CLECs	ILECs, CLECs	ILECs, CLECs	ILECs, CLECs, Internet	ILECs, CLECs	Cable company
Cost per line	Most	Most	Moderate	Low	Moderate	Low
Calling features	Usually cost extra	Usually built in	Usually built in	Built in	Built in	Usually none
Interface device to TDM system	Analog PBX trunk port	Analog PBX trunk port	T1 port or channel bank	VoIP Server	VoIP Server	VoIP Server
Interface device to VoIP system	ATA or analog trunk port	ATA or analog trunk port	T1 port or media gateway	Ethernet interface	Ethernet interface	Ethernet interface
Pros	Easy to use; available everywhere	Easy to use; extra calling features	Cheaper per-trunk cost when lots of channels are used	Can be used for voice and data without channelizing	QoS can be managed by the telephone company	Can be used for voice and data
Cons	Too expensive for connect points with a lot of trunks	Too expensive for connect points with a lot of trunks	Not always a good choice for small connect points (fewer than six lines)	Doesn't always provide QoS		Doesn't always provide QoS

How Many Dial-Tone Trunks Are Needed?

Whether you use POTS or your dial-tone is brought to you by way of SIP, you've got to make sure you've got *enough* trunks for your telephony application. If you're building a large hosted telephony application, like a call center, you'll probably need loads of them. If you're building a simple PBX in your home, you may require only one or two. In a medium-sized office, you might need a few dozen.

The place where your PSTN interfacing occurs is your *PSTN connect point*. Recall that, through VoIP, you can have voice conversations over your IP network, which may be a WAN that spans many, many miles. This will no doubt affect the location of your PSTN connect point, as well as the number of trunks required at that point. In Figure 12-1, you can see how a traditional wide area voice network might look. Its connect points are at every location where there's a PBX.

The case for fewer trunks

You'll use fewer PSTN trunks if you leverage your IP network. Say you've got three offices—one in New York, one in Detroit, and one in Kansas City. The one in

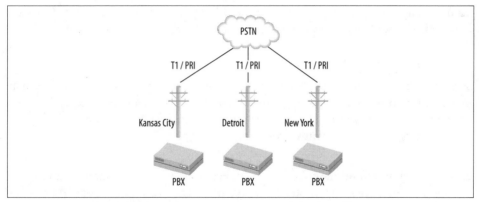

Figure 12-1. A traditional wide area enterprise voice network has lots of PSTN connect points

Detroit may be considered the central office because it acts as the corporate data center, the "hub," providing services for New York and Kansas City, the "spokes." It might make sense, if a lot of the spokes' voice traffic is long-distance traffic, to reduce the number of trunks at the spokes' PSTN connect points and increase them at the hub's PSTN connect point, as in Figure 12-2.

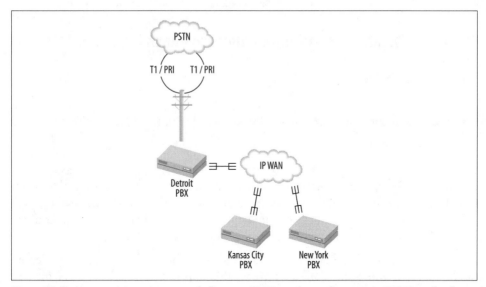

Figure 12-2. An enterprise voice network that uses VoIP trunks can leverage its WAN, and reduce the number of PSTN connect points, and often the number of PSTN trunks, it requires

This way, the organization could conceivably get a volume deal from the carrier providing dial-tone in the Detroit office. As a rule, the more bandwidth you have at one PSTN connect point, the less expensive, per trunk, that point becomes. You might also be able to trim the PSTN overall trunk count across the network because, when you consolidate PSTN trunks to a single connect point, you may find you just don't need as many trunks to support a centralized setup. (There are plenty of reasons *not* to centralize, too; Chapter 13 describes them.)

The case for more trunks

Some telephony applications, like conference calling, bridging, and hold queues may tie up PSTN trunks and require you to use more of them than you'd normally need. For instance, conference calling requires one additional trunk for every outside party participating in the call. If you've got three outside participants, then you need three trunks.

Call bridging and "find-me-follow-me" applications, which ring internal and external phones simultaneously in order to make it easier to reach the intended party, tie up PSTN trunks, too. If your application rings a user's IP desk phone and his cell phone simultaneously, keep in mind that a PSTN trunk is used to ring the cell phone. If the call causing this to occur originated from outside your network on the PSTN, then you've got another PSTN trunk tied up. That's *two* phone lines tied up.

In these situations, the only solution is to add additional trunk capacity. By using the Erlang B table in Chapter 4, you can properly size each of your PSTN connect points so that applications don't suffer busy signals due to inadequate capacity.

Project 12.1. Make It Easier for Callers to Reach PBX Users

What you need for this project:

- Asterisk
- Two X100P interface cards or a TDM400P card with two FXO modules
- One or more SIP phones
- LAN

In the early 1990s, phone companies began to introduce a feature called *call forwarding*. This feature allowed a subscriber to enter a short DTMF code that would cause the CO to ring calls destined for her phone number to another phone number of the subscriber's choosing. This way, you would "never miss a call."

Today, a newer telephony application exists to solve the same challenge in the enterprise. Called *find-me-follow-me* or FMFM, this app goes through a list of internal extensions and external phone numbers stored on the PBX and rings them each, simultaneously or in a certain order, in an attempt to track down the desired person when the PBX has received a call for his extension. Sometimes, the caller will hear a special recorded message or on-hold music while this "find" is proceeding.

More advanced FMFM setups allow users to log in to extension phones or otherwise notify the PBX where they are and how best to reach them, so users' extensions can "follow" them. Using Asterisk's AGI and Asterisk Manager socket API, for example, you could program presence features that contribute to an elegant FMFM solution.

FMFM and hunt groups are two different things. Hunt groups are statically defined and comprised only of endpoints or trunks associated with a certain group of users. FMFM uses a group of endpoints and phone numbers associated with a certain single user. In Asterisk, though, their extension programming is similar.

To create a simple FMFM app with Asterisk—one that doesn't use presence—you must program the dial-plan with the right steps to carry out the FMFM "find" when an extension is dialed. To do this, first pick out the three or four endpoints or PSTN phone numbers you want to use. Then decide if you want the find pattern to be simultaneous or sequential. Don't forget that for every simultaneous PSTN destination you put in your pattern, your PBX server will need an additional PSTN trunk. We'll start with a sequential pattern and do a simultaneous one next.

Now, make sure you have a cell phone and a couple of IP phones turned on and working. Make a note of the cell phone's number and IP phones' extensions. If they aren't configured, set them up now.

When creating an FMFM find pattern, what you're really doing is programming an extension to hunt through other extensions and PSTN numbers. So, in the following example, calling extension 601 will result in the sequential dialing of extension 201, then 202, then cell phone number 440-523-0555. Eventually, if there's no answer anywhere, the extension will result in a voice mail greeting.

Take a look at this sample extension from */etc/asterisk/extensions.conf*, which achieves just that:

```
exten => 601,1,Answer( )
exten => 601,2,Dial(SIP/201,15,m)
exten => 601,3,Dial(SIP/202,15,m)
exten => 601,4,Dial(Zap/1/14405230555,15,m)
exten => 601,5,Voicemail(201)
exten => 601,6,Hangup( )
```

So, when extension 601 is dialed, Asterisk answers the call and immediately attempts to connect it to channel SIP/201, assumed to be a SIP peer already configured in */etc/asterisk/sip.conf*. If, after 15 seconds, nobody answers at SIP peer 201, then Asterisk attempts to connect the call to SIP/202. If nobody answers, then Asterisk attempts to connect the call on a Zaptel channel, Zap/1, to 440-523-0555—the cell phone. If there's no answer signaled before 15 seconds elapses, the call is transferred to voicemail box number 201. Finally, Asterisk hangs up the call. The m options in each of the Dial commands will cause Asterisk to play a music-on-hold message that's based on your on-hold configuration and context (more on that in Chapter 17).

The options used with the Dial command, such as m in the preceding example, are case sensitive. m is used for music-on-hold, and M is used for macros, so watch those caps! For a complete list of Dial command options, check out the command reference in Chapter 17.

It only takes a few tweaks to the FMFM extension to improve its functionality. By concatenating the SIP channels into a single Dial command, we can ring them simultaneously, *then* continue the find attempt with the cell phone:

```
exten => 601,1,Answer( )
exten => 601,2,Dial(SIP/201&SIP/202,15,m)
exten => 601,3,Dial(Zap/1/14405230555,15,m)
exten => 601,4,Voicemail(201)
exten => 601,5,Hangup( )
```

In this case, the two SIP phones are rung simultaneously for 15 seconds before the find proceeds to the cell phone number. If you wanted to ring all three simultaneously, you could concatenate the Zap/1/1440... expression into the Dial command.

 Much more elaborate FMFM tricks can be accomplished with web-based or Asterisk Gateway Interface programming. You could create a CGI script that lets you tell the Asterisk database the extension at which you're currently available (i.e., cell phone, desk phone, home phone), and then set up your dial-plan to pull that extension out of the database when someone is trying to call you. Asterisk's database commands are described in Chapter 17.

Home phone call bridging

Now, suppose you wanted to build a simple server for your home that would ring both your home phones and your cell phone simultaneously whenever your home phone line receives a call. This is easy to accomplish with Asterisk. Bear in mind, you'll need two phone lines connected to your Asterisk server—one to receive the call and one to bridge the call to the cell phone. We'll assume for the following config that the analog channels are called Zap/1 and Zap/2 and that your home phones are SIP phones. (This could be done with analog phones too, if you had an additional Zaptel channel.)

In all dial-plans, the special extension s handles inbound calls on channels that are destined for the Asterisk server itself. So, when a PSTN caller reaches a trunk connected to Zap/1, the s extension tells Asterisk how to handle the call:

```
[default]
exten => s,1,Answer( )
exten => s,2,Dial(Zap/2/14405230555,35,m)
exten => s,2,Dial(SIP/201&SIP/202,35,m)
exten => s,3,Voicemail(201)
exten => s,4,Hangup( )
```

This config answers the incoming call on Zap/1, then bridges it to the cell phone by placing an outgoing call on Zap/2. It also attempts to bridge on the two SIP phones as in prior examples. Finally, if after 35 seconds, neither the cell phone nor the SIP phones answer, the voice mail module takes over.

Time-based context includes

You've already set up an FMFM find list to track you down wherever you go and even set up an Asterisk machine to just bridge everything to your cell phone. Both of these solutions are effective, but neither is especially elegant. That's why we're going to alter the home cell phone bridge dial-plan so that it bridges all calls to the cell phone only during the day, not at night.

Asterisk's include directives in *extensions.conf* can be used to specify context inclusion in a time-dependent manner. The format for include directives is:

```
include => context | times | weekdays | days of month | months
```

Since we want to prevent bridging of all inbound PSTN calls during non-business hours, we'll use time-dependent contexts to make it happen:

```
[default]
include => business|09:00-19:59|mon-fri|*|*
include => nonbusiness

[business]
exten => s,1,Answer( )
exten => s,2,Dial(Zap/2/14405230555,35,m)
exten => s,2,Dial(SIP/201&SIP/202,35,m)
exten => s,3,Voicemail(201)
exten => s,4,Hangup( )

[nonbusiness]
exten => s,1,Answer( )
exten => s,3,Voicemail(201)
exten => s,4,Hangup( )
```

In the preceding snippet of *extensions.conf*, the [business] context is included only during business hours, Monday through Friday, all days of the month, all months of the year. The [nonbusiness] context is also included. Though both contexts contain instructions for the s extension, Asterisk evaluates them at runtime in the order they're included in the dial-plan. So, [business] will be applied first, only during the time indicated in the include directive.

Now, suppose we wanted to do some additional exclusion—say, during holidays. Who wants to get a cell phone call when she's at Grandma's carving the Thanksgiving turkey? Or on New Year's Day, for that matter?

```
include => nonbusiness|*|*|1|jan
include => nonbusiness|*|*|31|may
include => nonbusiness |*|*|4|jul
include => nonbusiness |*|*|6|sep
include => nonbusiness |17:00-23:59|*|24|nov
include => nonbusiness |*|*|25|nov
include => nonbusiness |17:00-23:59|*|24|dec
include => nonbusiness |*|*|25|dec
include => nonbusiness |17:00-23:59|*|31|dec
include => business|09:00-19:59|mon-fri|*|*
```

```
include => nonbusiness

[business]
...

[nonbusiness]
...
```

Now you've got a fully functional gateway that answers your calls, rings your cell phone when and *if* you want it to, and records a voice mail message if you're unavailable.

More on Trunk Sizing

T1 and its Euro-equivalent E1 provide 1.5 mbps and 2.0 mbps of bandwidth organized into 24 and 30 channels called DS0s. Throughout this book, we've referred primarily to T1 circuits, since they are the most common form of digital dial-tone trunk. But there are much higher capacity circuits available from your friendly local exchange carrier that are based on the same technology. These hi-cap circuits are the preferred way of hauling dial-tone from the LEC because they are generally less costly for it and you to maintain. Still faster speeds are available using SONET optical carrier connections.

Unlike T1s, which are connected by two copper pairs, SONET connections are made on multimode fiber-optic cable.

Once you decide how much capacity is needed at a PSTN connect point, you can choose one of the T carrier or OC circuits described in Table 12-2.

Table 12-2. T Carrier and OC circuits for PSTN trunking

N. American	European	N. American capacity	European capacity
DS0	DS0	64 kbps (1 call)	64 kbps (1 call)
T1	E1	1.5 mbps (23 calls)	2.0 mbps (29 calls)
T3	E1 Third Level	44 mbps (~ 670 calls)	34 mbps (~ 480 calls)
OC-1	STM-1 or OC-1	51.8 mbps (~ 800 calls)	155.5 mbps (~ 2400 calls)
OC-12	STM-4 or OC-12	622 mbps (~ 9,600 calls)	622 mbps (~ 9,600 calls)
OC-48	STM-16 or OC-48	2,488.3 mbps (~ 38,000 calls)	2,488.3 mbps (~ 38,000 calls)

Connecting Trunks to Your Telephone Network

Whether your PSTN trunks are PRI-signaled T1s, POTS lines, or VoIP trunks, there are basically two ways to connect them to the VoIP network or PBX. Either you're going to have an interface in the softPBX server, or you're going to have an offboard gateway that is itself trunked to the PBX.

Cisco's media gateways are Cisco routers with VoIP firmware. They have modular expansion slots that allow them to connect T1s, optical connections, POTS lines, and Ethernet interfaces for IP-based trunking. In a Cisco CallManager setup, all PSTN trunking to and from the softPBX is facilitated by these media gateways, as shown in Figure 12-3.

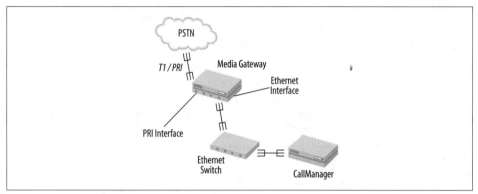

Figure 12-3. Cisco CallManager connects to the PSTN through a PRI-equipped media gateway router

Avaya's S8300 and S8700 softPBX chassis can connect to the PSTN in a more direct manner, using an onboard POTS or PRI interface. While the Avaya solution can certainly make use of a media gateway, say for trunking PSTN traffic back and forth to the softPBX over Ethernet, most Avaya chassis in the field are hosting PSTN trunks with onboard interface hardware. (You might have cause to use a gateway for PSTN trunks if the demarc is a great distance from the softPBX and you can use your IP network to overcome that distance.) (See Figure 12-4.)

An Asterisk server can also use either method. Digium's T1 and FXO/FXS interface cards give the Asterisk PC the ability to directly connect to PSTN trunks, while off-board gateway equipment lets you take a more distributed route. Figure 12-4 illustrates how an Asterisk server can use either approach. The dial-plan programming on the Asterisk server must be altered to support offboard gateways, of course, and those gateways must be programmed to route calls to the Asterisk server. For this reason, the preferred solution is to use onboard interfacing.

Channelized or Split-Use T1s

So far, we've talked about using T1s for voice dial-tone service from the CO, and we've talked about T1s for data: an IP point-to-point link for example. We've even covered getting your voice dial-tone service by way of VoIP on such a link. But what we haven't talked about is split-use service, which allows you to use a T1 for both.

Since a T1 circuit has 24 channels, most local phone companies will gladly group those channels into appropriately-sized pipes: one for Internet access service and the

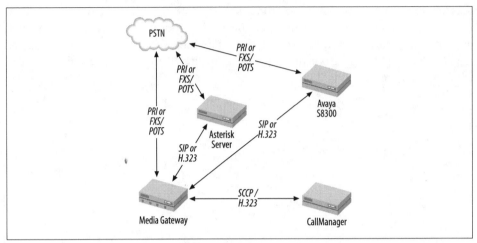

Figure 12-4. Asterisk and Avaya softPBX servers support direct PSTN trunking as well as media gateways

other for dial-tone service from the CO. So, it's possible to have a T1 trunking your PBX to the CO with, say, 12 channels and linking your IP network to the Net with the other 12. This could save you money, because most LECs are also ISPs, and they sell Internet access at a discount if you're willing to bundle their local dial-tone service on the same circuit.

 When connecting a T1 circuit from the smart jack to your DSU/CSU, router, or PBX, *don't use a CAT-5 cable!* Standard CAT-5 cables don't have the shielding needed by a T1. T1s will behave better if you use individually shielded twisted-pair cables instead. You can get these from any telecom supply catalog. Ask for a plain T1 cable.

Using the PSTN for Intraorganization Calls

Leverage Centrex groups

Calls between lines on the same Centrex group are usually cheaper than calls made to destinations outside the Centrex group. If you have a couple of offices in the same vicinity, close enough to be members of the same Centrex billing arrangement (the same area code and phone company), you can use Centrex signaling to your advantage to connect them seamlessly. Have your users reach one another by Centrex four-dialing, which is "in-group," rather than having them use seven-digit dialing, which is "out-of-group." This will save money.

Use dial-tone trunks to seamlessly route calls between PBXs in the same organization

Say we've got two separate offices, each with a PBX that has Centrex PSTN trunks. Let's also say that there's a ton of voice traffic between each of the offices because the workers are constantly calling one another. Now, let's say that the West office phones are numbered between 3400 and 3499, while the East office phones are all numbered 3000 to 3099 (see Figure 12-5).

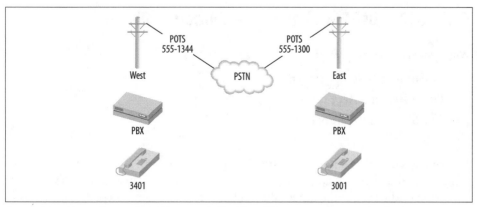

Figure 12-5. Two offices with PBXs connected to the PSTN

Ordinarily, if a West user wanted to reach an East user, he'd have to pick up his phone and dial the phone number of the East office, wait for an answer, and then request that user, either by speaking with a receptionist or by dialing that user's extension. This awkward process is shown in Figure 12-6. Direct inward dial could shorten this process, but the dialing user still wouldn't be able to reach his coworker using that convenient, four-digit extension.

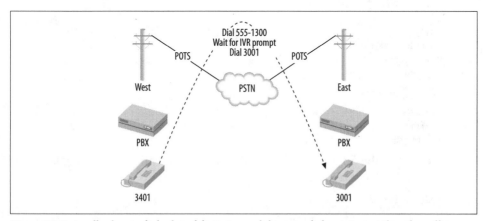

Figure 12-6. A caller has to dial a lot of digits to reach his intended recipient at the other office

But with a little bit of dial-plan setup on each of the two PBXs and a few PSTN trunks dedicated to the task, it's possible to allow users at the East office to dial users at the West office by their four-digit extensions. Each office will need a minimum of one PSTN trunk for this to work. Larger setups that require more concurrent calls will need more PSTN trunks. Project 12.2 explains how to set up four-digit dialing between PBX systems at different offices, making the convoluted process in Figure 12-6 disappear.

Project 12.2. Use PSTN Trunks in a Multioffice Dial-Plan

What you need for this project:

- Two or more Asterisk servers
- Two X100P interface cards (one in each server)
- Two SIP soft- or hardphones
- LAN

In keeping with the example illustrated in Figures 12-5 and 12-6, we can build a two-office unified dial-plan using two Asterisk servers. This way, users need dial only the extension of the user at the other office in order to reach her. Asterisk will then route the call to the other office's PSTN trunk, wait until it's answered, and dial the recipient's extension in order to complete the connection.

We're assuming here that we have two Asterisk servers—East and West. We'll use the same dial-plan extension numbering convention as in Figure 12-6. Phones at the East office will be 3000–3099; phones at the West office will be 3400–3499. Have one SIP phone register with the East server and the other SIP phone with the West server:

```
# East office sip.conf
...
[3001]
callerid="East User" <3001>
canreinvite=no
context=default
host=dynamic
mailbox=3001
secret=3001
type=friend
username=3001
```

The preceding is the SIP peer config for 3001 at the East office. Following is the SIP peer config for 3401 at the West office:

```
# West office sip.conf
...
[3401]
callerid="West User" <3401>
canreinvite=no
```

```
context=default
host=dynamic
mailbox=3401
secret=3401
type=friend
username=3401
```

With these first two configs committed, the SIP phones can now register with their respective Asterisk servers and place calls in the default contexts of each. But they still can't call each other without dialing a lengthy PSTN phone number, waiting for the autoattendant, and dialing the extension on the answering Asterisk system. To get around that, we can tell both Asterisk servers to route calls bound for the extension number range of the other office out through the PSTN and automatically dial the extension on the answering system, as follows. We'll start with the dial-plan config for the East office:

```
# East office extensions.conf
...
[default]
exten => _34XX,1,Dial(Zap/1/5551340,35,mD(${EXTEN}))
```

And a mirror of that config so that West office users can dial 30xx extensions:

```
# West office extensions.conf
...
[default]
exten => _30XX,1,Dial(Zap/1/5551300,35,mD(${EXTEN}))
```

Let's dissect this exten directive. First, _30XX is a wildcard expression that matches any number dialed that begins with 30. Next, the 1 is the extension priority. Then, the Dial command tells Asterisk to dial the number of the other office on the Zap/1 channel and wait for up to 35 seconds for the call to be answered. Then, the D(${EXTEN}) option tells the Dial command to send DTMF digits representing the extension number that was dialed by the user. ${EXTEN} is an Asterisk variable that always contains the extension number used for the current call. Finally, as with all Dial commands, the call will be connected after the DTMF digits are sent.

The net result of this config is that the users at West can dial 3001–3099 to reach the users at East, and the users at East can dial 3401–3499 to reach the users at West— all without any PSTN dialing or autoattendant interaction. Here, the PSTN trunks are used like private trunks to connect two switches while the dial-plan makes it easy for the users.

Controlling Caller ID when using PSTN trunks

In the previous example, the receiving PBX doesn't know the extension number of the party who is calling, because the calling PBX supplies the caller ID signals for the Zaptel channel and phone line being used, *not* the caller ID signals for the extension that originated the call. So, the receiving user will see that he is getting a call from the other office but won't know which user is calling.

Using some Asterisk dial-plan wizardry, you can preserve the original caller's caller ID information throughout the interswitch calling process:

```
# West office extensions.conf
...
[default]
exten => _30XX,1,SetCIDNum(${EXTEN})
exten => _30XX,2,Dial(Zap/1/5551300,35,mD(${EXTEN}))
```

In this case, Asterisk will supply the originating extension number as the caller ID number. SetCIDNum establishes the caller ID number for outgoing channels on the current extension. This config would, of course, have to be mirrored for the East office:

```
# East office extensions.conf
...
[default]
exten => _34XX,1,SetCIDNum(${EXTEN})
exten => _34XX,2,Dial(Zap/1/5551340,35,mD(${EXTEN}))
```

 Asterisk's built-in variables, like EXTEN, are case-sensitive! ${EXTEN} works fine but ${exten} does not. User-defined variables, on the other hand, are case-insensitive.

Routing PSTN Calls at Connect Points

In the previous section, we covered using PSTN trunks to facilitate users in the same organization calling each other. The following projects describe ways of using PSTN connect points for elegantly interfacing your PBX to the PSTN. This includes setting up outbound rollover groups, using automatic call distribution (ACD) to intelligently handle incoming calls, and leveraging common PSTN trunk features like distinctive ringing.

Project 12.3. Grouping PSTN Trunks

What you need for this project:

- Asterisk
- Two or more X100P interface cards (or other Asterisk-compatible TDM cards)

Just as PSTN trunks are grouped together for billing, like a Centrex group, or for aggregation, like a hunt group, your PBX needs to deal with trunks in groups, too. For instance, on a large system, it isn't practical to dedicate the same POTS interface to dial-plan functions that require an outbound line (like outbound calling or bridging). In the prior project, that's how we configured it: interswitch Dial commands used the Zap/1 channel.

Fortunately, Asterisk allows you to group analog interfaces together so that it can select an available PSTN line from the group when the dial-plan calls for it. This frees you from having to dedicate PSTN lines to a single purpose, as in Project 12.2. All commercial PBXs support similar functionality.

 Trunk groups are called outbound rollover groups, because they simplify programming of outbound dialing actions in the dial-plan.

In the Zaptel module's configuration file, *zapata.conf*, you can set up groups of channels that the Dial command can take as an argument where a single channel is required. The four ways the Asterisk dial-plan can address the trunk groups areas follow:

g   Use the lowest-numbered non-busy Zap channel in sequential order.

G   Use the highest-numbered non-busy Zap channel in sequential order.

r   Use the next higher channel than last time the group was invoked (sometimes called a rotary hunt group).

R   Use the next lower channel than last time the group was invoked.

Let's set up a sequential Zaptel group. Take a look at the following config bits from *zapata.conf*, which establish three Zaptel channels in a single group:

```
context=default
signalling=fxs_ks
usecallerid=yes
echocancel=yes
group=1
callerid="X100P 1"<(216) 524-1701>
channel=>1

context=default
signalling=fxs_ks
usecallerid=yes
echocancel=yes
group=1
callerid="X100P 2"<(216) 524-1702>
channel=>2

context=default
signalling=fxs_ks
usecallerid=yes
echocancel=yes
group=1
callerid="X100P 3"<(216) 524-1703>
channel=>3
```

With this config, group 1 is established using the three Zaptel channels that are provided by three separate X100P PCI cards. Once this is established, you can use the groups in the dial-plan using the `Dial` command. Here's an example that dials a PSTN-bound call on the group in sequential order (g):

```
exten => _XXXXXXX,1,Dial(Zap/g1/${EXTEN})
```

In the previous project, we set up a two-switch dial-plan that used the PSTN to automatically call between the switches. These calls were always placed on the same channel, but we could've just as easily used a group like the one set up in this project:

```
exten => _34XX,1,Dial(Zap/g1/5551340,35,mD(${EXTEN}))
```

Project 12.4. Create an Automatic Call Distribution (ACD) Scheme Based on Area Code

What you need for this project:

- Asterisk
- A PSTN trunk connected and working
- Two or more soft or hard SIP phones

Automatic call distribution is a way of delivering incoming calls logically based on the time of day, the availability of users to answer, or perhaps the number of calls already answered by a certain user. So, if a particular user hasn't answered as many calls as another user, she might be more likely to receive the call, if that's what your ACD policy calls for.

There can be very elaborate ACD policies that route calls based on the availability of users with certain skills or in certain departments. These ACD policies are often tied to IVR applications that allow the caller to qualify his need so his call gets to the right person. Sometimes, inbound call are routed by caller ID. Area codes can be used to route calls to a user who is responsible for a certain geographic region, for example.

In this case, we'll set up a simple app that routes calls based on their area codes. This project will deal with three area codes—216, 330, and 440, routing them to an appropriate extension. Consider the following configuration in *extensions.conf*:

```
exten => 2216,1,Dial(SIP/201,30)

exten => 2330,1,Dial(SIP/202,30)

exten => 2440,1,Dial(SIP/203,30)

exten => s,1,areacode=${CALLERIDNUM:0:3})
exten => s,2,Answer( )
exten => s,3,GoTo(2${areacode},1)
exten => s,4,Dial(SIP/200,30)
```

This configuration assumes SIP/201, SIP/202, and SIP/303 are already defined in *sip. conf* and working properly. The first exten directive establishes extension 2216, which will ring calls on the SIP/201 channel. The next exten directive establishes extension 2330, which will ring calls on SIP/202. Next, 2440 will ring calls on SIP/203. Notice that the extension numbers contain the area codes we're routing. This will make using the dial-plan to route them a snap.

In the next section, extension s, the default extension for incoming calls on Zap channels, is established. Its first priority is to store the area code in a variable called areacode. It does this by grabbing the first three digits of the caller ID number, which, in North America, is the area code. The syntax for this substring processing is very simple:

```
${VARIABLE:first character:length}
```

Once the area code is stored, the next priority (2) of the s extension, is to answer the call. Finally, priority 3 sends the call to the extension represented by the expression 2${areacode}. If the area code is 216, the extension taking the call is 2216. If it's 330, the extension is 2330, and so on. Note the use of the GoTo dial-plan command, whose purpose is to send the current call to the extension provided in parentheses. That extension consists of a 2 and the area code, but in another config, you could make it be 10 and the area code, or some other expression.

 Refer to Chapter 17 for the synax of the GoTo dial-plan command. You'll also find a more detailed description of Asterisk string processing.

Once the target extension springs into action, the appropriate SIP phone rings, and the call can be fielded. In large environments, this kind of caller ID–driven ACD app could send calls to ring groups instead of individual phones.

What if the call is from a different area code?

If the area code doesn't match any of the 2xxx extensions because it's coming from an area code other than the ones we've made provision for, the GoTo command in the s extension will fail, and the next priority for s will instead be used. In this case, give the call to SIP peer 200, which we'll say is the front-desk receptionist who could manually route the call. This functionality is important so that calls originating from area codes other that 216, 330, and 440 don't get categorically dropped by the Asterisk server.

What if the caller is using caller ID blocking?

The behavior will be as if the caller has provided an invalid area code, and the call will go to SIP peer 200.

Project 12.5. Use Distinctive Ring Detection

What you need for this project:

- Asterisk with verbose logging enabled
- A POTS or Centrex line connected and working
- Two or more soft or hard SIP phones

Distinctive ring is a feature offered by some phone companies that permits you to use two or three phone numbers with the same POTS line. Depending on which number was dialed, the ring signal will differ, causing the ring to sound unique for each number. This feature allows parents to avoid answering their teenagers' incoming calls. With a fax/voice ring switch device, distinctive ring can be used to inexpensively use a single line for receiving both fax and voice calls.

> Distinctive ringing is a legacy signaling solution. That is, it works only with POTS. On VoIP trunks, such functionality would be handled by out-of-band signaling.

With Asterisk, distinctive ring can be used to automatically route calls from the PSTN trunk to the right phones. Or the distinctive ring can just be passed through to the phones on the private network, so that incoming PSTN calls cause them to ring distinctively, too.

Each Zaptel channel can be configured to detect up to four different distinctive signals. The first thing we'll need to do is open *zapata.conf* and add this configuration to the section for the trunk in question:

```
usedistinctiveringdetection=yes
```

> Enabling distinctive ring on a Zaptel channel will cause a slight delay before Asterisk can answer incoming calls, because the distinctive ring signals can take up to 5 seconds for the Zaptel channel to detect.

The signals used by distinctive ring consist of analog electrical cadences—variations in voltage that cause analog phones to produce certain ring patterns. Asterisk uses the `dring` attribute in *zapata.conf* to describe the signals. Unfortunately, these signals vary from one regulatory jurisdiction to the next, and you'll have to figure out what value to give `dring` attributes yourself.

Here's how. When an incoming call is received on a POTS interface, Asterisk records the ring pattern in Asterisk's verbose logging output. (If you haven't already enabled verbose logging, see Chapter 10.) Use the `tail` command with its `-f` option to watch your logfile for changes as they occur:

```
tail -f /var/log/asterisk/full
```

While `tail` is following the logfile, call each of the numbers that cause distinctive rings on your POTS lines. When the POTS interface senses the ring pattern, a log entry will appear containing Asterisk's representation of it: a string of digits has three values separated by commas. Each digit represents a duration of ringing, such that each ring pattern could have up to three rings of varying length in a 1- or 2-second timespan. The pattern repeats at regular intervals until the call is answered.

The string representing this pattern is used to supply a value to the `dring` argument in *zapata.conf*. Repeat this process until you've identified the strings needed for each of the phone numbers associated with your POTS line. Here's a sample config in *zapata.conf* that describes two distinctive ring signals and assigns them different contexts in the dial-plan:

```
usedistinctiveringdetection=yes
dring1=325,95,0
dring2=95,0,0
dringcontext1=TedsCalls
dringcontext2=JakesCalls
channel =>1
```

 Distinctive ring features outside North America can use caller ID signaling instead of ring pattern signaling to indicate which phone number is being called. Check with your telephone company to see how it supports distinctive ring.

Routing based on distinctive ring

The Zaptel channel's configuration will tell Asterisk the context into which distinctively rung calls are sent. In this example, we've used a POTS line with two ring signals and two corresponding contexts. Now, we've got to create those contexts in the dial-plan. Here's a sample that accomplishes that in *extensions.conf*:

```
[TedsCalls]
exten => s,1,Dial(SIP/201,30)
exten => s,2,Voicemail(201)

[JakesCalls]
exten => s,1,Dial(SIP/202,30)
exten => s,2,Voicemail(202)
```

Timing Trunk Transitions

It's important to schedule any changes at the PSTN connect point in a way that minimizes possible negative impacts on production voice systems. These changes might be rewiring of a demarc to add additional capacity, or moving of a connect point from one location to another, from one floor to another, or from an old PBX to a new one. Usually, major changes like this coincide with the adoption (or move) of a new PBX. Since they can add to the success—or create the failure—of such a project, their timing is critical. Timing is especially important when "new" PSTN connections are displacing the old ones—that is, if you're replacing an old PRI with a new one in a different building or if you're consolidating a bunch of POTS trunks into a new PRI.

Major changes to trunk connections are often called *switchovers*. Once you've designed your VoIP topology, and you know the proposed location of every phone, every media gateway, and every PSTN connect point, you've got to map out a switchover plan. This is a schedule for your trunk moves from one PBX system to another, from one voice mail system to another, from one PSTN connect point to another, or possibly all of these.

If possible, coordinate the switchover with the phone company so that any major failures can be discovered immediately and the CO reverted back to its prior state until the switchover can be attempted again. (Third-party implementers of PBX systems are experienced at managing interaction with the phone company.)

Plan to do your switchover during off-hours—not an hour before the start of business on a Monday. Allow plenty of time for testing of the internal dial-plan *after* the switchover is done, too. Even if your shiny new PRIs are working right, there may be flaws in your call-flow logic, DID signaling, or inbound hunt groups that need to be worked out before users can start making and receiving calls.

Key Issues: PSTN Trunks

- The Public Switched Telephone Network provides connectivity to legacy telephony systems such as switched long-distance carriers, residential telephony subscribers, and TDM-based business phone systems. Connections between a private voice system and the PSTN are called PSTN trunks or dial-tone trunks.

- POTS, Centrex, and T1/PRI are the most common technologies for PSTN trunking. VoIP-based links, ATM, television cable, and fiber links are higher-capacity trunk technologies that are supported by some telephone companies.

- TSPs are companies that provide VoIP-based dial-tone services but don't own the network infrastructure on the "last mile."

- While VoIP can help you reduce the number of PSTN trunks you need, there are certain applications, like cell phone bridging, that increase the number of trunks you need.

- Optical carrier (OC) links offer vast amounts of calling capacity for extremely high-density applications.

- PSTN trunks are connected to the softPBX in one of two ways—either through gateway devices or through interface hardware contained in the server chassis.

- The PSTN can be used to link multi-PBX systems via DTMF signaling so that dedicated, sometimes costly, private trunks aren't needed.

- Asterisk permits you to control caller ID signals for better integration of PSTN-connected switches in the same organization.

- Automatic call distribution (ACD) is a technique that employs a logical algorithm or interactive data-collection routine (like an IVR app) to determine who in the organization should receive a call from the general public.

- Carefully manage the timing of any changes to your PSTN connect point, and do all changes during off-hours.

CHAPTER 13
Network Infrastructure for VoIP

In earlier chapters, we've talked about dial-plan design, PSTN trunks, and dial-tone services. We've covered some of the equipment used to link VoIP-based networks with legacy telephony systems: equipment like media gateways and ATAs. Signaling protocols like SIP, media protocols like RTP, and quality of service have been described. Each of these elements is dedicated to a specific, tactical duty in your telephony system.

But the "big picture" that contains and unifies all of these disparate technology objects together into a functional system is *topology*: the lay of the telephony land. At the base of this functional system is infrastructure. The transport, security, and directory services elements enabling your voice applications on the IP network are your VoIP infrastructure. Topology includes geographical issues, too—the physical locations of voice resources and connectivity maps of your wide area network.

There's quite a bit to take into account. In this chapter, we'll talk about methods for building a reliable VoIP infrastructure. We'll cover general IP WAN layouts, the use of private trunks to link PBX systems, disaster recovery and survivability, choosing a location for PSTN connect points, optimizing VoIP WAN links, and directory services for telephony.

You could think of VoIP servers and private trunks as Tinkertoy hubs and sticks. The old Tinkertoy hubs had six holes and could therefore host six sticks, which could connect to other hubs and so on, creating a network of Tinkertoys. This network could be straight and long, it could be shaped like a hexagon or star, or it could be a complex mesh, depending upon how the hubs and sticks were arranged. Such is the case with WAN topology, and particularly enterprise VoIP.

In the context of an IP WAN, the Tinkertoy hubs are routers (or switches), and the sticks are data links like frame-relay PVCs (permanent virtual circuits)or Ethernet switches. In the higher-layer context of a VoIP network, the Tinkertoy hubs are analogous to softPBX servers, and the sticks would be physical (T1 carrier) or logical (IP-based) voice trunks.

Legacy Trunks

Legacy trunks are links that connect private voice switches using a traditional technology like FXO/FXS or T1. There may be many reasons why you use legacy trunks, even if your heart's desire is to move totally to VoIP:

- Legacy trunks are the only way to integrate legacy PBXs with VoIP systems.
- Existing legacy trunks, especially point-to-point T1s, may be subject to service contracts that require you to maintain them for a while—sometimes as long as seven years from the date of inception.
- Legacy trunks are known for quality. Since they derive from the circuit-switched PSTN, legacy trunks have intrinsic bandwidth guarantees. If you can't justify (or convince other decision-makers on) the merits of packet-switched QoS across your private network, then legacy trunks provide a solid connection between voice switches.

Private Analog Lines

If two PBXs are in the same building or on the same campus, they can be connected by analog copper and FXO/FXS interfacing. This technique is effectively the same as that in Project 12.2 that linked two softPBXs using a POTS line. The difference in this case is that you aren't using the PSTN; you're just using a standard telephone cable with RJ11 connectors on both ends. One end connects to an FXO/FXS port on each PBX. Then, the PBXs' dial-plans are programmed to route calls appropriately between them.

Leased lines

If the two PBXs are not within the same campus or building, then the PSTN should be enlisted to provide analog (or TDM) connectivity between them. If the PSTN connection is going to be utilized very heavily, the phone company can provide a dedicated, monitored connection called a *leased line*.

Leased lines were very popular for data connections before the advent of cheap T1 circuits. They can provide the same data throughput as a POTS line (around 56 kbps) but with a much higher level of reliability.

Dry lines

Almost all connections provided by the phone company cross through its network—the PSTN. But *dry lines* don't. They are copper loops that begin at one customer's premises, route through the CO without entering the local exchange switch, and terminate at another customer's premises. Dry lines can be used to link PBXs via FXO/FXS. Also called *dry pairs*, these lines were once commonly used to connect security system monitoring companies with their customers. Dry lines can be used only to link sites that are served by the same CO. This means the sites must be within a 3- or 4-mile radius in urban areas. In rural areas, dry lines may not be practical because of distance-imposed attenuation problems.

 Not all telephone companies offer dry lines because, if you use the right hardware, dry lines can compete directly with their other, more expensive services such as POTS and DSL. Other telephone companies may have a policy of not selling dry lines to customers who plan to use them for voice applications. Check with your telephone company to be sure.

Private Digital Trunks

T1s and ISDN BRI connections are used to connect PBXs that have the appropriate digital interfaces, though BRI's practicality in this role has evaporated. T1s are also used to connect groups of TDM phones to the PBX by way of a device called a *channel bank*. This permits 24 TDM phones to be used with a single T1 port on the PBX.

In order to connect two PBXs by T1, a DSU/CSU (data and channel service unit) device is required at both ends of the T1. The DSU/CSU's role is to provide low-level signaling and diagnostic feedback for the T1 data link. DSU/CSUs are used in both legacy voice and data encapsulation scenarios on a T1 (such as TCP/IP and IPX/SPX). They are used to configure how the T1's 24 channels are broken into voice and data functions.

If two PBXs are located in the same building, a T1 cable can be used to directly connect them. If not, the LEC provides point-to-point T1 service that uses the PSTN to provide a completely unswitched connection between the two PBXs. This means that no dial-tone services or local access signaling is included with the T1. These services aren't needed for a private trunk, because the PBXs are going to signal directly to each other without using PSTN switching.

 Private T1 trunks don't need a local signaling technology like PRI or DID, so you need pay only for the loop, and not for dial-tone.

For trunk connections of hundreds of miles, a recipe of long-haul technologies is required to use T1 for point-to-point private trunking. Many network carriers provide T1 access across large SONET links that encircle entire metropolitan areas. IXC network uplinks haul T1 connections even further—from coast to coast, if necessary. Using VoIP (rather than legacy T1 voice signaling) for trunking is more appealing as the distance increases, because (unlike T1), VoIP trunks aren't paid for by the mile.

If you have T1 links that have very long contract periods remaining, you might have to keep them in place even if your endgame is to replace them with VoIP-based trunks. Some phone companies will assess a hefty termination liability fee if you cancel your T1 lease before it's up!

VoIP Trunks

As you know, a VoIP trunk uses digitized voice in IP packets to link two PBX servers. VoIP trunks can replace legacy trunks only when the two PBXs being linked are VoIP-enabled. A VoIP trunk could run between two IP-enabled PBXs, like the Avaya S8700 and a Nortel Meridian, or between two legacy PBXs that have outboard VoIP media conversion. Outboard conversion devices, like Cisco media gateways, allow legacy PBXs to connect to the local Ethernet network and perform the voice packetizing for the VoIP network since the PBX can't on its own. (The cost of outboard conversion often helps build the case for using VoIP-enabled or native VoIP softPBXs instead.)

Like other IP protocol families, VoIP can be tunneled within VPNs and GRE (generic routing encapsulation)point-to-point tunnels. By now, you know it can be routed, switched, and load-balanced. These qualities, inherited from the Internet-like networks that came before it, give VoIP's trunks greater flexibility than legacy trunks. You'll connect voice systems using VoIP trunks when one or more of the following conditions exist:

- Two or more PBX systems on a private network are IP-enabled.
- Two or more legacy PBX systems on a private network have outboard media conversion (Ethernet interfaces) to link them using a VoIP trunk running on the IP network.
- The cost of a legacy trunk is prohibitive, especially in long-distance scenarios.
- IP WAN links exist between sites that have PBX systems.
- Two sites have broadband connections to the Internet, which can be used as a transport for IP telephony applications.

Unwanted Effects of Load Management

Whether you use a completely private WAN or a mix of ATM, frame-relay, and VPN connections, you may employ a few different techniques to preserve or balance traffic between those links. You might use complex multipath routing, or you might use load-splitting or multilink bundling.

The challenge these techniques pose to VoIP is simple. Wherever two physical network paths to the same destination exist, there are likely to be differences in latency and jitter between those two paths. That is to say that, if you have two T1s side by

Squeezing More VoIP Out of VoIP Trunks

You would think getting a couple of 64 kbps voice calls over a 128 kbps link would be a simple matter, but not so in VoIP. IP packetization, RTP, and the framing on the data link itself all introduce overhead—so much overhead that a single G.711 call can all but swamp such a slow link.

Aside from using low-bandwidth codecs, here are some tips to ensure your IP-based pathways provide the highest possible capacity for VoIP calls:

- Use SigComp (Signaling Compression) if it's supported by your VoIP devices. SigComp is described in RFC 3320.
- Use IP header compression over low-capacity links.
- Enable silence suppression and voice activity detection to stop the transmission of packets when nobody is speaking.
- If trunking a large number of calls between two systems, use IAX2 for the trunk. IAX2 multiplexes much more efficiently than SIP or H.323.

---

side between point A and point B, they won't always run at exactly the same error rate day in and day out.

Traffic diversion (failover circuits)

To illustrate this, let's say that two WAN links between the same point A and point B exist. Let's say one is a full point-to-point T1, while the other is a 512 kbps frame-relay PVC. Ordinarily, the frame-relay connection isn't used. But it provides some geographic diversity and a good outlet for bandwidth overflow. So, when the T1 is maxed out (or down), traffic is diverted across the PVC using a router we'll call an *overflow valve*. This may be a fine solution for non-real-time IP traffic.

Frame-relay PVCs are *not* a good backup for point-to-point T1 trunks since they're far more jittery, though they're sometimes cheaper than T1s in long-distance situations.

But frame relay is a much more latent, jittery technology than T1, so during times when the overflow valve is diverting IP traffic to the PVC, things will be noticeably slower and, quite possibly, more jittery than they would ordinarily be. This could result in a situation in which, most of the time, VoIP media channels function fine, but suddenly, once the overflow or diversion point is reached, phone calls start sounding bad. This is a basic example of a precautionary topology decision having a potentially destructive effect in the world of VoIP.

Load-splitting

A similar problem can occur with simple load-splitting. Let's say you have two or more point-to-point links that all start and end at the same locations. Let's also say you use IP routers to split the traffic load equally (or unequally) across them. You can use BGP (Border Gateway Protocol) and other routing protocols to accomplish this, but be careful of the potential for variances in jitter and delay—especially if the links run at different speeds or use different data link technologies (like a radio link side by side with an FSO link).

Multipath jitter

Jitter that's incurred by complex routing and/or load-balancing can be minimized. Here are three things you want to avoid when setting up WAN links to support VoIP trunks:

- Avoid using a multipath routing setup for parallel IP links that use differing transport technologies (i.e., point-to-point T1 and a VPN). While it may be fine to use one or the other as a backup link, daily use will sabotage the consistency of phone calls.

- Avoid terminating any one end of a call path on more than a single router. This will create jitter. If you want to use multiple routers for disaster preparedness reasons, then take steps to make sure each RTP media stream (in both directions) is being handled by only one of them.

- Don't do load-splitting across two links of differing latency (like an 802.11b link and an 802.11g link). This exacerbates the jitter problem.

Multilink PPP

Multilink PPP bundles allow a single router to bond multiple interfaces, so that two or more data links can act as a single cohesive pathway. If four T1s ran from point A to point B, and a router with 4 T1 interfaces existed at each end, then those four T1s could be bonded into a multilink PPP connection. The result would be four times the bandwidth across a single logical link—similar to the simple load-splitting example before, but with a lower risk of jitter (especially if the four links are of nearly equal speed and latency).

TCP/IP as a Transport for Voice Trunks

So far in this chapter, we've been talking mainly about the physical wide area network connections, or trunks. This is physical and data link layer stuff. But VoIP infrastructure happens at the network layer, too. This is where TCP/IP replaces analog lines and T1 signaling as the voice carrier. The IP carrier can take many forms—plain-old, insecure UDP datagrams, VPN connections, GRE tunnels, even SSH tunnels. Some are more well-suited to enterprise voice apps than others, as we'll discuss.

Insecure, unencrypted UDP

One of IP telephony's key advantages over traditional telephones is that of security. While it's nearly impossible to secure the access perimeter on the PSTN (see Chapter 10), it's even tougher to secure analog phone conversations against clandestine surveillance—the old-fashioned "phone tap." Legacy telephony's glaring security issues have always been a key selling point for VoIP. Yet, if you want to use IP telephony in an insecure way, it's easy to do—just don't encrypt anything.

 Many IP phones now support Secure RTP (SRTP), a protocol that allows them to use encrypted media streams.

If the VoIP network were to replicate the insecurity of the PSTN, it couldn't enroll any secure transport technologies or encapsulation. This means avoiding the use of VPNs and encrypted tunnels. Using unencrypted signaling and media channels means clear-text authentication and eavesdroppable audio signals. So, a G.711 phone call across the Internet between two endpoints that don't support media encryption is quite easily monitored by a third party. These calls are vulnerable to man-in-the-middle attacks, allowing a remote observer to record, and possibly even hijack, them.

VPN

Virtual private networks created encrypted connections across the Internet between two private IP networks by encapsulating private traffic into public traffic and sending it between two routers (LAN-to-LAN VPN) or from a remote user to a VPN gateway router ("road warrior" VPN).

The two most common VPN technologies in use today are PPTP and IPSEC. While both can be used for LAN-to-LAN and road warrior setups, it's more common to see IPSEC as a part of high-density or mission-critical solutions because it implements IETF standards. Despite their differences, PPTP and IPSEC are both excellent for securing traditional, non-real-time traffic, and both are poor for securing VoIP traffic, because:

- VPNs introduce packetization delay, from 5 ms to 50 ms.
- When established across the Internet (which is almost always), VPNs are subject to typical Internet traffic delays, making them less suitable for VoIP.
- The devices used to connect VPN clients (VPN servers and gateway devices) sometimes don't have enough processing power to support a large number of media channels.

But these challenges don't rule VPN out totally. If you're using a VPN connection between two offices that both receive Internet access from the same provider's network, then it's likely that the provider can manage the VPN or guarantee the maximum latency of traffic that doesn't have to leave its network, such as the traffic between the two offices, will always be treated with high priority. Here are some tips for successful VoIP over VPN:

- Try to keep VPN traffic between remote locations on the same backbone network to keep the number of router hops down and minimize end-to-end latency. Keeping traffic on a single network may also allow the provider to guarantee you a certain level of service.

- If using a dedicated device for VPN termination, such as a specialized VPN gateway router or "concentrator," be sure it can tag priority traffic *after* it's encapsulated into VPN traffic. This way, the CoS information recorded by the telephone endpoint in each LAN packet won't get lost inside the encapsulated VPN packet it ends up in as it travels over the Internet. If your VPN solution doesn't account for CoS, VoIP quality may suffer.

 A great source of information about implementing VPNs is O'Reilly's *Virtual Private Networks*.

GRE tunnels

A simple but effective way of linking two disparate networks securely over the Internet is the use of a GRE tunnel. GRE is a simple IP protocol used to encapsulate most kinds of VPN traffic. GRE can be used to tunnel directly between two routers, providing the secure, encrypted transport of a VPN without the need to support a VPN appliance or pricey VPN server. Cisco routers that support 3DES encryption and have the Cisco IP Firewall IOS firmware can create a highly secure GRE tunnel.

 Even if your Cisco router can support 3DES encryption and IP Firewall firmware, it may not have enough flash memory to support them both at once. At a minimum, a 16 MB flash memory card is needed.

Consider the following Cisco IOS configuration, which establishes a tunnel using 3DES encryption over an Internet-facing serial interface:

```
crypto isakmp key crypt0k3y address 10.10.10.2
!
crypto ipsec transform-set 3des ah-md5-hmac esp-3des
!
crypto map mymap local-address Serial0
!
crypto map mymap 10 ipsec-isakmp
 description VPN to Satellite Office
 set peer 10.10.10.2
 set transform-set 3des
 match address satellitevpn
```

```
  qos pre-classify
 !
interface Tunnel0
  description ---------- Tunnel to Remote Office
  ip address 192.168.1.1 255.255.255.252
  tunnel source Serial0
  tunnel destination 10.10.10.2
  crypto map mymap
 !
interface Serial0
  description ---------- T-1 to Internet Provider
  ip address 10.10.9.2 255.255.255.0
  ip nat outside
 !
ip access-list extended satellitevpn
  permit gre host 10.10.9.2 host 10.10.10.2
 !
access-list 150 remark ACL for PBR route-map
access-list 150 permit ip 10.0.0.0 0.0.0.255 10.0.1.0 0.0.0.255
access-list 150 deny ip any any
 !
route-map PBR permit 10
  match ip address 150
  set interface Tunnel0
  set ip next-hop 192.168.1.2
 !
ip route 10.0.1.0 255.255.255.0 192.168.1.2
 !
```

The crypto commands at the beginning of the configuration establish the encryption parameters the tunnel will use, called a crypto map. These parameters describe what type of encryption techniques will be used to secure the link and which networks can use the link. In this example, satellitevpn is the name of an access list that defines the address of the remote host allowed for this tunnel.

 If you'd like further study in cryptographic features on Cisco routers, and other Cisco knowledge, refer to O'Reilly's *Cisco IOS in a Nutshell* and No Starch Press's *Cisco Routers for the Desperate*.

The interface called Tunnel0 is a logical interface established to distinguish tunneled, private traffic from public, unencrypted traffic to and from the Internet. The remote router must have a similar logical tunnel interface to operate the other end of the tunnel (see Figure 13-1).

The configuration for this router establishes a tunnel to a peer router at 10.10.10.2, and sets up an access list that enables it to pass traffic through this router. In this case, the IP Firewall firmware is performing NAT on the Serial0 interface, which puts a NAT firewall between hosts on the remote end of the tunnel and the Internet. (Serial0 is the interface facing the ISP.)

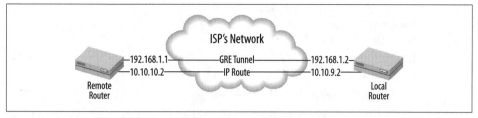

Figure 13-1. A GRE tunnel between two routers is a relatively easy VPN link to use for VoIP trunking

Once the link is established and able to pass normal IP traffic through the tunnel (tracerouting across the tunnel won't show any intermediate hops other than the local and remote routers), you can put any combination of IP phones and IP/PBX equipment on either end of the tunnel, as long as the capacity and end-to-end latency are acceptable to you. So, like a point-to-point T1 with IP, this tunnel could be used as a trunk between two PBX servers or as a way of connecting several dozen IP phones to a PBX on the other end of the tunnel.

Incidentally, all of the Internet-bound traffic from the remote site will also traverse the tunnel rather than going, unencrypted and directly, to the Internet.

IP addressing schemes

You may have a desire not to burden the tunnel with non-voice traffic, but rather tunnel only voice traffic between the two sites. An easy way to distinguish between voice and non-voice traffic is to organize all voice traffic into its own subnetwork and then simply route the non-voice traffic outside of the tunnel. (This has implications for the whole enterprise WAN, not just tunneling.) For example, if your private network address is 10.0.0.0/24, then you could subnet that address range in order to organize by mode of service (data or voice) and location:

```
10 . location . mode of service . 0
```

So, 10.1.1.0/24 would be for Detroit data, while 10.1.2.0/24 would be for Detroit voice. 10.2.1.0/24 would be for Chicago data, while 10.2.2.0/24 would be for Chicago voice and so forth. Using network addresses to "label" things is a simple convention for managing routing rules and making the address structure admin-friendly. But in a very large network, using a "labeled" address scheme could cause you to run out of address ranges. If you wanted to better preserve available IP addresses, you could do some fancy subnet slicing:

Detroit: 10.1.0.0/255.252.0.0
    Detroit data: 10.1.0.0/255.255.255.0
    Detroit voice: 10.1.2.0/255.255.255.0
    Detroit future growth: 10.1.2.0/255.254.0.0
Chicago: 10.1.4.0/255.252.0.0
    Chicago data: 10.1.4.0/255.255.255.0

Chicago voice: 10.1.5.0/255.255.255.0
Chicago future growth: 10.1.6.0/255.254.0.0
Cleveland: 10.1.8.0/255.252.0.0
Cleveland data: 10.1.8.0/255.255.255.0
Cleveland voice: 10.1.9.0/255.255.255.0
Cleveland future growth: 10.1.10.0/255.254.0.0

This way, instead of each location using up 65,000 addresses, they'll use only 512. In fact, this example allows growth of 512 more addresses in the future.

Even if you aren't tunneling traffic—differentiating the mode of service by using different subnets for voice and data is still a good way to organize your network.

Class-of-service prequalification. In the Cisco configuration sample, you can see the cos pre-qualify command. This tells the router to recognize CoS tags within packets inside the tunnel and then tag the encapsulating packets appropriately. This is crucial for VoIP performance. Without prequalification, the layer 2 and 3 class of service tags normally carried by each packet would be encrypted into the tunnel, no longer legible to routers that are handling the tunnel. The effect would be that those routers, which cannot see "inside the tunnel," would think these packets have a regular priority class of service like any other traffic. Prequalification ensures that the tunnel packets retain layer 2 and 3 tags.

Use prequalification and keep your VPN within a single service provider's network, as in Figure 13-1, to avoid delay problems that occur when traffic hops from one Internet provider to another.

Supporting VoIP Road Warriors

A *VPN concentrator* is a dedicated server or appliance that allows many users to use a VPN for direct connection to a private corporate network—one PC at a time rather than one network at a time, as in the prior examples. The main duty of VPN concentrators is giving road warriors secure access to the main office.

That could mean email access, file and print services, or even monitored web browsing. VPN simplifies the traveling user's configuration experience. The user plugs in his laptop, and once the networking software on the laptop obtains its TCP/IP configuration from DHCP, the user can "dial up" the VPN connection—preconfigured by an administrator. The user types his password, and off he goes, just as if he were in the office sitting next to the file server.

The notion of "it works anywhere" is available for IP phones, too. Suppose a user accessing the VPN had a softphone running on his laptop. As long at that softphone can register and make calls over the VPN, then he can:

- Receive phone calls at the same E.164 number no matter where he is physically, as long as his PC can access the Internet. SIP (or H.323) does the job of signaling incoming calls to the phone once it has registered with the VoIP server through the VPN.

- Make use of the private dial-plan—i.e., three-, four-, or five-digit extension dialing and autoattendant features normally used only *inside* the office.

- Originate calls from the corporate call center instead of from his hotel. This is useful for traveling salespeople who don't want the hotel's caller ID showing up on their customers' phones when they call. (Or for those who don't wish to pay exorbitant hotel phone fees.)

To VPN or not to VPN?

Now, to do all three of the great things listed, you don't really need a VPN. The VPN merely provides security and ease of access for the end user. Setting VPN's niceties aside, you really just need your softphone to be able to register and communicate with the softPBX over the Internet. As long as your softPBX can be accessed from the Internet without a VPN, then you don't need to use one. This probably isn't the most secure idea in the world, however.

In telephony, the VPN trade-off is black and white. With VPN, you gain security. Without it, you (may) gain quality.

TSPs don't use VPN to support their customers because of its negative impact on quality of service and because it's more difficult to maintain. But, in a corporate environment, it's probably a much safer idea to provide VPN's protective wrapper for your telephony app. As long as call quality is acceptable to your road warriors, then VPN does the trick.

Soft- or hardphone in the mobile office?

If you use VoIP for your mobile workforce, then you'll have to choose whether each user will be using a hardphone or a softphone. Aside from convenience and preference, the choice is affected by whether you employ a VPN to connect your mobile users.

Since a VPN client runs on a PC, the VPN tends to encapsulate any softphone traffic from that PC. But hardphones are more difficult to use with a VPN. They don't have a built-in VPN client as the PC softphone often does, so the only way to use them with a VPN is to connect them to an offboard VPN router. Try convincing your salespeople to do *that*. If you opt not to use VPN and decide to support your road warrior with a direct, VPN-free connection, you afford her the option of using a hardphone in her mobile office.

When an IP hardphone is connected to an Ethernet segment, it behaves like a networked PC that's booting up. It contacts a DHCP server, gets its IP configuration, and, if configured for a VoIP network, attempts to register with a VoIP server. It doesn't care where the VoIP server is. It could be in the next room over, or it could be in Manitoba, Canada. Such is the nature of TCP/IP. The point here is that you can set the behavior of the IP hardphone to be identical no matter from where it connects. In theory, if you provide for a consistent registration path from any spot on the Internet, the phone will behave consistently regardless of where it's sitting. (This is what Vonage and BroadVoice have done.)

Like your corporate web site, which can be viewed by mobile users anywhere in the world, it's possible for you to allow your telephony users to access your voice services from anywhere in the world.

But there are a few gotchas to this Utopian road warrior's dream:

- There ought to be firewall(s) between the road warrior and his VoIP network. Depending on your network's requirements, you may need to allow TFTP, RTP, IAX, SIP, H.323, SCCP, and the host of VoIP-related traffic (port numbers and protocols can be referenced in Chapter 10).

- Don't bet on hotels and other road warrior hotspots offering inline power. Make sure the road warrior carries an AC power adapter for his hardphone.

- In order to be truly transparent to the end user, you must set up your VoIP network so that road warriors' IP phones can be configured once and not repeatedly—regardless of how many stops they're making or how many days a week they are in the corporate office. Road warriors don't want to have to remember to pop in a certain DNS setting when they're on the road and another when they're in the office.

If these issues are making you think twice about skipping VPN, consider sticking with VPN and using a USB handset. These devices provide the look and feel of a real hardphone, but they connect to the PC's laptop so they can be used as a UI element for the softphone. This can save frequent callers the cumbersome task of dialing 11-digit phone numbers on a laptop keyboard, because the USB handset has a 12-key dial-pad. For more information on USB handsets, visit *http://www.pc-telephone.com/ usb-phone.htm* and *http://www.eutecticsinc.com/products*.

Project 13.1 Use Dial-Plan to Connect to Multiple VoIP Networks

What you need for this project:

- Asterisk
- Internet access

While we're on the subject of VoIP trunks and using the Internet to extend your voice network, let's tackle a project. This time around, we'll use the Internet as a trunk carrier to connect the local Asterisk server to a third-party VoIP network to route select calls. The logical trunk we'll set up is really just an IAX protocol client connection to an Asterisk server on that remote network.

Inter-Asterisk Exchange protocol Version 2, or IAX2, isn't yet embraced heavily by the large commercial telephony manufacturers. But its design goals are very noble: to be the most lightweight, flexible signaling and media streaming protocol for telephony. This goal has been largely achieved, and a number of commercial voice networks have begun to support IAX endpoints. Others, like IAXTel, are exclusively reliant upon IAX. This project describes how to establish a dial-plan that makes use of a logical VoIP trunk for handling calls to and from IAXTel.

The Inter-Asterisk Exchange Telephone (IAXTel) network

IAXTel is sponsored by Digium, the creators of the Wildcard TDM interface cards and Asterisk. The purposes of IAXTel are to use to connect Asterisk systems over the Internet, to promote the adoption of the IAX protocol, and to provide free calling for the Digium-sponsored IAX softphone, Gnophone.

IAXTel's service is free. You can use any IAX2 endpoint to terminate and originate calls on the IAXTel network—this includes IAX analog telephone adapters, IAX IP phones like GnoPhone, and of course, Asterisk itself.

Other networks, including Free World Dialup, Vonage, Packet8, and VoicePulse, employ IAX for direct, Internet-based trunking of calls between their subscribers and those of IAXTel. Free World Dialup (FWD), like IAXTel, is free. VoicePulse, Packet8, and Vonage are commercial TSPs that also provide PSTN calling, but IAX-Tel is one of the only TSPs that terminates calls to its subscribers from callers using IAX.

In this case, we'll be creating an Asterisk dial-plan that routes outbound calls to these different networks based on extension pattern matching. One way we can do this is by area code. This simulates a least cost routing (LCR) that uses the private network or low-cost third-party peer networks to connect calls in the cheapest possible manner.

To make sure this project remains totally free to you, we're going to use only toll-free area codes: 800, 888, 866, 877, and 700. (700 isn't a traditional toll-free area code, but it can be used to dial calls, free of charge, between IAXTel and FWD. More on this later.)

We'll also create an LCR context for the other IAXTel-participating networks so that when calls are placed to their subscribers from our VoIP network, the calls will be routed via IAXTel rather than the PSTN. This benefits us in two ways: no long-distance charges, and our POTS line isn't tied up during those calls.

Since IAXTel is a free service, it is intended for experimental use only. Nobody promotes it as a reliable, "always-on" service, but it's a good way to learn about VoIP-over-Internet trunks and IAX2 channels.

Get access to IAXTel

In order to place calls using the IAXTel network, you'll need to obtain an IAXTel user account and password. To do this, visit IAXTel's sign-up site at *http://gnophone. com/directory/createAccount.php*. Here, you can obtain a username and password that will be used to authenticate you when you attempt to originate calls into the IAXTel network. Asterisk will have to be programmed with this username and password; this configuration is discussed later, so read on.

Set up the dial-plan for regular PSTN calls

First off, we've got to make sure that local and toll calls (not toll-free calls) are routed out the local PSTN interface—which we'll assume is an X100P card. To get an "outside line," we'll have Asterisk pick up a POTS line on Zap/1:

```
[default]
exten => _NX.,1,Dial(Zap/1/${EXTEN})
```

Route toll-free calls to the Internet

Next, outbound toll-free calls will go to IAXTel, which supports the IAX protocol for call termination. First, we'll route all calls starting with 1700 to IAXTel.

1700 is the customary prefix used to reach IAXTel users and not a PSTN area code, per se. In fact, if you dial 1-700-*IAXTEL#* from VoicePulse or FWD, the call will be routed from these networks to IAXTel.

You'll need to plug in your own IAXTel username and password here:

```
[iaxtel]
exten => _1700NXXXXXX,1,Dial(IAX2/user:password@iaxtel.com/${EXTEN}@iaxtel)
```

Now, we can send 800 and 866 calls to IAXTel, as well. We'll put them in their own context, too:

```
[tollfree]
exten => _1866NXXXXXX,1,Dial(IAX2/user:password@iaxtel.com/${EXTEN}@iaxtel)
exten => _1800NXXXXXX,1,Dial(IAX2/user:password@iaxtel.com/${EXTEN}@iaxtel)
exten => _1888NXXXXXX,1,Dial(IAX2/user:password@iaxtel.com/${EXTEN}@iaxtel)
exten => _1877NXXXXXX,1,Dial(IAX2/user:password@iaxtel.com/${EXTEN}@iaxtel)
```

So, at this point, you can reach IAXTel subscribers at their 1700 numbers, and you can reach toll-free destinations, all using only the Internet, without using the PSTN.

Now, we'll use IAXTel to connect our system's calls to users of Free World Dialup. All FWD subscriber numbers are six-digit numbers. This makes it easy to handle them in the dial-plan. Since six-digit numbers aren't generally used with the PSTN, we'll just route them all to FWD via IAXTel. This can be done by tacking a 17009 onto the beginning of the number that's sent to IAXTel. The IAXTel servers recognize the extra 9 as a token indicating the call should be routed to FWD for termination:

```
[fwd]
exten => _NXXXXX,1,Dial(IAX2/user:password@iaxtel.com/17009${EXTEN}@iaxtel)
```

FWD and IAXTel aren't the only TSP services that have dial-plan tokens for routing calls to each other. Try experimenting with the ones in Table 13-1.

Table 13-1. Dial prefixes for calling between TSPs using the Internet

Call an IAXTEL user from FWD	*-1-700 and the 7-digit IAXTel number
Call a Vonage user from FWD	**-2431 and the full 11-digit Vonage PSTN number
Call a FWD user from Vonage[a]	0110393 and the 6-digit or 5-digit FWD number
Call a FWD user from Packet8[a]	0451 and the 6-digit FWD number or 5-digit FWD number
Call a Packet8 user from FWD	**898-1 and the full 11-digit Packet8 PSTN number
Call a VoicePulse user from FWD	1-700-900-0000 and the full 11-digit VoicePulse PSTN number
Call an IAXTel user from VoicePulse	1-700 and the 7-digit IAXTel number
Call a FWD user from VoicePulse	1-700-9 and the 6-digit FWD number or 1-700-99 and the 5-digit FWD number

[a] See warning, next.

 There's not yet a way to register an Asterisk server as a SIP client on Vonage or Packet8, so the prefixes in Table 13-1 will work with those services only when dialed through the TSP-provided gateway equipment/ATA.

Allow incoming calls from IAXTel

Routing calls to IAXTel is a function of the dial-plan. But in order to receive calls *from* them, we've got to set up an IAX2 registration peer. In other words, the Asterisk server will become an IAX2 client registered with a server on the IAXTel network. To do this, add a directive such as this to the [general] section of *iax.conf*:

```
register => username:password@iaxtel.com
```

The registration tells the IAXTel network how to reach you—specifically, to contact your Asterisk server's IAX2 client whenever a call is bound for your IAXTel number. Next, we'll need to set up the IAX2 peer in *iax.conf* that tells Asterisk the context in which to place calls when they come in from IAXTel:

```
[iaxtel]
type=friend
host=iaxtel.com
context=from-iaxtel
```

Route incoming IAXTel calls

At this point, we've got IAXTel calls coming to the `from-iaxtel` context, but this context doesn't exist yet in the dial-plan. So, in *extensions.conf*, we can use this context to tell Asterisk what to do with the calls. In this case, we'll just have them ring a Zaptel channel and then dump to voice mail box 100 if there's no answer after 45 seconds:

```
[from-iaxtel]
exten => s,1,Dial(Zap/1,45)
exten => s,2,Voicemail(100)
```

Monitoring registrations

When you set up VoIP trunks from the Asterisk server to various providers, what you're really doing is registering Asterisk as a VoIP client. To monitor the status of each trunk's registration, use the `sip show registry` and `iax2 show registry` commands. Here's a SIP trunk registration in progress:

```
pbx*CLI> sip show registry
Host                         Username       Refresh State
access1.voicepulse.com:5060  s00227788          120 Auth. Sent
```

Here's an IAX trunk registration:

```
pbx*CLI> iax2 show registry
Host                 Username    Perceived       Refresh  State
69.73.19.178:4569    tedwalling  <Registered>         60  Registered
```

You can also follow VoIP trunk registrations in Asterisk's detail log output.

WAN Design

The layout of your WAN has huge implications for VoIP, particularly when it comes to failover ability, disaster preparedness, and latency. The high-level topology model—distributed or client/server—will affect where you place gateways, registrars, and PSTN connect points. Meanwhile, your network's topographic layout—hub and spoke, meshed, or peered—will affect how well your network survives local outages and how latency impacts your telephony apps.

Distributed Versus Client/Server

As in network computing, there are two essential models for voice applications: distributed and client/server. In the distributed model, processing resources are on smart endpoints and servers are spread out over the network, which acts as an unintelligent transport between them. In distributed computing, each client agent has a substantial processing role in the application, and the server may or may not. IP phones working with a softPBX to connect users over an Ethernet network is a decidedly distributed idea.

Client-server (also called *server-based*) *computing* is a more consolidated, centralized approach to information processing. In client-server models, the endpoints have very little processing capability and serve a narrow role in the neighborhood of the OSI presentation layer. Meanwhile, processing-intensive tasks and network transport are managed by a central server that works on behalf of all the dumb endpoints. Mainframes that use serial terminals to display and collect data—say, in a POS (point-of-sale) environment—are a good example of client-server computing. So are TDM phones connected to a legacy PBX.

Just as in client-server computing, TDM endpoints are often referred to as *terminals*.

Consider the difference between a SIP phone and TDM phone. A SIP phone is somewhat functional apart from the softPBX, since it has both a user agent (UA) and server agent (SA), but the TDM phone can't be used at all on its own. In client-server models, the endpoints or clients are slaved to the server. On the other hand, SIP phones, which have memory and onboard programming, are smart enough to be configured to work with other SIP-compatible PBX systems. This difference will affect the placement of telephony resources within your network: SIP phones can work in a distributed fashion, while TDM and analog phones cannot. SIP phones all around the Internet could use your PBX server back at HQ, but TDM phones have no such luxury.

Both models are important in IP telephony. Even though "native" VoIP (i.e., no TDM or analog connections) uses a distributed model, client-server objects may still need to be integrated into your VoIP network. TDM phones may be necessary in offices where adequate cabling isn't possible to support IP phones or where IP phones are not cost-justified. With the right interface hardware, VoIP can be used to trunk between these legacy resources on the WAN.

It's also possible to force certain IP phones to become completely dependent upon server resources in order to boot and be usable. This makes them more like TDM phones, taking on a client-server guise. Most IP phones can be configured to obtain their VoIP configuration from a TFTP server. This allows a group of IP phones to be maintained centrally. All of their configuration files can be stored on a single server, just as a TDM phone receives its configuration from a central PBX.

The Cisco 7960's SIP firmware can be configured to retrieve the phone VoIP configuration from a TFTP server. After unlocking the configuration options, enter the IP address of the TFTP server in option 7 on the Networking Options menu. Or, better yet, have your DHCP server assign the TFTP server address.

WAN Layout

Hub and spoke

If your WAN is a hub and spoke layout, you most likely have a single (or a few), centralized data center where all of your business locations' connections with the rest of the network are made. Hub and spoke networks, like the one at the lower left in Figure 13-2, allow for easy central network administration and often have lower facilities costs than other layouts. In hub and spoke networks, all remote locations connect directly to the data center without any intermediate hops.

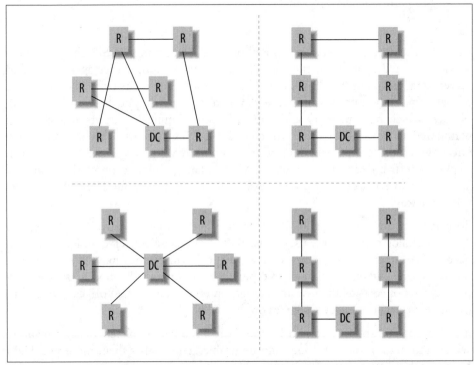

Figure 13-2. WAN layouts, clockwise from top left: partially meshed, circular peered, linear peered, hub and spoke

The benefit of hub and spoke layouts to VoIP is that they minimize latency because there's only one hop between the remote office and the data center (where the soft-PBX and primary switching equipment are). Other layouts may contribute more to latency than hub and spoke does. If there's a drawback to the hub and spoke approach, it's that there's a single point of failure on the WAN—the data center.

Peered

Sites in a peered network form a chain along which data travels. The data hops from one site to the next until it reaches its intended destination. The sites furthest from the destination use the ones that are closer to the destination as their "next hops" along the path, getting the data closer to its destination as it travels through each consecutive site.

The benefit of this technique is cost reduction. Peering keeps point-to-point connectivity costs down by decreasing the mileage of each link. In a hub and spoke layout, you might use six remote site links with an average length of 300 miles apiece. But, if the distances between each site and its closest peer are all lower than the distance to the data center, a peered layout can drastically lower that average distance and the associated cost. This is because each site is connecting to nearby peers, which are much closer than the HQ, which is further away. In a peered layout, only one peer needs to be connected to the HQ.

But there is a big drawback to this layout. Peered networks (like the lower-right example in Figure 13-2) are prone to cascading failure patterns. When an office that supplies a route to the HQ for other offices experiences an outage, so do the offices that rely upon that route. An ingenious way to eliminate this problem, if the geography is suitable, is to make a circular peer layout, like the upper-right example in Figure 13-2. When dynamic routing is employed, circular peer layouts are very resilient to isolated failures.

Exclusively peered layouts impose a higher number of router hops on call paths that traverse the WAN than a hub and spoke or meshed layout would. This might be OK for data applications, which aren't as sensitive to delays, but for VoIP, excessively peered networks can be showstoppers. If you had a 10-hop chain of peers with T1s between them, you could easily have 50 ms of latency from end to end, not accounting for higher-layer sources of latency. Making a high-quality VoIP phone call in this scenario just wouldn't be possible.

Meshed

In meshed networks, remote sites may be connected to one another, like a peered network, *and* connected to the data center, like a hub and spoke network. Or a single remote site might connect with two or more other remotes. This provides diversity for the transport to the data center and reduces the risk of widespread outage.

Depending on *how* meshed the network is (how much redundancy it offers), the threat of network downtime can be nearly eliminated. The Internet is a highly meshed network in the sense that many of its backbone carriers connect to multiple other backbone carriers.

In VoIP, a meshed network provides the greatest protection against unwanted system downtime. Meshed layouts, however, are the most expensive and aren't always practical. They might be needed in very demanding scenarios, such as a call center with offices in multiple states or highly redundant military or intelligence applications.

Quality-of-service measures must be more sophisticated when used in a highly-meshed network. Indeed, one thing RSVP is designed to do is negotiate the best-quality path through the network for a phone call when there are multiple potential paths. This is quite different from dynamic routing alone, which chooses paths across the network based on policies that have no direct relationship with the voice application.

Layout and PBX Placement

Most networks employ a combination of distributed and client-server models for different applications and a combination of different layout techniques for different balances of cost and resilience to failure. As long as the total number of remote locations is under 10 or so, a circular peer network is a great solution to the downtime issue, because if an upstream route fails, the other arc of the circle is still up and also leads back to the data center. This functionality is handled by—you guessed it—IP routing. So it takes two points of failure on the circle in order to bring the voice transport down for any one site, as illustrated in Figure 13-3. If a particular arc on the circle can't handle the traffic being thrown at it by the voice application, then it's a matter of adding a mesh link between that arc and the DC.

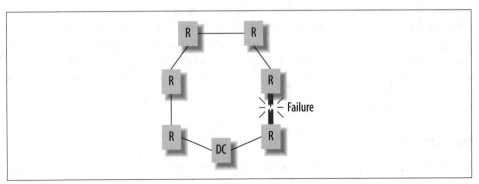

Figure 13-3. Even though a failure has occurred on one of the WAN links, all the remotes on this circular peer network can still reach the data center

The old saying from the real estate business, "location, location, location" is also true in IP telephony. Indeed, your choice of WAN layouts and computing models is always tied to the locations of your application servers and user groups. Since you can't move your users around to meet the needs of your network design, you must strategically locate VoIP resources (like PBXs and PSTN gateways) to meet *their* needs.

Locate to conserve network availability

It would seem that it's ideal to take the existing WAN and just pick the best locations within it for all of these elements—but that's not always the right approach. The WAN footprint may need to change shape to optimally support the VoIP network you're about to overlay upon it.

For example, the places where large amounts of traditional network traffic (say, database traffic) are transported may not always be the places where huge amounts of phone calls travel. Look at Figure 13-3. If a majority of voice traffic travels between the remote sites at the top of the diagram, it might make sense to put a PBX there—rather than in the existing data center at the bottom of the diagram. The last thing you want to do is decrease network availability to existing apps in order to add voice. This is exactly what you'd be doing if you unnecessarily overlaid a voice pathway onto an already-busy data pathway.

Locate to save money

There may also be geoeconomic reasons to place a telephony resource at an otherwise unlikely location. Consider international call centers. Lately, it's become very fashionable for insurance, mortgage, and collection companies to house big groups of low-cost outbound telephone operators in countries such as India and Mexico. These English-speaking employees call American households on behalf of American companies.

It would be very expensive for all of these calls to traverse the international LD network from India to the U.S. Instead, these companies may use VoIP to trunk calls over a comparatively low-cost international WAN to a PSTN connect point in the United States. Calls that originate inside the U.S. PSTN are much cheaper when destined for U.S. destinations. So, depending on your line of business and the needs of your particular application, locating a PSTN connect point a great distance from your call center, or perhaps a great distance from your PBX, might be a good idea.

Locate for capabilities

The location of telephony equipment is often dictated by the equipment's purpose and interfacing capabilities. For example, PBX servers with built-in PSTN interfaces may need to be in the same building as the PSTN connect point. But a PBX server with an outboard PRI chassis could be located several hops away, and perhaps miles away, from the connect point. The PRI chassis would need to be near the connect point, but, WAN bandwidth notwithstanding, the PBX server itself could be anywhere on the private network.

Voice-mail servers, like Cisco's Windows-based Unity, may need to communicate with an email server if you're going to use integrated messaging. In this event, very high amounts of bandwidth between the Unity server and the email server will likely be required. It's quite common for voice mail and email servers to be installed in the same rack or to be running on the same PC.

Since we're talking about layout, all of these issues must be taken into account when looking at how your VoIP network will overlay your IP network layout. Is there enough bandwidth to support the necessary loads between all endpoints? Would adding a new connection solve a capacity problem imposed by VoIP, or would it be better to place a PBX somewhere to solve the problem? Which solution would be more cost-effective? Is your vision of Voice over IP even feasible given today's load on the network? If it's a peered network, would the outlook for VoIP be better with a hub and spoke layout or a few new mesh links? How would such a change affect other network systems?

Don't locate for convenience

The VoIP network ought to drive the IP network design. A PBX shouldn't be placed in a particular location because "that office already has a server rack and a UPS" or because "that's the office where the old phone system is." Ultimately, while these issues are an influence, they can't be deciding factors in locating VoIP resources. The VoIP network's design must not be retrofitted around the current network's preexisting topography. If this were a good way to approach IP telephony, then VoIP-over-Internet would long ago have replaced the PSTN. Clearly, the existing layout of the Internet isn't suitable to replace the PSTN (yet).

If it makes sense to have a PSTN connect point in another country with VoIP trunking back to your PBX, build your WAN to accommodate that. The bottom line is this: the IP network you have in place today probably won't be the IP network you'll have in place when migration to VoIP is complete.

Disaster Survivability

The ability of a network to survive isolated equipment and link failures is called *survivability*. It's a subject that, like networking itself, is addressed at every layer of the OSI model. Backup power supplies, dynamic routing, and remote-survivable dialplans (groups of phones that can call each other even when their access to the PBX is cut off) operate at the physical, network, and application layers, respectively.

Surviving Power Failures

The most fundamental survivability measures occur at the physical layer, starting with backup power. Without backup power, your site's primary source of electricity may go dead, taking your phone system with it. Whether you use standalone battery systems or a combination of batteries, a generator, and a transfer switch, backup power is a requirement in all data centers and at all crucial network connection points.

Unlike analog phones on the PSTN, VoIP systems don't get their power from the PSTN. VoIP systems will fail during power outages unless they have adequate backup power.

Multiphase power

Most small offices and residences receive their AC power in the form of a single 120/240-volt connection. This connection feeds a circuit-breaker block that distributes individually limited power circuits throughout the premises. When the power fails at the breaker block, the power fails for the entire premises.

But when power is delivered in multiphase, it can create redundancy. Multi-phase power means that the same connection to the electric company can deliver two or three AC supplies to the subscriber's premises. The supplies are connected to sections of the breaker block or to different breaker blocks. So, when a single phase fails, the other phases are still intact, and equipment on the failed phase can be moved to them. This won't eliminate all failures, but it can protect you against certain kinds of failures that occur within the electric company's facilities.

Uninterruptible power supplies

In order to survive a power failure, all of your network equipment must remain running—switches in phone closets, servers at the data center, and IP phones themselves. This means you either have to back up every device individually, using uninterruptible power supplies (UPS) or create a centralized power distribution system. One way to do this is to place a backup switch with battery and/or generator at a central location and then pull AC wiring from the backup system to each of your phone closets. This way, each critical phone closet has an AC source that's backed up centrally.

For IP phones, use PoE, and make sure the powered switches or injectors are backed up, too. The moral is this: it won't do any good to have your Linux PBX server on a high-quality backup power system if your phones and switches aren't on one, too.

Surviving Network Link Failures

Redundancy is your best defense against network link failures—those that affect only an individual link like a single T1 or an Ethernet switch. If a network link is absolutely critical, there should be, if at all possible, a redundant alternate link that provides an identical logical path.

Point-to-point T1s can be made more resilient to failure by bonding them together into multilink bundles. This way, one of the T1s can fail without totally downing the networking pathway. Moreover, two T1s running through two different providers' networks are more resistant to failure than a pair that runs through only one network.

But redundancy costs money. It may be tough to justify a completely redundant network and even tougher to manage one so that, when failures occur, it behaves as originally envisioned. Moore's Law infers that whatever capacity you make available, your application will become dependent upon it and grow to exhaust it—even if it's placed there for backup reasons to begin with. So, even if you have double the capacity needed for every link—in the name of redundancy—you may still find yourself in a state of panic when that capacity is merely *reduced*.

PSTN trunk failures

Some types of network links are easier to make redundant than others. IP links can be automatically failed over using dynamic routing at the network layer, but voice T1s and phone lines aren't so simple. A PRI, for example, may go down—and when it does, all of its DID numbers and inward signaling configuration will become unavailable to the PBX. Even if a second PRI exists that the PBX can use for outbound calls, some emergency switch at the telephone company will have to occur in order to reroute inbound calls to the second circuit.

The same is true of POTS and Centrex lines. If you have 10 POTS lines in a hunt group and the line with the published number (the lead line) experiences a failure, you'll have to contact the phone company to get all calls to that line forwarded to the next line in the group, until the problem with the first line is resolved. Phone companies do offer high-availability solutions for these scenarios at your expense, so contact your local phone company to see what it offers.

Hot failover—instant, user-transparent switching from one telco circuit to another—is difficult to achieve. Some trunk bypass switches can redirect private trunks from one T1 to another, but this can create challenges for DID, caller ID signals, and call routing. Plus, it isn't exactly cheap to maintain backup PRIs merely for the sake of failover.

Here's what to do if your PRI or POTS trunk goes down:

- Have the phone company forward calls from your lead number to a backup line.
- If the failure is in a POTS hunt group, have them "busy out" the failed lines so calls will roll to the next line in the group, which is presumably still working fine.
- Some phone companies let you manage your Centrex groups by software or web interface. Make the appropriate changes yourself.

Remote Site Survivability and ALS

While dynamic routing and good network design can make it less likely for link failures to disrupt your voice or videoconferencing application at remote sites, such measures are often too expensive or complex. Moreover, they're low-level measures that don't address the needs of telephony applications during a link failure.

Minimizing the Havoc of a PBX Crash

Few disaster scenarios are more frightening than the loss of a single, critical server… except, perhaps, a single critical *PBX* server. In the age of distributed computing and PC components, the PC chassis is becoming the new home of the private branch exchange.

The PC brings its well-known characteristics to telephony: cheapness, modularity, extensibility, and, unfortunately, instability. Better PC servers equal better stability, of course, but PC backplanes will never be constructed with the untouchable reliability of old-school PBX systems.

The mere fact that PC servers rely upon hard disks means that PC-based PBXs have a pretty good chance of a downtime-inducing crash. So what can be done to prevent your next-generation dial-tone from dying unexpectedly?

- Back up your dial-plan regularly and have a standby server ready to go in the event of a failure.
- Use redundant, mirrored hard drive arrays on your softPBX servers or a central, redundant network-attached drive array to eliminate the threat of hard disk failures.
- If you use a commercial VoIP platform like Meridian or Avaya Media Server, invest in failover equipment. The biggest advantage of commercial systems over open source ones is that they have reliable, well-tested automatic failover ability.
- If using Asterisk, you can create emergency contexts in a secondary server's dial-plan—one that matches the active dial-plan of a primary server. This way, when the primary server goes down, you can "promote" a secondary just by including the emergency context in its dial-plan. If you wanted to get fancy, you could use the Asterisk Manager API (described in Chapter 17) to trigger the failover automatically and notify an administrator by doing a Dial( ) to his cell phone.
- Use a distributed call-switching technology such as DUNDi (discussed later in this chapter) to minimize the effect of a single PBX server's downtime.
- Use IP-based connections to the PSTN rather than PRIs or POTS lines. If a PBX crashes, it's easier to redirect IP-based connections than it is PRIs or POTS lines.
- If you're using a PRI attached to a crashed PBX, you can automatically redirect it to a secondary PBX by way of a mechanical T1 failover switch, also called a trunk bypass switch.
- If an H.323 gatekeeper crashes, it's easy to fail over to a backup. When using multicast locate requests from IP phones, you can configure a backup gatekeeper to listen for requests on the same multicast address as the primary gatekeeper but enable it to respond to those requests only when the primary server has failed.

Fortunately, many IP telephony manufacturers have stepped up with solutions at the application layer that are effective guards against the symptoms of link failures that won't break the bank.

The survivability problem

At a remote office location that has eight IP phones linked together by Ethernet, a single WAN link might provide a connection to the PBX server at HQ for all eight phones. If that link were to fail, the phones would lose their centralized signaling, directory, and call-switching server. The IP phone users wouldn't be able to call other users in the organization, much less PSTN users. They wouldn't even be able to call each other!

To solve this problem, one could put a PBX at this site, but that would be an expensive and wasteful solution. It's also hard to justify putting in a redundant WAN link for such a small office. And what if there were an emergency at the site and public safety services were needed in a hurry? An emergency 911 call couldn't be originated from the remote site without access to the PBX at HQ.

The survivability solution

The solution to all of these problems is application layer survivability, or ALS, a generic term that describes a number of technologies that allow sites without full PBX services on-site to survive an unexpected disconnect from the PBX and remain functional until WAN access is restored. Usually, this means redirecting PSTN-bound calls to a locally connected POTS or Centrex trunk. It could also mean redirecting calls that are ordinarily transported over the WAN to the HQ's PBX over the PSTN instead. These on-the-fly changes are handled by a device often referred to as a remote site gateway. This device may be incorporated into the site's WAN access router or IAD, or it could be a standalone, VoIP-oriented Ethernet switch.

 ALS isn't a concept that's exclusive to VoIP. Non-telephony systems have been programmed to survive network failures for a long time. Resumable FTP is one example.

Multitech, Avaya, Cisco, Zoom, and others provide integrated devices for small remote offices that can handle ALS. Zoom's X5 device, for example, provides a WAN router, a place to connect to the PSTN, and a place to connect both analog and IP phones. It can also bypass any VoIP trunks and send 911 and other local calls right to an attached POTS line.

Cisco's embedded media gateways can be programmed for just about any ALS scenario using IOS voice commands. If you use analog voice interface cards to hook up POTS lines to these boxes, they can do local 911 bypass, too. Plus, they integrate very easily with Cisco's softPBX, CallManager.

Each device handles ALS in a proprietary way, though all ALS solutions apply the distributed computing model to solve the remote survivability problem. Those that are integrated with a central PBX make a local copy of the dial-plan so that local users can still call one another when the link to HQ is broken. Those that have an FXS port allow it to be used for emergency 911 calls regardless of the state of the central PBX.

Some vendors have special nicknames for ALS. Cisco calls its ALS technology SRST (Survivable Remote Site Telephony), and Avaya calls it LSP (Local Survivable Processor). Most ALS solutions include:

- A local cache of the dial-plan so that phones at the ALS-equipped remote site can still call each other without the central PBX server.

- Instructions on how to handle calls to the private voice network from the remote site during times when the IP path to HQ is down. These instructions might mean diverting the calls over the PSTN and into the HQ site's PSTN trunks.

- Parameters on how long to wait after an outage occurs before attempting to give call control back to the central PBX at the HQ.

- Parameters that describe when and how often to replicate the local cache of the dial-plan.

- The ability to route 911 calls to a locally connected POTS line or PRI, even when the upstream link is working correctly.

- The ability to act as the last private signaling point in an LCR call path. For example, if Cleveland users want to call PSTN destinations local to Miami, the WAN can trunk calls from a Cleveland PBX to the Miami remote office, and then the ALS-equipped gateway device can dial these calls on the local POTS line, so they can be local PSTN calls instead of expensive LD calls. (LCR features aren't necessarily tied to failures. Most ALS-equipped gateways also enable LCR with or without regard to survivability.)

Metro-Area Links

The use of MAN (metropolitan area network) technologies is increasing, as the quality and reliability of radio and wireless optical systems is rising, and the need for campus-scale networking is driving adoption of 802.11 radio Ethernet. Free space optics (FSO) is a wireless bridging technology that uses infrared and laser band transmissions. These technologies are described in Chapter 4. WiMax is a newer radio bridging solution that deals with many of the shortcomings of 802.11.

When it comes to survivability in a voice scenario, these systems exhibit the same characteristics as they do in traditional networking. 802.11 is a data link layer technology family, so it has no QoS capabilities. FSO is a physical layer technology, and it has the same shortcoming. This makes it hard to tell when an 802.11 or FSO link is at capacity in order to make a prioritization decision or bandwidth reservation. Consequently, these notions just don't exist in wireless Ethernet.

Failing over from one 802.11 link to another is also complicated. Using 802.11 (or FSO) by itself provides no way of knowing *when* an automatic failover is necessary, because there are no built-in alert mechanisms that tell you when a link is down. One possible solution is to place routers on either end of an 802.11 link pair, and then use load-splitting or dynamic routing to deal with one of the links going down. Beware of the potential for substantial jitter on 802.11 links, too. For a much more elegant (and detailed) discussion of wireless networking, check out O'Reilly's *802.11 Wireless Networks: The Definitive Guide.*

Firewall Issues

In most networks, a firewall provides an access gateway to network resources that need to be restricted and secured. The primary method firewalls use to secure a resource is limiting network traffic to and from the resource by port number, by protocol, or by network address.

Another key use of firewalls is network address translation (NAT), which provides an exchange point for data transmitted between privately and publicly addressed IP networks. NAT keeps track of outbound sockets (those from users on the private network to the Internet) and performs the necessary packet translation as each packet traverses the firewall. This way, hosts with private addresses can access the services of hosts with public addresses, like World Wide Web servers. A side effect of NAT is that protocols that must use both outbound and inward sockets (like SIP and H.323) simply don't work.

Since firewalls are a reality of doing business on IP networks, they aren't going to be cast aside merely for the sake of SIP compatibility. As a result, they're an important topological consideration when locating VoIP resources on your network. Figure 13-4 illustrates some firewall scenarios you may encounter with VoIP.

We'll illustrate the twofold problem posed by NAT using SIP as a context. Typically, when SIP requests are sent from a privately addressed client to a publicly addressed VoIP server, their headers contain the private address of the SIP client device. If this address is being NAT-translated to some other address, the VoIP server will send its response to the wrong address (the private one, not the NAT-translated one), and the operation will fail. If the client uses the NAT-translated address (which some clients can be statically configured to use), you would still have to program your NAT firewall to forward all inward SIP traffic (ports 5060 and 5061) to the private host running the SIP client. This way, the VoIP server's responses would reach the client. The effect of this is that you'd only be able to use one SIP client at a time. So, port-forwarding doesn't really solve the problem.

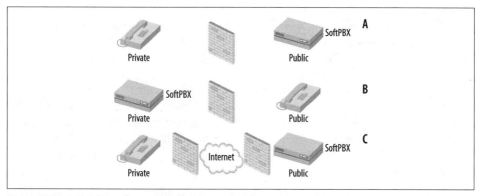

Figure 13-4. NAT firewall scenarios

The second aspect of the NAT problem is the RTP media channel, which, like the signaling conversation, is also bidirectional. Connectionless UDP packets are sent to the VoIP server from the client and from the VoIP server back to the NAT firewall that stands between the client and server. Simply port-forwarding RTP traffic is fine—for a single client. But a more elegant solution is needed to support a group of SIP endpoints behind a common NAT firewall.

In order to get VoIP phone calls to work through NAT firewalls, both signaling transmissions *and* media streams must be able to flow in both directions, to multiple clients simultaneously. Also, the endpoints must signal the appropriate IP address during call setup—an IP address bound to the NAT firewall, an address that's publicly routable.

Neither of these issues has been resolved within the confines of SIP, MGCP, SCCP, or H.323, although IAX2 has solved both of them because it requires just one socket for all communications on a single call. Unfortunately, since there isn't much commercial support for IAX yet, you'll most likely get stuck dealing with the NAT allergies of the other protocols. There are several ways to tackle NAT, either by eliminating it or by cooperating with it. Read on.

DMZ Eliminates the Need for NAT

In a setup like example A in Figure 13-4, your IP phones won't be able to receive signaling or audio from IP endpoints outside the firewall. An easy way to solve this problem is to place the softPBX on a DMZ, where it can still be firewalled, but without having to have traffic from the private network be translated via NAT.

Traffic from the private network to the Internet is still translated, of course, but since the softPBX is on the DMZ, it can act as a media proxy for VoIP phone calls between the private network and the Internet. The softPBX can communicate without NAT in the way. A DMZ such as this is illustrated in Chapter 10.

Placing your VoIP server on a DMZ solves the NAT problem for all signaling protocols—SIP, H.323, SCCP, and MGCP.

Using a DMZ requires that you have access to more than one IP address. You'd have to obtain, at a minimum, three public IP addresses from your ISP—one for the soft-PBX, one for the DMZ interface on the firewall, and one for the Internet-facing interface on the firewall. If you are unable to get a block of IP addresses, then you'll have to consider another solution. But if you are able, placing VoIP server resources on a DMZ is the way to go.

Setup B in Figure 13-4 is what you may be considering if you plan to have some road warriors or Internet-based subscribers accessing voice services through a server on your private network. DMZ solves the NAT problem here, too.

But there may be situations in which the Internet-based phone must be behind a NAT firewall, and there's nothing the user can do about it. Setup C in Figure 13-4 illustrates this idea. Residential broadband routers have built-in NAT firewalls, and so do the firewalls at hotels and in some coffee shops and public access points. Since you're relying upon the policy of other organizations for your voice transport, you're at the mercy of what their infrastructure permits—or doesn't.

Fortunately, there are solutions to the NAT problem that don't involve DMZ.

IAX Eliminates the Need for NAT

The IAX2 protocol is a great alternative to SIP in telephony applications, as long as you require only endpoint signaling and trunking. That's because SIP implements a whole lot more than these two things. IAX2 is centered around these two roles, and it does them very well. So well, in fact, that IAX2 can traverse NAT firewalls without any special configurations. IAX2 uses a single UDP socket for all signaling and media, and is therefore NAT-proof. If you can, consider using IAX as an alternative to SIP where NAT is concerned.

STUN Allows Coexistence with NAT

Simple Traversal of UDP NAT (STUN) is a simple protocol that allows applications to discover the presence of NAT firewalls. It also tells these applications the public IP address(es) allocated to them by the NAT firewall. STUN requires no special configuration on the part of the NAT firewall. Consequently, STUN is easy to integrate into your existing network.

But STUN does require that the client application that uses NAT traversal be equipped with a STUN client. Adoption of STUN among many IP phone vendors has been slow, to say the least. Most commercial IP telephony platforms advocate placing a media proxy on the DMZ, while others suggest avoiding the Internet altogether as preferable to using STUN. But this advice is overblown. STUN works great—when it's there.

STUN is defined by RFC 3489. A free, open source implementation of the client and server is available from *http://www.vovida.org/applications/downloads/stun*. Vovida also operates two public STUN servers that you can use in small-scale applications and for your own STUN client development. Try configuring the X-Lite softphone to use Vovida's STUN servers, either 128.107.250.38 or 128.107.250.39, to place SIP phone calls through a NAT firewall.

VPN Allows Coexistence with NAT

There's another solution that can help you avoid VoIP-through-NAT, if you are the proprietor of both the VoIP server and the endpoints it supports. Virtual private networks allow VPN clients on the Internet to use the same IP network address as the VPN server. Therefore, if a VPN server and a VoIP server are on the same LAN, VPN clients connecting to that VoIP server from the Internet needn't be concerned with NAT. Of course, you have the normal headaches of a VPN, such as latency and the lack of QoS. But if your VPN is good enough, it saves you from the NAT problem.

Peer-by-Peer Codec Selection

Just as your VoIP network is affected by the design of the IP network it overlays, the IP network will be affected by the decisions you make at the higher layers—especially codec selection. Using a bandwidth-conserving codec on a WAN link is often mandatory to avoid swamping the link with excess voice utilization. In the same way, the less utilization you impose with each call, the more calls you can "squeeze" onto the link.

But that doesn't mean your VoIP network will use a highly thrifty codec like G.729A in all circumstances. Processing requirements and sound quality degradation are prime reasons to use bulkier, noncompressed codecs instead of a compressed codec. On a fast Ethernet segment, where there's usually a glut of bandwidth, you can safely use G.711 (or G.722 for even higher quality) as your standard codec.

The VoIP servers responsible for enforcing codec policy must therefore be programmed to select certain codecs for use in certain calls. Different vendors have different ways of doing this, but they have one thing in common: they all associate a list of allowed and prohibited codecs with each endpoint, port, or trunk. This way, you can force a group of phones in a remote office with limited WAN bandwidth to use a low-bandwidth codec, while still allowing local phones on the LAN to use a high-quality codec.

Controlling Codec Selection in Asterisk SIP Peers

Asterisk's *sip.conf* file, in */etc/asterisk*, contains entries for each SIP endpoint or trunk that can communicate with the Asterisk SIP server agent. Each entry can have multiple allow and disallow keywords, which indicate what codecs are to be used for each peer. The following configuration sets up a group of LAN phones and a group of remote, WAN-connected phones:

```
[general]
disallow=all
allow=ulaw
allow=g279
allow=speex

[301]
; phone on same LAN as
callerid="Jake" <301>
context=Cleveland
host=dynamic
type=friend
username=301
secret=browns

[402]
; phone in remote location, connected by WAN
disallow=all
allow=g729
callerid="Maddie" <402>
context=Maui
host=dynamic
type=friend
username=402
secret=aloha
```

 You can refer back to Project 6.1 for more details about per-peer codec selection on Asterisk.

Directory Services

Directory services transform your VoIP network from a transport apparatus into a set of user-friendly voice applications. Without directory services, the network can do very little. Sure, you can make IP-to-IP phone calls, and you might not mind memorizing IP addresses instead of E.164 numbers or SIP URIs. But, of course, there's a better way to get hold of people.

On the PSTN, that better way is the phone book: that yellow-paged tome that often doubles as a booster chair for toddlers or as a flattener of accidentally crumpled papers. But since transistors are cheaper to make than paper, the phone book has its digital equivalents. Many digital PBX systems add a centralized phone directory that can be accessed from their telephones. Those that offer voice mail also (usually) offer an interactive IVR directory that lets you find a person's extension number by keying in the dial-pad equivalent of the person's name.

Directory services: resolution and advertising

There are two components to directory services in VoIP: resolution and advertising. *Resolution* translates already-known resource identifiers (like SIP URIs and E.164 numbers) in network layer identifiers (IP addresses and port numbers) so VoIP endpoints can communicate. *Advertising* tells callers what SIP URIs and E.164 numbers to use when they want to call somebody. It associates the resource identifier with the name of the party to be called.

So *resolution* translates IDs into protocol-friendly addresses, like 10.3.4.10:5060, and *advertising* translates IDs into human-friendly addresses, like Bob and Pat. The reason for the distinction between these two concepts is so that you can implement them separately or together. Resolution and advertising can be codependent, as in a legacy PBX phone directory, or they can be independent, as in SIP URIs, wherein resolution is accomplished using DNS.

Most legacy PBX systems force resolution and advertising to be codependent because the possibility of globally aware phone directories never existed before IP telephony. The PBX always relied upon the PSTN to properly route calls bound for foreign exchanges (resolution), and businesses always relied upon the phone company to publish a phone book for those foreign exchanges (advertising). In the same way, the private exchange publishes its own directory and routes its own calls independently of the rest of the world. Now that the Internet and private global networks are a viable alternative to the PSTN, global directory services are important, and resolution is separating from advertising.

In IP telephony, resolution is the responsibility of DNS or some other distributed mechanism, leaving advertising as the true function of the PBX's directory, an LDAP (Lightweight Directory Access Protocol) server, or some other directory authority appointed by you, the VoIP adopter. This directory advertising service is the digital phone book, where your fingers do the walking.

Resolution of SIP URIs and E.164 numbers

Different VoIP protocols handle resolution differently. For example, SIP provides for DNS resolution of its URIs, like HTTP. H.323, conversely, relies upon a gatekeeper for resolution of dialed numbers. A DNS tool like Dig can be used to resolve SIP URIs, while only an H.323 client can allow you to resolve E.164 numbers. A SIP URI can reach you anywhere in the world you're registered via the Internet, but an E.164 number is always tied to a particular H.323 zone.

SIP URIs are globally unique, but H.323's extension numbers make no such provision. This poses an engineering challenge for H.323, especially on extremely large networks that are managed by several groups with independent authority over their own numbering within the greater network. Fortunately, large commercial implementations of H.323, like Avaya's MultiVantage and Nortel's Meridian platforms, do offer an in-phone directory that can be centrally maintained and object-based security measures for that directory. Still, SIP offers an open solution to a common problem—use DNS for distributed name resolution, and you don't have to maintain a closed, proprietary directory in order to let people locate one another. For H.323, another standard can provide similar functionality: ENUM.

ENUM

ENUM is the IETF's protocol for resolving E.164 telephone numbers via DNS. It can be used with all kinds of IP telephony endpoints, not just H.323. Its usefulness with H.323 is clear: it's the only way to globally resolve E.164 numbers using DNS. This frees H.323 from the PSTN, allowing it to be the basis of a global, Internet-based voice network. ENUM is described in RFC 2916.

DUNDi

Distributed Universal Number Discovery is Digium's protocol for failure-resilient, decentralized resolution of phone numbers on VoIP networks. DUNDi uses a peer-to-peer protocol to dynamically discover how to reach users. A reference implementation of this newly developed protocol is available from Digium's public CVS server for Asterisk Versions 1.0 and higher.

Advertising

In VoIP, user resources like IP phones aren't callable unless you know what their resource identifiers or phone numbers are. You key the phone number into the IP phone, and then it (or your PBX) performs the resolution of the resource identifier. Once resolved, your IP phone knows how, at the network layer, to send and receive data to/from the called endpoint.

But how do you know the resource ID of the person you're trying to call? The answer, as in the Yellow Pages, is simple: advertising.

On the Web, you won't get far without a URL to begin your browsing. Without a URL, you can do nothing. But go to *http://www.google.com*, and suddenly the world is at your fingertips. Even though SIP URIs don't work in precisely the same fashion, the basic idea is the same: Google advertises the existence of URLs, making them browsable by you. In effect, Google is a phone book of the Internet.

Building your own directory lets you advertise your voice resources within your private network. You may also advertise them to the outside world. It's possible to use

LDAP as a directory server for large VoIP networks. You could allow anonymous access to your LDAP server from foreign networks so callers can browse your directory, but beware of the security risks.

 LDAP can solve some administrative headaches for Asterisk. If you've got a find-me-follow-me user with three cell phones, a home phone, and two office phones who's already listed in your LDAP directory, it may be easier to link that user's extension to the LDAP directory than it would be to program all of these phone numbers into the dial-plan. Asterisk offers LDAP integration in the dial-plan using the LDAPGet command, which is configured in */etc/asterisk/ldap.conf*. You can download the LDAP Asterisk module from *http://www.mezzo.net/ asterisk*. With LDAPGet and Asterisk's IVR commands, you could even make Asterisk audibly read the data from the LDAP server to callers.

Commercial directory services for VoIP don't tend to interoperate, but they are usually very slick and work well as long as you stick with a single vendor's PBX and phones. Cisco's AVVID architecture implements a directory system that is included with the CallManager softPBX. It uses XML to drive the displays and command buttons on the 7960/7970 SCCP phones to make an interactive, searchable directory of users and endpoints registered with the CallManager. Most high-end PBX vendors offer something similar.

Web-Based Corporate Phone Directory?

Once the phone system is networked and soft-driven, the promise of integration with other systems—like the Web—brings with it some really cool novelties. One of them is a web-based corporate phone directory. A useful add-on to any corporate intranet, this web tool would allow users to easily and quickly look up extension numbers. The Asterisk Management Portal (AMP) is a PHP web interface for Asterisk that you could easily modify to provide an online phone directory for your end users. For more on AMP, refer to Chapter 17.

Project 13.2. Build an Interactive Directory on a SIP Display Phone

What you need for this project:

- A Cisco SIP display phone (7940, 7960, etc.)
- Asterisk
- An HTTP server like Apache

Cisco's 7900-series SIP phones can access a web service to drive interactive menus and application frontends right on the phone. These applications use the phone's display and soft buttons and are often tied, via a web server, to a backend database or PBX system like Asterisk. In this project, we'll build a simple, static user directory that you can browse with a 7900-series SIP phone.

 Ever wonder what the difference between a 7960 and a 7960G is? The G stands for *global*. G phones have icons instead of words printed on their buttons: a mail envelope instead of the word messages and so on.

Cisco's IP phones use XML markup to describe the user-interface elements the phone is capable of constructing—simple UI elements like images, button labels, and lists. The 7970G phone is even capable of displaying color graphics. (A fully functional web browser is a standard feature of the SCCP version of the 7970G.) The XML files are downloaded by the phone from a web server, so they can be static text files or CGI- or servlet-generated output.

On the 7960 phone, there is a Directories key that is used to trigger an XML application on the phone's display. Despite its label, this key can be programmed to do anything, because you control the XML file that it loads. The URL used to find the directories file is found in the `directory_url` setting of the phone's SIP configuration file, which is loaded by TFTP when the phone starts up.

TFTP and Cisco SIP Phones

All Cisco IP phones support the use of TFTP to load configuration files, and most Cisco IP phones also support configuration using the phone's built-in display and buttons. If your Ethernet segment has a DHCP server that can be used to get the phones on the IP network, then TFTP is the key to doing mass configurations at once. Your DHCP server can tell the SIP phones what your TFTP server's address is.

To mass-configure 7900-series phones, start with Cisco's default configuration file (there's an example at *http://www.cisco.com/cgi-bin/tablebuild.pl/sip-ip-phone7960*) and customize it as needed. Once you've got it set up the way you want, copy it to your TFTP server's root directory and rename it according to the convention:

    SIP<mac address of the SIP phone>.cnf

Everything in the filename, including the MAC address of the SIP phone, should be uppercase, *except* for the extension *.cnf*. Make one copy for each of the phones you need to configure, until you've got a file for every SIP phone. At this point, when the phones boot up, they'll get their configuration from the TFTP server.

Once the config file is loaded by TFTP, the directory_url defines where the XML will be loaded from. The next step is to put that XML file at the path of the URL. Here's a sample of what a simple directory file might look like:

```
<CiscoIPPhoneDirectory>
<Title>Phone Directory</Title>
<Prompt>Asterisk Extensions</Prompt>

<DirectoryEntry>
<Name>Wallingford, Ted</Name>
<Telephone>105</Telephone>
</DirectoryEntry>

<DirectoryEntry>
<Name>Wojtowicz, Susie</Name>
<Telephone>106</Telephone>
</DirectoryEntry>

<DirectoryEntry>
<Name>Gray, Sue</Name>
<Telephone>107</Telephone>
</DirectoryEntry>

<DirectoryEntry>
<Name>Wallingford, Jayne</Name>
<Telephone>108</Telephone>
</DirectoryEntry>

</CiscoIPPhoneDirectory>
```

So, your directory URL could point to a static XML file, or it could point to a dynamic Perl, PHP, or VB script. In any event, make sure that the proper headers are sent to the SIP phone, or you won't have good results. Here are the headers the 7900-series phones need:

```
Content-type: text/xml
Connection: close
Expires: -1
```

If you're using PHP for your dynamic directory app, you should make a script with a name like *79xx-lib.php* that lets you output these headers (and other 79xx-specific stuff) easily. A PHP header script for Cisco SIP phones would contain this:

```
$header79xx="Content-type: text/xml\nConnection: close\nExpires: -1\n\n";
```

Now, the $header79xx PHP variable contains the headers, and you can place it at the start of your PHP directory script. The following script grabs a list of usernames and extensions from a fictitious MySQL database and outputs them as a 7900-series phone directory that can be accessed by pressing the Directories button.

```
<?
include "79xx-lib.php";
print $header79xx;
```

```php
$mysql_conn = mysql_connect("localhost","asterisk","");
mysql_select_db("directory",$mysql_conn);
$rowset = mysql_query("SELECT username,extension FROM extensions", $mysql_conn);

print("<CiscoIPPhoneDirectory>\n");
print("\t<Title>Giant Motor Corporation</Title>\n");
print("\t<Prompt>Office Phone Directory</Prompt>\n");

while($row = mysql_fetch_row($rowset))
{
    print("\t<DirectoryEntry>\n");

    print("\t\t<Name>");
    print($row[0]);
    print("</Name>\n");
    print("\t\t<Telephone>");
    print($row[1]);
    print("</Telephone>\n");

    print("\t</DirectoryEntry>\n");
}
print("</CiscoIPPhoneDirectory>\n");
?>
```

In the preceding PHP example, when the user presses the Directories button on the Cisco phone, the HTTP client downloads the output of the directory_url setting (which is driven by the PHP script), interpret it, and display the directory on the phone's LCD. The user can then place a call directly to any of the users in the directory.

Cisco IP Phone Hacks

Cisco's IP phones, through their XML Services firmware, allow you to use highly sophisticated web-based apps to enhance your telephony offerings. They are accessed via the Services button and from the URL stored in the services_url setting.

These apps can access a wider spectrum of functionality than those accessed as directories. You could have your OpenOffice or iCal calendar show up on your phone's display, or you could turn the phone into a data-collection device, a time clock, a stock ticker, or an instant messaging tool.

A host of commercial third-party applications have been created to take advantage of this functionality, too. Cisco publishes a developer's guide for IP phone XML Services. Once you've gotten your feet wet with XML Services, check out *http://asterisk.edihost. co.uk/am-web*, a site that offers an XML application that gives you administrative control over an Asterisk PBX from a Cisco 7960.

Key Issues: Network Infrastructure for VoIP

- The topology of a VoIP network rides atop that of an IP network.

- Your IP network layout affects where you locate voice resources on the network and how resilient voice services will be to failure.

- Hub and spoke networks offer little remote redundancy but result in cheap connectivity.

- Meshed networks offer redundancy, but result in more expensive connectivity.

- Peered networks offer the least-expensive connectivity but are often the least capable of yielding positive results with IP telephony because call paths across peered networks are prone to latency.

- When choosing a location for a voice resource like a PBX server or media gateway, locate to conserve bandwidth, to save money, and to gain long-term capabilities, not because of superficial or short-term conveniences.

- The best way to survive a power failure is battery backup, but using multiphase power from the electric company can somewhat reduce the risks of power loss.

- PSTN trunk failures almost always incur some manual failover, whether it's calling the phone company to get a line forwarded or physically juggling trunks around to compensate for a downed trunk.

- IP-based trunks (VoIP trunks) can be more easily failed over than PRIs and POTS lines.

- Remote site survivability, the ability to keep telephony applications at remote sites working when disconnected from the central PBX, is accomplished using application layer security. ALS techniques and failover protocols are proprietary to each IP telephony platform.

- Metro area bridging technologies like FSO and 802.11a/b/g don't offer automatic failover and must be used in combination with dynamic routing in order to accomplish it.

- NAT firewalls impose a communications breakdown on protocols like H.323 and SIP that require multiple sockets. To get around this problem, you can employ a DMZ or use the IAX protocol (which requires only one socket per client), STUN, or a VPN.

- Peer-by-peer codec selection is necessary if you want to use bandwidth-conserving codecs across certain links and not across others. In Asterisk, this is configured in */etc/asterisk/sip.conf*.

- Directory services are a crucial aspect of VoIP infrastructure. Using an XML-generating web site, you can create dynamic directories and other applications for Cisco IP phones that are tied to a database backend.

## CHAPTER 14

Traditional Apps on the Converged Network

When first designed, landline phone service was intended to carry sound signals, and its uses as a carrier of data were years away from realization. It's ironic that the technology that predated the telephone was itself a data transport technology: the telegraph. This device carried encoded messages from terminal to terminal across the 19th-century equivalent of a peer-to-peer network.

A lifetime later, in the 1960s, sound-encoding devices emerged, and, very soon, computers were able to send data, represented as sound, across the telephone network. Those devices were modems, and later fax machines—the descendants of the telegraph. Modems, fax machines, voice mail systems, emergency 911 service, and a slew of other messaging tools evolved around the international telephone network.

Today voice and data networks converge and VoIP begins to replace Bell's brainchild. IP telephony has the same fundamental goal as legacy telephony: facilitate human interaction at a distance. But, since IP telephony goes about this goal differently, not all of the specialized devices that evolved around the old system work with the new one. Fax machines, modems, and voice mail systems aren't necessarily compatible with VoIP, because they grew into a mold that was shaped by the *old* network.

In this chapter, we'll cover some of the great legacy technologies we've come to rely on and discuss ways of migrating their functionality to the converged network.

Fax and Modems

Fax machines and modems encode digital data into analog sound signals for transport across the PSTN. At the local CO, these analog sound signals are digitized into PCM digital signals at 64 kbps. After being transported over the network, the CO at the opposite end of the link decodes the PCM signal and plays it back for the analog receiver, whose job it is to reassemble it into a facsimile of the original digital data. Since the maximum rate of transmission on a POTS line is 64 kbps, faxes and modems can't transmit data very quickly (one twenty-fourth the speed of a T1). Even with compression techniques, which introduce signaling overhead, most modems will never transmit data faster than about 54 kbps. They are limited by the digital resolution of the sound pathway they use.

Sending a fax through an encoded audio channel such as those provided by VoIP and the PSTN is called in-band faxing.

The challenge posed by VoIP is this: codecs like G.729A distort the sound signal by compressing it using lossy vocoder algorithms such as CELP. When a modem or fax's analog sound signal is encoded and decoded using vocoders, it becomes distorted, such that the device on the receiving end of the transmission is receiving a different analog signal than the one that was sent. The side effect of compressed VoIP codecs is that faxes and modems simply don't work. So devices that rely on modems, such as some burglary alarm systems, TiVo consoles, and maybe that old-fashioned Amiga 1000 computer, have a hard time getting along in a VoIP network. (Series 2 TiVo devices can work using the Internet instead of a modem—check out *http://www.tivo.com/adapters*.)

Terminology Recap

POTS
: Plain old telephone service, the basic analog phone line most folks have in their homes.

Fax
: Facsimile, a method of sending a copy of a paper document over an analog telephone line.

Modem
: Modulator/demodulator, a device that connects computer systems or fax machines by transmitting signals using analog signals.

CO
: Central office, the phone company's switching center that provides POTS service to the surrounding area.

PCM
: Pulse ccde modulation, a way of storing an audio signal as a bit stream.

CELP
: Code-excited linear prediction, a compression codec used in advanced VoIP codecs. CELP reduces the number of bits necessary for successful transmission of the sound, but it does so by degrading the audio signal somewhat.

There are a few ways to tackle this problem. The first and most obvious is to avoid lossy codecs and use only non-lossy codecs like G.711 and G.722. It won't distort the analog modem signals as long as jitter is under control. But avoiding lossy codecs may not be possible, especially on bandwidth-starved WANs.

Fortunately, the ITU has two recommendations to aid in migration of fax to VoIP. Both are discussed in this chapter.

There is no real answer to the matter of modems, however. The ITU's V.150.1 standard describes how to relay modem signals over a converged network, but no open reference implementations exist yet. One could argue that it's a bit arcane to use a modem in a converged network, since any data sent by the modem could be sent hundreds of times faster using Ethernet. In other words, if you have an Ethernet connection, use that instead of the modem.

 Thanks to the insistence of the U.S. Department of Defense, Version 4.1 of CallManager, Cisco's venerable softPBX, does indeed support modem pass-through using V.150.1.

T.30, T.37, and T.38

The ITU's T.30 recommendation describes how fax devices should work. Just about all legacy fax machines (and fax software) are T.30-compliant.

The T.38 protocol is the ITU's recommendation for sending faxes over data networks. T.38 software runs on a server that can perform T.30 fax signaling in order to send and receive fax transmissions over analog lines to fax machines. The server encodes the fax signals into packets and sends them to a T.38 server that can decode the fax signals on behalf of a receiving fax machine. This circumvents reliance on in-band faxing and increases the efficiency of fax transmission over the network. T.38 has a few nicknames, including Group 3 faxing and FoIP (Fax over IP).

The T.37 protocol is another ITU recommendation. It also allows faxes to be sent over the IP network, but T.37 doesn't package fax signals into its own packets. Instead, it creates SMTP mail messages using MIME-encoded TIFF (Tagged Image File Format) files. These messages can be sent directly to an SMTP mail server for delivery to email users, or they can be sent to another T.37 server for delivery to a fax machine. T.37 has a few advantages over T.38, among them its ease of integration with standards-based email. T.37 is a store-and-forward protocol, meaning it can hold onto messages if they aren't deliverable at the moment. T.38 has no such capability.

ATAs and fax/modem support

To support a fax machine on the VoIP network, you might think all you have to do is plug it into an analog telephone adapter (ATA) and configure that adapter to only use G.711. But this won't always work. Other VoIP devices on the network may alter the sound encoding somewhere along the fax machine's call path, possibly crippling it.

There are three ways to ensure faxes can traverse your VoIP network:

- Guarantee that G.711, and only G.711, will be used for all possible fax call paths, and eliminate all jitter.
- Implement T.38 servers at locations with fax machines.
- Get a T.37-compatible fax machine or connect fax machines to T.37 servers.

There's only one way to positively ensure modem traffic can traverse your VoIP network:

- Guarantee that G.711, and only G.711, will be used for all possible modem call paths, and eliminate all jitter.

To make fax and modem communications over VoIP more reliable, decrease the baud rate of the fax or modem connection to 9600 or lower. While this will slow the connection down, it will allow you to overcome minor jitters, which will at least keep the connection intact.

Project 14.1. Turn Your Linux Box into a Fax Machine

What you need for this project:

- Asterisk running on Linux
- A POTS line connected to a Zaptel-compatible interface card

Asterisk offers a built-in fax-detection mechanism. This allows you to handle faxes that are sent to your Asterisk PBX on a POTS line. Asterisk's Answer( ) command triggers the fax detection. If an incoming fax is detected, Asterisk automatically transfers the call to the special extension called fax, if it exists.

In order to use this special extension, you'll need to compile and install the SpanDSP package. Download the latest version from *ftp.opencall.org*, and unzip the file into */usr/src/spandsp*. To compile it, issue these commands:

```
# ./configure --prefix=/usr
# make; make install
# cp app_rxfax.c /usr/src/asterisk/apps
# cp app_txfax.c /usr/src/asterisk/apps
# cp Makefile.patch /usr/src/asterisk/apps
# cd /usr/src/asterisk/apps
# patch <Makefile.patch
```

If you're worried about the security concerns associated with compiling as root, you can use a non-root account to compile SpanDSP.

These commands compile the SpanDSP package, which provides a source code patch for Asterisk. Thus, you'll need to recompile Asterisk now:

```
# cd ../asterisk
# make clean ; make install
```

The next step is to enable fax detection on the Zaptel channel you want to use for faxing. This doesn't stop the channel from being used for normal voice calls; it just enables the channel to discern fax calls from normal calls. To enable this function, be sure that the channel's section in */etc/asterisk/zapata.conf* has this entry:

```
faxdetect=both
```

The valid parameters for the faxdetect option are incoming, outgoing, both, and no. By default, fax detection is disabled.

Receiving faxes

Now, consider the following snippet from a dial-plan:

```
[incoming-local]
exten => s,1,Answer
exten => s,2,Dial(SIP/202,45,rm)
exten => s,3,voicemail(202)

exten => fax,1,SetVar(TIFFILE=/var/spool/faxes/thisfax.tif)
exten => fax,2,rxfax(${TIFFILE})
```

In this context, the Answer( ) command triggers fax detection. If there's no fax, the dial-plan calls for a call to SIP peer 202. If there is a fax, the fax extension takes over, saving the fax image into a TIFF file. Another script could then process the file however you see fit, perhaps printing it immediately, like this:

```
exten => fax,1,SetVar(TIFFILE=/var/spool/faxes/thisfax.tif)
exten => fax,2,rxfax(${TIFFILE})
; dump the fax file to the default printer and remove the fax file
exten => fax,3,System('tiff2ps ${TIFFILE} | lpr')
exten => fax,4,System('rm ${TIFFILE}')
```

 Tiff2PS is a utility provided in the *libtiff* distribution, a library for dealing with TIFF files. It's a standard part of many Linux distributions, Red Hat included.

Sending faxes

Receiving faxes with Asterisk is quite a bit easier than sending them. This is because, when receiving them, the work of scanning them into a digital form is already done. This is the part neither Asterisk nor SpanDSP addresses. These packages *can* very easily fax a TIFF file. So it's up to you to get that TIFF file in a path where SpanDSP can grab it.

This can happen any number of ways. You could create a simple web interface that allows you to upload TIFF files to the server, or if you've got the right software, you could just scan them directly using the Linux machine. I don't recommend either of these approaches, however, because neither of them provides a straightforward way of telling Asterisk where to send the fax. Fortunately, the Hylafax package provides a solution.

Project 14.2. Build an Inbound Fax-to-Email Gateway

What you need for this project:

- Asterisk running on Linux
- SpanDSP
- Sven Slezak's LDAPGet module
- A T1 interface with an active PRI attached or multiple X100P cards with POTS lines attached

In Project 14.1, we built a configuration to direct all incoming faxes from Zaptel channels to a file, which, in turn, we could have automatically printed. But if the server were working on behalf of many possible fax recipients, we would have to rely on the incoming fax's cover sheet to know which recipient it's destined for. Worse still, we'd have to go to the printer, pick up the fax, and hand-deliver it to the correct person.

There's a better way, of course: email. It's just as easy to email that TIFF file to somebody's inbox as it is to print it:

```
exten => fax,1,SetVar(TIFFILE=/var/spool/faxes/thisfax=${CALLERIDNUM}.tif)
exten => fax,2,rxfax(${TIFFILE})
; email the fax file to the receptionist and then delete it
exten => fax,3,SetVar(EMAIL=receptionist@oreilly.com)
exten => fax,3,System('mewencode -e ${TIFFILE} | mail -s fax ${EMAIL}')
exten => fax,4,System('rm ${TIFFILE}')
```

This configuration receives the fax, MIME-encodes it using the mewencode command (a standard part of most Linux distributions), and emails it to the email address stored in the ${EMAIL} variable. This is a catchall solution—it sends every fax that's received to the same recipient, who can then forward (and screen if necessary) to the appropriate person based on the content of each fax message.

Automatic fax routing

In order to have the Linux server automatically route each fax to the right recipient (instead of a certain email user doing it), we must have a way of associating each fax with the correct recipient. We'll have to associate a certain line (or a certain DID) with each user, so that, whenever a fax is received on that line (or DID), we'll know where to route it. Each phone line's number, or each DID number if we're using a PRI, will become a single user's fax number.

Fax routing with DIDs

In order to associate a DID with a particular user's email address, we can use an LDAP inquiry. An LDAP client library for Linux, *openldap*, provides applications the ability to access LDAP servers and perform such inquiries. Your LDAP server may be an Exchange or Lotus Domino server where directory information is stored. The Red Hat distribution includes *openldap* and its developer packages.

But you'll need more than just the LDAP client library. You'll also need an actual LDAP client for Asterisk, like Sven Slezak's LDAPGet package. Download it from *http://www.mezzo.net/asterisk* and unzip it into the Asterisk source directory, */usr/src/asterisk*.

Next, as root, copy the *app_ldap.so* file into */usr/src/asterisk/apps*. Then, use a text editor to add *app_ldap.so* to the list of applications that begin with APPS= in */usr/src/asterisk/apps/Makefile*. While you've got the make file open, also add the following rule just below above the app_voicemail.so line:

```
app_ldap.so : app_ldap.o
    $(CC) $(SOLINK) -o $@ $< -llber -lldap
```

Now, LDAPGet is ready to be compiled. Save your changes in the text editor and exit back to the shell, where you'll issue these commands to compile the package:

```
# cd /usr/src/asterisk
# make; make install
```

If Asterisk isn't currently running, start it. Then, go to the Asterisk command line load the LDAPGet module (alternatively, you could just restart Asterisk):

```
pbx*CLI> load app_ldap.so
pbx*CLI> show application LDAPget
```

The show application command will confirm the module is installed and loaded by showing you a brief description of the LDAPGet dial-plan command. Now, you can set up the LDAP inquiry your dial-plan will use to get email addresses based on the DID provided by ${EXTEN}. To set this query up, open */etc/asterisk/ldap.conf*. It may not exist yet, since you've only just compiled the LDAP module. Create an entry like this in *ldap.conf*:

```
[mailfromdid]
host = ldap.oreilly.com
user = cn=root,ou=People,o=oreilly.com
pass = jarsflood
base = ou=Addressbook,o=oreilly.com
filter = (&(objectClass=person)(|(fax=%s)))
attribute = email
convert = UTF-8,ISO-8859-1
```

This configuration will cause the LDAP inquiry to ldap.oreilly.com asking for an object of the person class with the attribute fax equal to the value of the %s token (which will be replaced with the DID at runtime). The attribute setting tells the inquiry which attribute from the object to return as a value to the dial-plan's variable. This may seem confusing right now, but it should be clearer once you see how the LDAPGet command is used in the dial-plan.

In the context where your incoming PSTN calls begin (specified in *zapata.conf*), you can capture the DID from the ${EXTEN} variable, and use it to supply an argument to an LDAP inquiry. If the inquiry is successful, Asterisk's LDAP client will return the email address to which the fax number (i.e., DID) is associated, as in this snippet of *extensions.conf*:

```
[incoming-pstn]
exten => s,1,SetVar(DID=${EXTEN})
exten => s,2,Answer
exten => s,3,Ringing
; allow 4 seconds for the fax detection
exten => s,4,Wait(4)
; if no fax, send this call to be handled elsewhere
exten => s,3,GoTo(incoming-voice)

; here's the fax handling extension, which sends the call to the
; 'inc-fax' context
exten => fax,1,Goto(inc-fax,1,1)

[inc-fax]
exten => s,1,SetVar(TIFFILE=/var/spool/faxes/${DID}.tif)
; The 'mailfromdid' LDAP inquiry is defined in Asterisk's ldap.conf file.
exten => s,2,LDAPGet(EMAIL=mailfromdid/${DID})
; If the LDAP inquiry succeeds, priority will be 2+1.
exten => s,3,rxfax(${TIFFILE})
exten => s,4,GoTo(105)
; If the LDAP lookup fails, priority will be 2+101.
exten => s,103,SetVar(EMAIL=receptionist@oreilly.com)
exten => s,104,rxfax(${TIFFILE})
; Now, email the fax file to whichever email address was decided upon.
exten => s,105,System('uuencode ${TIFFILE} uuenc | mail -s fax ${EMAIL}')
exten => s,106,System('rm ${TIFFILE}')

[incoming-voice]
; non-fax calls are handled here
```

The sum of all this compiling and config-tuning is that different email recipients now have assigned fax numbers on the PRI (or assigned POTS lines for their exclusive use as inbound fax lines). When you send a fax to Todd's fax number, Todd receives the email. When you send it to Susie's, Susie receives the email, and so on. Of course, it's up to you to populate your LDAP server with the right information and to make sure the inquiry config in *ldap.conf* matches your LDAP server's schema.

Don't have an LDAP server? You could use Asterisk's built-in database commands to resolve DIDs to email addresses, instead. Chapter 17's dial-plan command reference covers these commands.

For those who prefer PDF files

In the last project, we used the `tiff2ps` command to create a printable version of the fax, but with a few extra steps, we can turn it into a PDF file, too. PDF may be preferable to TIFF when using email as we are in this project. Consider the following dial-plan changes to the [inc-fax] context:

```
exten => s,105,System('tiff2ps -2eaz -w 8.5 -h 11 ${TIFFILE}| ps2pdf \
>${TIFFILE}.ps')
exten => s,106,System('uuencode ${TIFFILE}.ps uuenc | mail -s fax ${EMAIL}')
exten => s,107,System('rm -f ${TIFFILE}*')
```

Now, instead of just encoding the TIFF file and emailing it, the file is converted to a PostScript file, then to a PDF file, before being uuencoded and emailed to the appropriate recipient.

Simulating T.37

This project can be taken one step further, to crudely simulate T.37—that is, using SMTP to trunk faxes back and forth between email systems that have analog fax endpoints connected to them. In a scenario with three offices and multiple fax machines at each office, you could build a fax bank that routes the faxes to the right machines, anywhere within those three offices. Each participating server would be configured to package incoming faxes into mail messages, as before, but they'd also need to be able to automatically decode incoming email, save the TIFF file attachment to a temporary directory, and fax it to the correct fax machine using the SpanDSP txfax( ) dial-plan command.

FoIP Solutions

Cisco AVVID

Cisco media gateway routers can host analog telephone ports. Cisco implements T.37 and T.38 protocols on their analog telephone interface cards, so that, once you provision T.37 or T.38, you can connect a fax machine and have reliable faxing over IP. IOS commands are provided for administration of fax functionality on firmware loads for the Cisco 2600XM, 2800, 3700, and 3800 series routers. More information is available by searching Cisco's web site at *http://www.cisco.com*.

Avaya Media Servers

The Communication Manager software platform for Avaya's Media Server chassis provides support for fax machines connected to analog ports. Avaya supports both T.37 and T.38, though its T.38 support is older and more refined. Only Media Servers with analog ports can host fax machines. More information is available by searching Avaya's web site at *http://www.avaya.com*.

Other solutions

Nortel, Alcatel, and other, smaller vendors offer T.38 support. Chapter 16 contains a listing of web sites for commercial PBX vendors so you can investigate commercial solutions further.

Commercial soft fax solutions

Since fax is a largely legacy technology that has been replaced, for better or worse, by email, you may want to consider the option of dropping the use of fax machines and switching to a 100% soft environment for faxing. Commercial software products like Interstar's Lightingfax and ACCPAC's faxServe can turn a Windows server into a fax concentrator that can route incoming and outgoing faxes to and from users' desktops. For a Macintosh-compatible commercial solution, try 4-sight fax from Soft Solutions.

Hylafax

Without a lot of hacking, Asterisk just doesn't make a good day-to-day, occasional-use fax server for outbound fax transmittals. There are better solutions to this need already. One of them is Hylafax, a freely available fax server for Linux and BSD operating systems. Hylafax can use standard fax/modems, which also makes it cheaper to implement than Asterisk with (comparatively expensive) Digium voice cards. Hylafax can be obtained from *http://www.hylafax.org*.

Hosted Fax

If all that dial-plan configuration seems a bit over the top, consider using a hosted fax service provider instead. Some service providers host fax servers with lots of PRI circuits, allowing subscribers to send and receive faxes from a desktop computer using the Internet. This is a stress-free alternative to implementing FoIP. Here is a partial list of hosted fax service providers:

- BroadFax (*http://www.broadfax.com*)
- DataOnCall UniFAX (*http://www.dataoncall.com*)
- eFax (*http://www.efax.com*)
- Extreme FAX (*http://www.extremefax.com*)

- InnoPort (*http://www.innoport.com*)
- MyFax (*http://www.myfax.com*)

Fire and Burglary Systems

Many fire detection, sprinkler, and burglary alarm systems require the use of one or more POTS lines. This line is used to notify the monitoring company in the event of an alarm trip or a sprinkler head break. A modem inside the alarm system's CPU dials a computer at the monitoring station, and the alarm dispatcher's computer tells her information about the alarm trip.

These systems use an analog data modem that won't work reliably in a pure VoIP environment (see the earlier section, "Fax and Modems"). There are a few ways to overcome this problem. You could replace the system with a newer, Internet-aware system. Some alarm installers offer CPUs with Ethernet interfaces that can replace the modem for notification. But, if necessary, you may have to keep a POTS line (or two) around until you can upgrade.

Surveillance Systems and Videoconferencing

As you move to a converged network, you may want to factor video into your plans. Just as the converged network replaces the legacy voice network, it can also replace the legacy surveillance and videoconferencing networks.

Camera Surveillance

Most traditional surveillance networks consist of meters and meters of expensive, inflexible coaxial cable reaching from a low-resolution monochrome monitoring station out to a handful of analog surveillance cameras. The videocassettes used to record surveillance videos are bulky and prone to failure, and the quality of most surveillance video is poor.

With digital, IP-enabled surveillance cameras, a converged network, and good centralized surveillance software, you can eliminate all those shortcomings. Indeed, as you prepare your network for VoIP by implementing high-speed switching and quality of service, you're also preparing your network for video. Contrary to popular belief, it's surveillance, not videoconferencing, that ranks as the number one application for video in business.

Like VoIP endpoints, video surveillance endpoints (cameras) are fully self-contained, have a microprocessor, an Ethernet interface, and a variety of features. Some cameras have a web-based interface that gives you pan, tilt, and zoom control so you can point a stationary camera, zoom in and out, and so forth. More sophisticated cameras add infrared night vision, multicast streaming, and other cool features.

 To check out some surveillance cameras that are inadvertently available for your perusal on the Internet, trying Googling:

```
inurl:"ViewerFrame?Mode="
```

Video can eat up a lot of network bandwidth, 5 to 25 times that of a typical G.711 phone call, depending on the video codecs, resolution, and color depth employed. Some developers, like DIVR Systems (*http://www.divrsystems.com*) offer remote video surveillance solutions that ride on the same network as VoIP. In order to keep from having video swamp the network and break down phone call quality, your video surveillance apps need to play nice with QoS. This means making sure that surveillance traffic is treated with a lower priority than phone call traffic.

Videoconferencing

Video meetings are closer to telephony than video surveillance, but their requirements are a bit different. First off, while video surveillance needs reliable delivery to ensure that every frame of video is recorded, videoconferencing is more like a phone call; if a frame gets dropped here and there, no problem. So reliable packet delivery, a la TCP, isn't necessary. Also, the need for multicasting (that is, having many viewers watch a single video source) is more prominent in video conferencing than in surveillance. Finally, videoconference transmissions tend to be bursty and fairly short—under a few hours usually. But surveillance video tends to be steady, and round the clock.

While not necessarily inclusive in VoIP, videoconferencing has implications for the VoIP network, just as surveillance video does—in QoS mainly. But, since videoconferencing is indeed a call-switching application, it is often implemented in, or with, phone systems. Some TSPs even offer videoconferencing service with special, camera-equipped screen phones. Packet8's VideoPhone plan is one such service. It allows Packet8 customers to place VoIP/video calls to other Packet8 customers.

Cisco's AVVID IP/VC solution allows video-equipped H.323 endpoints to participate in videoconferencing. SIPQuest Corp. offers a commercial SIP-based videoconferencing server that runs on Linux. Sony offers a SIP-based videoconferencing solution (Sony PCS) that works with any video-compatible SIP-switching server, such as NEC's Univerge SV7000. Technically, if SDP is used for call setup (as it is in just about every SIP implementation), adding videoconferencing to a call is merely as simple as supplying a SIP video phone on both ends.

As such, it's quite easy to create a videoconferencing solution with Asterisk. Start by enabling H.263 and H.261 video codecs in *sip.conf*. Video softphones you can try include LinPhone for Linux and FreeBSD, Windows (MSN) Messenger for Windows desktops, EyeBeam for Mac OS X and Windows, and iFon for PocketPC. SDP decides the parameters of the video channel during call setup.

Voice Mail and IVR

Voice mail and interactive voice response systems have always been soft-driven, because they embody business logic. They need to be customizable in ways that a traditional phone switch doesn't. Consequently, voice mail and IVR have been at home on the microcomputer server for a lot longer than call switching has been. Early IVR and voice mail software packages ran on Solaris, OS/2, and Windows NT.

The voice mail/IVR server would be connected to the PBX by dedicated trunks, usually a T1 interface or a handful of FXO/FXS links. The PBX would assign those trunks special extensions, and when a caller needed voice mail or IVR function, his station would be bridged with the trunk connected to the voice mail/IVR server. In some cases, especially on very high-end chassis, the voice mail/IVR and PBX would be housed in one unit.

Today, the IVR and PBX systems are both soft-driven and can therefore be maintained at the same place. Dedicated trunks are no longer needed to link PBX to IVR as long as you choose a platform that integrates them on the same server, unless you design it so they're housed separately for load-management or logistical reasons.

Asterisk offers built-in IVR functions, which are issued in *extensions.conf* or through the Asterisk Gateway Interface. Probably the most oft-used IVR application in Asterisk is its remarkably powerful voice mail system, Comedian Mail.

Project 14.3. Build a Custom Voice Mail System

What you need for this project:

- Asterisk

One of Asterisk's most compelling features is its voice mail server. In fact, you could use Asterisk voice mail as a transition piece as you move your current PBX to VoIP. Chances are good that Asterisk offers more voice mail functionality than your current voice mail system, anyway. Once you transitioned your voice mail to Asterisk, you would be only a dial-plan away from having a full call manager online and ready to handle calls.

But since we've covered call management quite a bit already, let's concentrate solely on voice mail. In Asterisk, the voice mail system is configured by the */etc/asterisk/voicemail.conf* file. Like the dial-plan in *extensions.conf*, the voice mail configuration file establishes contexts. Each voice mail box exists within a certain context, allowing logical separation of subscriber groups, say, on a system that hosts voice mail for several different businesses.

Like *extensions.conf*, *voicemail.conf* can use include directives to encompass the content of other files, as long as they contain settings that are valid for the voice mail server—and not some other module. As is the convention elsewhere in Asterisk, the *voicemail.conf* file begins with a [general] section that defines settings that apply to the entire voice server.

In this project, we'll be customizing Asterisk's voice mail config in order to make it email newly recorded voice messages to different peoples' inboxes. We'll also be customizing the content of the voice mail announcement message that is sent when new voice mails are received.

To get started, let's establish a couple of voice mailboxes in *voicemail.conf*:

```
[default]
201 => 201,John Lennon,jl@oreilly.com,,attach=yes|tz=mountain
202 => 202,George Harrison,gh@oreilly.com,,attach=no|saycid=yes
```

Now, take a look at the format that defines the syntax of this config:

```
mailbox => pin,user_name,user_email[,user_pager_email][,option(s)]
```

The alphanumeric *mailbox* must be unique within the voice mail context but can be duplicated elsewhere in the voice mail configuration. The *pin* is the DTMF string that will be required in order for the subscriber to retrieve messages by phone. The *user_name* and *user_email* settings tell the server how to address email-forwarded messages. The *user_pager_email* setting gives the server a secondary email address that's intended for use with email-interfaced paging and cell phone text messaging networks. Finally, the *options* setting is a string of pipe-limited options that are outlined in Table 14-1.

Table 14-1. Asterisk voice mail box options

Option	Description	Possible values
tz	Defines the time zone for this subscriber so that timestamps can be localized	Any time zone configured in */usr/share/zoneinfo*
attach	Specifies whether or not to attach the sound file of the voice mail message when notifying by email	yes; no
callback	Specifies the dial-plan context to which callers are returned after they've left a message for this subscriber	default
dialout	Specifies the dial-plan context in which subscribers exist if they dial an extension from within the voice mail application	default
operator	Specifies whether or not the user can return to the operator extension by dialing 0 (zero) within the voice mail application	yes; no
review	Specifies whether or not callers to this subscriber can review (listen to and approve or rerecord) messages	yes; no
saycid	Specifies whether the caller ID number of the person who left the message should be announced when the message is played by the subscriber	yes; no

Turn on email notification

Enabling notification of new voice mail messages by email is as simple as three steps. First, edit the `servermail` and `fromstring` settings in the [general] section of *voicemail.conf* so that notification messages will come from an appropriate email address like *voicemail@yourcompany.com*. Second, make sure the local mail routing agent (sendmail) is operational on the voice mail server. Third, enter the email address of each subscriber into her mailbox declaration in *voicemail.conf*.

Attach voice mail sound files to email notifications

Adding `attach=yes` to the options string in a mailbox declaration will cause recorded messages to be forwarded (as attachments) to the email addresses associated with that mailbox. There's really only one thing you can customize here, and that's the format in which the recorded sounds are encoded. You have three choices—`wav49`, `wav`, and `gsm`. WAV49 is highly compressed and compatible with most client desktops, so it's the recommended format for emailing. Unless you're doing some further processing with an SMTP agent after the message that requires it is sent, there's no reason to use WAV or GSM.

Customize the email notification message

The template for the notification email is stored in the `${emailbody}` variable, defined in *voicemail.conf* or one of its includes. You just have to define it. Feel free to mix in any of these system variables while you're at it:

VM_NAME
>   The name of the subscriber for whom the message was left

VM_DUR
>   The duration, in seconds, of the message

VM_MSGNUM
>   The number of the message relative to the subscriber's mailbox

VM_MAILBOX
>   The mailbox number of the subscriber

VM_CALLERID
>   The caller ID string for the caller who left the message

VM_DATE
>   The date and time the message was left, relative to the time zone specified in *voicemail.conf* for this subscriber

Consider the following emailbody declaration:

```
emailbody=Hello ${VM_NAME},\n\nThere is a new voice mail message in mailbox number
${VM_MAILBOX}. It was left on ${VM_DATE} from ${VM_CALLERID}. To listen to this
message, dial 1000 from your desk phone or 216-524-0071 from an outside phone.
\n\nThanks,\nThe voicemail System
```

This message would appear more or less as follows, when used to send a voice mail notification:

```
Hello John Lennon,

There is a new voicemail message in mailbox number 201. It was left on 7/3/05 12:42:
42 from (440) 365-1964. To listen to this message, dial 1000 from your desk phone or
216-524-0071 from an outside phone.

Thanks,
The Voicemail System
```

The \n token, common in Unix, starts a new line. If the recorded message were actually attached to the email, we would need to use a different message template, indicating that.

Connect the dial-plan to voice mail

Once the voice mail server is set up the way you want it, connecting it to the dial-plan is a snap. First, establish an extension where subscribers can retrieve their messages and manage their mailboxes. This example attempts to plug in the extension number of the caller in order to forgo a prompt to enter their mailbox number, assuming the extension and mailbox numbers match:

```
; extensions.conf
[default]
...
exten => 1000,1,Voicemailmain(${CALLERIDNUM})
```

If the caller ID doesn't match any mailbox, the subscriber will have a chance to enter his mailbox number. Next, make sure that some extensions trigger the voice mail application:

```
; extensions.conf
[default]
...
exten => 201,1,Dial(SIP/201,30)
exten => 201,2,Voicemail(201)
```

In this case, extension 201 will attempt to ring SIP peer 201 for 30 seconds. If there's no answer, the voice mail greeting for mailbox 201 will be played and the caller will have an opportunity to record a message.

Web access to voice mail

Asterisk includes a Perl CGI script (*vmail.cgi*) that you can use to give your voice mail subscribers access to their mailboxes via the Web. This is great for road warriors who may not have direct telephony access to the phone system while they're out and about. By placing the script in Apache's *cgi-bin* script path and making it executable by the web user, this script works fine for a plain-vanilla install of Asterisk. Figure 14-1 shows what it looks like from a web browser.

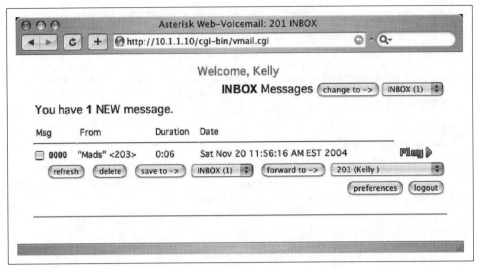

Figure 14-1. Asterisk's vmail.cgi allows web-based access to voice mail

Project 14.4. Create Custom Sounds to Interact with Callers

What you need for this project:

- Asterisk
- A sound recording application (optional)
- SoX (*http://sox.sourceforge.net*)

Interactive voice response is the backbone of voice mail, but there's so much more you can do with it. You could replace pen-and-paper time cards and punch clocks. You could use it to perform billing functions, such as collecting a credit card number and storing it in a database. But in order to make all these things cohesive to the caller (the user), your user interface will have to be customized. This means using, and making, sounds.

Use a prerecorded sound for a simple paging routine

We'll start with a quick little dial-plan tool that allows us to send quick emails. This will allow us to use one of Asterisk's prerecorded sounds. In this example, we'll build a basic numeric pager. It will screen an incoming call, record the caller's phone number, and email it to an SMTP recipient with instructions to "please call" the number.

```
exten => *89,1,Read(PhoneNo|allison7/enter-phone-number10|10)
exten => *89,2,System('echo Please call ${PhoneNo}. | mail user@oreilly.com')
exten => *89,3,Playback(thank-you)
exten => *89,4,Hangup( )
```

You may also notice output like this in Asterisk's verbose console log when the Read( ) command captures the dialed digits, in accordance with the first priority of extension *89:

```
VERBOSE[1122993344]     --User entered '5552345678'
```

The Read( ) command plays the *enter-phone-number.gsm* file and then stores the 10 digits entered in the ${PhoneNo} variable, which is used at the next priority in a shell command (via the System( ) command) that sends a brief email message to *<user>* *@oreilly.com*.

Use your own sounds in IVR

If you can record WAV or AIFF files, you can make custom IVR prompts. These formats must be converted into GSM-encoded files, though, to work with Asterisk. It's better to record the prompts directly into GSM encoding rather than recording in WAV and then encoding to GSM. But not all sound recording applications support GSM out of the box, so you might need to record in one step, and convert to GSM, using the Unix application SoX, in another step.

In Windows, you can use the trusty Windows Sound Record application to create WAV files. On the Mac, you can use a sound recorder like Audiocorder (http://*www. blackcatsystems.com*) to create AIFF files. Either way, once you record the sounds, you'll have to convert them using SoX, the Unix SOund eXchange application.

Converting sounds using SoX

Using SoX for this task couldn't be any easier. This example converts an AIFF file called *announce.aif* into GSM encoding, merely by specifying the file extension (.*wav*) in the output filename:

```
# sox announce.aif announce.gsm
```

SoX can do plenty of cool things. It can change the sampling rates, equalization, length (of time), and encoding of the output, too. In fact, a full-rate, high-fidelity sample (say, 44,000 Hz) wouldn't be a very good idea in an IVR app, so you could trim it down to size by resampling it into 8 kHz:

```
# sox -r8000 44KHzfile.aif 8KHzfile.gsm
```

There also wouldn't be much sense in using stereo samples, so those can be made single-channel sounds:

```
# sox -c1 stereo.aif mono.aif
```

Finally, you could boost the volume of a quiet sample with the -v option, followed by a decimal value that represents an increase in volume if it's greater than 1, or a decrease in volume if it's lower than 1:

```
# sox -v1.5 quiet.gsm louder.gsm
```

Use soxmix to put background music into an announcement

The soxmix application can mix two sound files into a single file. This could be useful for combining background music with a voice recording. The mix of sound levels between the two source files is determined using SoX's "average" algorithm. Before mixing the two files, you can adjust their volume levels, as shown earlier. Here's a simple example of soxmix:

```
# soxmix music.gsm holdmessage.gsm combined.gsm
```

Sounds for Business and Pleasure

"One moment please..." (*one-moment-please.gsm*)

"That party is busy; please hold..." (*busy-please-hold.gsm*)

"Enter an extension..." (*enter-ext-of-person.gsm*)

Also included are a few that are suitable for, well, *nonbusiness* use:

"All your base are belong to us!" (*all-your-base.gsm*)

"This is not the extension you're looking for..." (*jedi-extension-trick.gsm*)

"We're off gambling..." (*gambling-drunk.gsm*)

Record IVR announcements using Asterisk instead of a sound recording app

You can use Asterisk to record GSM files by configuring the dial-plan itself to record your channel into a sound file. This is accomplished using the Monitor( ) command. Consider the following dial-plan snippet:

```
exten => *50,1,SetVar(pathname=/var/lib/asterisk/sounds/)
exten => *50,2,Read(FileName|beep|4)
exten => *50,3,Monitor(wav,${pathname}${FileName}.wav)
exten => *50,4,Read(foo|beep|1)
exten => *50,5,StopMonitor( )
exten => *50,6,System('sox ${pathname}${FileName}.wav-in.wav ${pathname}${FileN$
exten => *50,7,Playback(${FileName})
exten => *50,8,System('rm -f ${pathname}${FileName}.wav-*')
```

This novel little snippet is designed to record your voice when you dial *50. It stores the recording in a WAV file, converts it immediately to GSM using SoX, plays it back for you, and then deletes the WAV file originally used to record the call. To use it, call *50, listen for the beep, and enter a four-digit DTMF string. This will be used for the filename. As soon as you enter four digits, you'll hear a second beep. Now, begin speaking. When finished, dial any digit. The file is automatically saved and converted, and you should hear it play back immediately. Once finished, the unconverted files used to record the call are deleted.

You'll won't hear any DTMF tones in the recording as long as you use a phone with out-of-band signaling, such as SIP Info.

Emergency Dispatch/911

E911, or Enhanced 911, is the common American PSTN's public safety protocol for emergency situations. PSTN subscribers dial 911 and are connected to the fire department or police dispatcher at the closest PSAP (public safety answer point). Between the subscriber's CO and the PSAP are dedicated trunks that carry ALI (automatic location identification) signals from the phone company to the PSAP, so that emergency personnel can know the geographic location of the emergency taking place. This is particularly useful for fires and emergency medical responses.

E911 on the PSTN

The federal government mandates that all ILECs are required to provide 911 signaling service by way of dedicated trunks. In turn, the local public safety authorities collect a tax, usually through phone bills, to fund operation of the system. Public Utilities Commissions enforce the compliance of 911 operations, and the FCC approves standards that are used in the system.

Competitive LECs that don't have switching facilities can use those of the ILECs in order to route 911 calls.

E911 on wireless providers' networks

In addition to LECs, wireless voice carriers are now required to provide Enhanced 911 services with ALI signaling. All the major carriers have chosen the Global Positioning System (GPS) as a method of providing accurate location information to the PSAP during emergencies.

E911 on the converged network

As long as your PSTN trunks come from a LEC and the entire premises served by those trunks are at a single postal address, the default E911 service will work fine. Just make sure that your dial-plan permits 911 dialing. If it doesn't, it could literally have fatal consequences. There are essentially two ways to handle 911 calls from a VoIP network in accordance with FCC and public safety standards:

- Attach a LEC's POTS line or PRI circuit to each office location on the network for local routing of 911 calls, rather than handling them centrally in another location.

- Use a PBX-attached database server that can pinpoint the locations of users who dial 911, and pass that information to the PSTN.

A third option exists when direct PSTN connectivity just isn't possible, such as a VoIP network whose only outlet to the PSTN is a VoIP trunk that has no ALI service attached:

- Use local dial-plan configuration to route calls to "extension 911" out to the local authority via a standard 10-digit E.164 phone number.

911 on large VoIP networks

In situations in which the private premises is very large—say, a college campus, a high-rise building, or several floors of a skyscraper—E911 is more tricky to implement. If you had a VoIP network with two offices that were 50 miles apart and a single PSTN connect point served both offices, public safety workers could be dispatched to the wrong office. This is because the E911 ALI signals would reference the postal address of the connect point, even if the emergency occurred at the other office!

 Routing all 911 calls in an Asterisk context to a POTS trunk is very easy. Just make an extension called 911 and use the Dial() command with an outbound trunk group:

```
exten => 911,1,Dial(Zap/g1/911)
```

Fortunately, there are a few ways around this problem. Probably the easiest (though not necessarily the most economical) way to deal with E911 is to put a LEC's POTS line at each remote location and reserve it for use only when somebody dials 911 at that location. That way, the PSAP would get the call via the local CO closest to that location. You could also use this POTS line as a failover line if the WAN link from the remote office failed; this way, people there could still make PSTN calls.

Another way of dealing with the E911 issue is to use database servers that store location data about each endpoint or TDM gateway. This way, when somebody dials 911, the PBX can pull the location of her endpoint from the database and supply it, via ALI signals, to the municipality. Such a solution would mean being able to send ALI signals through the LEC's network, rather than having them originate there. This is an expensive proposition that makes sense in only the largest environments. If you had 10 floors of a skyscraper, 3 floors of the skyscraper across the street from it, and 1,400 users, then you would benefit from this technology. Cisco makes such a solution. They call it the Cisco Emergency Responder.

Private Switch ALI

If Duane works in Lexington half the time and Lousville the other half of the time, he may tote a WiFi SCCP phone back and forth to both offices. This SCCP phone may register with two different CallManager servers—so who knows who will be reached when Duane dials 911 on it? Aside from statically linking endpoint assets to locations, Cisco Emergency Responder (CER) also provides a simple way for mobile users like Duane to use the same phone to reach the closest PSAP from many geographically disparate locales.

CER relates geography and floor plans, not just to endpoint assets, but to network identifiers that aren't mobile, like MAC hardware addresses. The Cisco Discovery Protocol (CDP) is enlisted to find out where on the network each endpoint is at any given moment, then CER correlates that network location with a physical postal address. Finally, the call is connected to the PSAP using a PRI channel whose ALI address matches that which exists in the CER database. Cisco calls this technique Private Switch ALI, or PSALI.

Mapped ALI

Several companies that hail from the computer-aided dispatching software business offer software tools for Mapped ALI, an ALI solution that integrates geographic information systems (GIS) technologies such as GPS. Telecontrol Corp., Telecommunications Systems Inc. (*http://www.telecomsys.com*), and Contact One (*http://www.contactone.com*) each offer Mapped ALI solutions for service providers and public safety organizations. The approach of Mapped ALI is different from that of PSALI, because it doesn't require any stationary infrastructure like Ethernet switches. This makes it ideal in wireless applications or in hosted PBX situations in which you, as the service provider, don't have control over the customer's network. Not all PSAPs are able to use Mapped ALI technology.

POTS pass-through

Of course, all of these solutions work only when there's power. If the lights go out, so do the IP phones, the PBX, and the CER server, if you have one. Now, you wouldn't put in a phone system without a battery backup, but even a top-quality UPS can fail. In such a situation, you need to be able to reach an emergency dispatcher without the use of any of your VoIP gear—because none of it will be operational.

To solve this problem, connect an analog telephone to a pass-through port on the PBX server. A pass-through port is a connector on a POTS trunk interface that allows a standard phone to be used with the trunk directly, circumventing the PBX—*even when the PBX has no power*. Digium's X100P card has a pass-through port that you can use to make calls on the attached POTS trunks even when the host PC is turned off. So, the POTS pass-through phone, which draws its power from the CO through the POTS trunk, becomes your link to the PSAP during a total blackout power failure.

Dial-around

In networks with no POTS lines or PRIs connected to a LEC, such as a VoIP network that gets its PSTN connectivity by way of VoIP trunks, there's no reliable way of sending ALI signals to the PSAP. Some VoIP TSPs are working on this problem; others aren't planning on offering a solution. This doesn't mean you can't reach an emergency dispatcher with VoIP. It just means that the dispatcher won't know your location.

Still, dialing 911 won't work with most TSPs. You'll have to reprogram your on-premises VoIP gateway or IP phone to "translate" 9-1-1 into the full E.164 telephone number of the PSAP. To make matters worse, it's not always easy to find out what this number is. With the ubiquity of 911, these numbers aren't generally publicized. You may be able to get only the local fire department's or police department's administrative phone number—and in an emergency, this *isn't* who you're going to want on the phone. Do the research and find out how to get through to the emergency dispatcher (and not some receptionist) via a 10-digit phone number. Then, you can program your VoIP devices to connect 911 calls to that number. An example of such a configuration is given in Project 14.5.

Project 14.5. Use VoIP Dial-Around to Connect 911 Calls

What you need for this project:

- Asterisk
- VoIP trunk service from an Asterisk-friendly TSP like VoicePulse

 This is very much a hack! The FCC and your local public safety offices would much prefer you to use a solution that supports ALI, like those discussed earlier. But if you absolutely cannot support ALI by providing a POTS line or PRI, this project shows you how to route 911 calls using only a VoIP TSP and Asterisk.

If you use a service such as VoicePulse Open Access, then you are permitted to have a single SIP call ongoing at any time. If your VoIP trunk has this restriction, then dialing 911 from one phone while that trunk is in use won't result in a connection to the emergency dispatcher. Instead, you'll just get a busy signal. That wouldn't be good.

Fortunately, Asterisk allows you to "hijack" channels in priority situations such as a 911 call. The dial-plan command that does the job is SoftHangup( ). You supply it with the channel name, and it hangs up the channel so a call can be originated on it. So, a very rough hack to hang up the line and grab it to make a 911 call would be as follows:

```
exten => 911,1,SetVar(LocalPSAP=440-361-9000)
exten => 911,2,SoftHangup(SIP/VOICEPULSE)
exten => 911,3,Dial(SIP/VOICEPULSE/${LocalPSAP})
```

The second priority in this example hangs up the channel, while the third priority places a call to the local public safety answering point, whose phone number has been stored in ${LocalPSAP} with a SetVar command in priority 1.

This is fine when you have only one SIP trunk channel at your disposal for 911 calls, but if you have more than one, you wouldn't want to automatically hang up a channel based on some proprietary setting in the dial-plan. The ChanIsAvail() command will help us determine which of several SIP trunk channels can be used to place the call—kind of like an outbound hunt group. Consider this dial-plan snippet:

```
exten => 911,1,ChanIsAvail(SIP/VOICEPULSE1&SIP/VOICEPULSE2)
exten => 911,2,SetVar(LocalPSAP=440-361-9000)
exten => 911,3,Dial(${AVAILCHAN}/${LocalPSAP})
exten => 911,4,Hangup( )
exten => 911,102,SoftHangup(SIP/VOICEPULSE1)
exten => 911,103,Dial(SIP/VOICEPULSE1/${LocalPSAP})
```

In this example, priority 1 checks to see if either of the two named SIP channels is available (these correspond to SIP peers in *sip.conf*). If one is available, its channel name is stored in ${AVAILCHAN} and processing continues at the next priority, 2.

Priority 2 sets the phone number of the public safety dispatcher as in the previous example. Priority 3 dials that number to connect the 911 call on the available SIP channel.

If no channels are available in priority 1, the priority becomes 1+101, or 102. Since no channels are available, priority 102 hangs up a channel arbitrarily so priority 103 can place the emergency call.

Administrator Tools

In legacy telephony, the tools of the system administrator ranged from a primitive serial terminal all the way up to a dedicated desktop software application that allowed administrative control over the phone system. Traditional PBX CPUs, such as those made by old-schoolers Isoetec, AT&T, and Executone, used a 2,400-baud serial connection to an RS-232 port for management of channels, trunks, and subscribers.

Later model phone systems offered dedicated Windows-based software for managing these things. Most recently, the administrator tools have become web-based. Administering phone systems has become easier and easier, even as PBX systems have grown increasingly feature rich. Commercial softPBX platforms extend telephony administration further than ever before by using desktop and web-based applications with CTI to give administrators point-and-click power.

Asterisk's administrative interface

Asterisk doesn't come with any administrative applications other than the (rather limited) Astman. As a result, if you want to administer Asterisk, you're going to be making a lot of visits to the command line and a host of text-based config files, just as we've done in many of the projects herein.

This makes Asterisk administration more akin to an old-school, fridge-sized PBX than to a next-generation VoIP telephony platform. Fortunately, some excellent third-party solutions exist to fill the void. Some are simple command-line tools, while others are XML applications that run on softphones. The are even full-blown, turn-key web-administration solutions for Asterisk.

AMP

Probably the most ambitious one of these is the Asterisk Management Portal, or AMP. This package is a set of Perl and PHP scripts that not only automate administration of the Asterisk PBX via a web interface, but truly transform it into a turnkey, entry-level PBX product suitable for deployment to offices with a single inbound trunk group. When you load AMP onto an Asterisk PBX system, it uses MySQL to store the system configuration and dial-plan, and a group of custom Perl scripts to mirror that stored configuration with the active configuration of Asterisk.

 Even though Asterisk no longer supports MySQL for CDR, because of license conflicts, AMP still uses MySQL to store its dial-plan

AMP turns the Asterisk PBX into a basic, small-office phone system that supports any number of SIP or IAX phones and any number of Zaptel channels. The Zaptel channels are all considered "inbound" and the SIP/IAX phones are all considered private extensions. Extensions can be grouped together in ring groups very easily using the web interface, as can simple IVR call flows and call-parking zones. You can download AMP from *http://sourceforge.net/projects/amportal*.

AM

Asterisk Manager (not Astman and not Asterisk Manager API, which are different pieces of software) provides a simple, Perl-driven web interface for configuring Asterisk's dial-plan and SIP peer configurations (see Figure 14-2). Intended to deal with adds, moves, and changes of private extensions, AM gives you more flexibility in programming the dial-plan than AMP. But AM is more of a convenience tool for an administrator who's already familiar with Asterisk than it is a total turnkey solution. You can get AM from *http://astman.sourceforge.net*.

AstConsole and Asterisk setup assistants

These packages provide those running Asterisk on Mac OS X the ability to maintain SIP extensions and monitor the PBX in a friendly Aqua interface. AstConsole can also connect (via Asterisk Manager API) to remote Asterisk servers to manage and monitor.

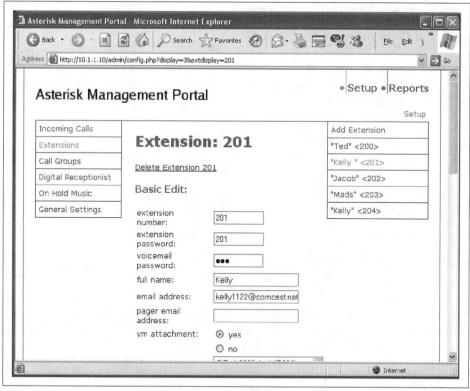

Figure 14-2. The Asterisk Management Portal web interface

The VoicePulse Connect Assistant provides Mac system operators with an easy GUI for configuring trunk connections to VoicePulse's long-distance termination service. These software packages can be downloaded from *http://www.sunrise-tel.com/*.

Open H.323 administrative GUIs

A number of GUI add-ons for Open H.323 exist. A simple Java-based gatekeeper status monitoring tool that allows you to disconnect registered users is available from *http://unix.freshmeat.net/projects/gkgui*. For the GnuGK, there's a Windows-based console snap-in that allows remote observation and control of the gate-keeper. It's called GnuGK Control Center, and you can download it from *http://www.gnugk-cc.com*.

Other administrator GUI tools

For more sources of tasty, GUI admin goodies, take a look at the supplier references in Chapter 16.

Key Issues: Traditional Apps on the Converged Network

- VoIP protocols and standards don't always make turnkey provisions for "old reliable" telephony applications like fax, modems, E911, voice mail, and administrator interface software.

- T.30 faxes and modems don't work over non-real-time voice pathways such as VoIP networks, unless a lossless codec is used and no jitter occurs.

- Lower modem baud rates are more likely to be successful if you inexplicably decide to run modems over VoIP anyway.

- Two ITU standards define your options for implementing Fax over IP, or FoIP. They are T.37 and T.38.

- T.37 is the store-and-forward, email-based Fax-over-IP solution.

- T.38 is the real-time, transport-inclusive FoIP solution.

- The SpanDSP module package for Asterisk gives you the ability to build fax servers that use Digium's Zaptel TDM hardware.

- Hylafax is a freely available (and more cost-effective than Asterisk with SpanDSP) package for building Linux-based fax servers. It uses cheap, abundant fax/modem cards instead of Digium cards.

- For small to medium businesses, hosted fax services can be a money saver compared to supporting an in-house fax server.

- Most fire and burglary systems have a modem that requires a POTS or Centrex line. For this reason, you're pretty much stuck supporting a legacy trunk until more security systems support IP networking for central office monitoring.

- IP-enabled video surveillance will eventually replace many analog camera systems. If you're thinking about moving to IP-based surveillance (or heavy use of videoconferencing), consider the impact of digital video on your converged network as you plan your transition to VoIP.

- Asterisk voice mail can perform email notification and forward sound attachments of the recorded messages to email recipients. Asterisk also supports web-based voice mail retrieval.

- It's very easy to create custom IVR greetings for Asterisk call flows using a basic sound recorder and the Unix-standard SoX sound conversion application.

- Alternatively, you could use the Asterisk Monitor( ) command to transform your phone into a voice recorder for IVR greetings.

- soxmix is used to merge two sounds together, such as background music and a foreground voice greeting.

- Emergency dispatch/911 services are not handled by VoIP protocols. You must build an adequate solution to handle emergency calls.
- POTS lines and PRI channels have built-in automatic location ID (ALI) signals that can alert the public safety dispatcher to a caller's location, but VoIP trunks usually don't.
- There are a number of open source administrative user-interface tools for Asterisk, including Asterisk Management Portal (AMP).

## CHAPTER 15

What Can Go Wrong?

As system designers, integrators, and geeks, we do our homework to become masters in a subject fully before we go headfirst, implementing it in a production environment, right? Well, we aspire to, anyway. But experience says you can never know a hundred percent of what there is to know about a subject... particularly with Voice over IP, which is still changing and evolving. So, once you've equipped yourself with the book knowledge, you hit the field. It's only then that you'll get a chance to discover whether your VoIP project plan has any remaining flaws. But, before you take that leap of faith, there are a few things you should know about: What can go wrong when implementing VoIP in the enterprise?

Common Problem Situations

The people you call complain about echo

Generally, echo is at its worst when end-to-end latency is high. If end-to-end latency is below 150 ms, echo should be nearly imperceptible. Remove echo by removing latency. Remember that using bigger packet sizes, which are often used with low-bandwidth codecs, can increase latency. If capacity is stopping you from removing latency, increase the capacity on the links that are causing the latency. Steer away from frame-relay and VPN if the link is critical—these technologies provide the slowest links.

The phone rings, but callers cannot hear you

This problem can occur when a NAT firewall exists between the caller and the recipient.

When placing a call using a SIP phone, your phone establishes the call after determining the recipient's socket through DNS and your (or your provider's) SIP registrar. The local socket used to communicate with the recipient is the local IP address and port used by your phone's RTP agent: i.e., your phone's IP address and RTP port. Consider Figure 15-1.

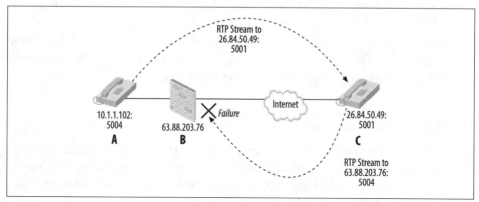

Figure 15-1. Since voice data is transported using a two-way protocol, a NAT firewall causes call setup to break

In Figure 15-1, phone A is able to send audio data to phone C, but phone C is unable to send audio data to phone A because the firewall (B) doesn't know what to do with the RTP packets coming from phone C. This is a problem created by NAT (network address translation), which cannot keep track of connectionless applications that use more than one socket pair, like a two-way phone conversation.

To solve this problem, add a SIP proxy server between phone A and firewall B. Configure the SIP phone so that it places all calls through the SIP proxy. The SIP proxy knows how to handle the RTP data sent to, and received from, many SIP endpoints simultaneously. In this instance, the SIP proxy would need to reside on a DMZ between A and B, so that it can have a publicly routable IP address. The SIP proxy could also reside on the same host as the firewall, if public IP addresses aren't abundant.

Another way to approach the NAT problem is to use a STUN (Simple Traversal of UDP) NAT server. This server assists endpoint devices in figuring out what sockets to use in signaling a VoIP call setup, so that UDP NAT traversal can occur without a DMZ. STUN is described in RFC 3489. More solutions to the NAT problem are presented in Chapter 13.

SIP registrations don't work through a firewall

Like the problem shown in Figure 15-1, NAT firewalls are prone to breaking two-way packet conversations. SIP is no exception. When a privately addressed SIP endpoint (or a voice server registering as SIP client) registers to a public SIP registrar, it writes its private address into the SIP REGISTER method it sends. This is the address used by the SIP registrar to send its 200 or 400 response, depending upon whether or not the registration attempt was successful. Since private addresses aren't routable on the public Internet, the response never gets returned to the registering endpoint, which interprets a timeout. Registration fails and the SIP endpoint can't make any phone calls.

The solution to this problem is also to use a SIP proxy that runs parallel with the NAT firewall. But the trick is that the SIP proxy must have a public address so that the remote SIP registrar's responses have somewhere to go.

The IP phone can't make any calls

Assuming the phone's IP configuration is correct, this problem normally occurs when the phone is unable to register or log on to the SIP registrar or H.323 gatekeeper server. Without an outlet to place calls, which is what these servers provide, the IP phone is unable to call anybody.

In SIP setups, this condition results in a 403 (and sometimes 401) SIP response, indicating registration failed. If the SIP setup uses a proxy, then response 407 occurs. Either way, the authentication of the IP phone has failed, so it can't place calls. (Chapter 11 demonstrated a failed SIP registration using Ethereal.)

To resolve the issue, be certain the phone's user ID and password match those stored in the registrar or softPBX. If MD5 is used, be sure the secret key matches as well. Since some phones don't support MD5, try removing the MD5 requirement for this phone to see if it can register without it.

Past a certain number of simultaneous calls, quality breaks down or calls are disconnected

Levels of utilization that result in a performance loss, or *performance limits*, are a result of improper planning or provisioning. This can mean a number of things:

- The RAM and processor power of servers along the call path may be too low to handle a high number of simultaneous calls. There's no hard and fast rule for provisioning these. But the test lab is the best place to establish the spec for your servers.
- The bandwidth availability on a data link in the call path, such as an Ethernet segment or frame-relay virtual circuit may not be great enough. An old-fashioned calculator will suffice for bandwidth projections
- The wrong type of codecs are being used, resulting in unnecessary processing load. For example, running G.729A over Ethernet isn't really necessary. (See the later section, "Calls across a wide area call path have dropouts in the audio.")

Pay attention to the obvious...

While writing and editing of this book, my editor and I were communicating by email about a VoIP TSP service. The editor was about to try it and was hooking up the shiny new ATA the TSP had just shipped him. He plugged his analog phone into the ATA, plugged in his Ethernet cable, and powered the ATA up. After a few minutes, he lifted the receiver, hoping to hear the dial-tone, but instead all he heard was silence.

So he looked over the lights on the front of the unit. They indicated nothing abnormal. He felt all the connections to make sure they were firmly in place and restarted the unit, but to no avail. After getting nowhere with the tech support department at the TSP (they said, "It should just work!"), he emailed me to see if I could help.

With the hope of troubleshooting the problem, I asked a bunch of questions about the editor's firewall, if the ATA was using DHCP, whether the ATA could be pinged from a nearby PC, and if the editor could do a packet capture of the ATA's registration process to be emailed to me for examination. So the editor did a packet capture using Ethereal and emailed it.

After receiving the packet capture, I saw that the ATA's SIP registration was being sent successfully and that the service provider's server was sending the appropriate 200 response. The ATA was clearly communicating with the TSP without a problem. So why was there no dial-tone? Resigned to the idea of advising the editor that his brand-new ATA was broken, I instead decided to sleep on it.

The next morning, I awoke to a message from the editor in my inbox. After all that troubleshooting and Q&A, the true culprit was revealed. The ATA had two RJ11 ports, one for the analog phone and one for a POTS line for 911 calls and the like. The analog phone had been connected to the wrong port on the ATA all along! Just plugging it into the other port resolved the problem immediately. The moral of the story— simple problems tend to have simple solutions.

You lose the dial-tone every few days or so, and you can't receive any calls from the PSTN

This problem usually occurs if the access router that connects to your Internet service provider is configured for DHCP, and its IP address has changed because the router's DHCP lease expired. The net effect is that the ATA's signaling socket is broken, and the ATA has no way of knowing it (remember, UDP is connectionless). So, the ATA must be rebooted in order to reestablish a valid socket. The best way to eliminate this problem is to get a static IP address from your ISP, or figure out a way to power-cycle your ATA daily during off-hours. Some TSP-provided ATAs have firmware that works around this problem. Contact your TSP to be sure.

Dialed digits work to place calls but not to interact with IVR prompts

The network can transmit dialed digits in two ways: in-band and out-of-band. This problem occurs most often when in-band signaling of digits is used. Because in-band signaling uses audio (dual-tone multifrequency) signals to represent each dialed digit, it is possible that the signals can be distorted during encoding, transport, and decoding. If the distortion is significant, IVR systems may not be able to interpret the audio signals they receive. Sometimes holding down each digit for an extra second or two can improve the IVR system's recognition, but not always.

To avoid this issue, configure your SIP endpoints and ATAs to use an out-of-band DTMF signaling approach such as the SIP INFO method. Generally speaking, the only time in-band DTMF signaling can be reliably used all the time is on legacy, non-packet-based voice links.

Callers sound robotic, or they say you do

Like in-band DTMF digits, the human voice can be distorted when many subsequent digital/analog conversions or encoding/decoding operations occur. When a sound stream is encoded digitally, decoded into an analog signal or transcoded, and then reencoded into a different codec, degradation of the original signal almost always occurs. It's like making copies of an old analog cassette tape and then making copies of the copies. By the third or fourth generation, the recording on the duplicates starts sounding awful.

This phenomenon is easy to demonstrate with a SIP ATA and a TSP like BroadVoice or VoicePulse. Try calling a digital cell phone, which probably uses the GSM codec to carry the sound stream from the phone to its CO, from an Internet-based TSP like BroadVoice or VoicePulse, which probably uses G.729A to carry the sound stream from the ATA to its SIP server. This call will sound somewhat robotic. Now, try using the TSP service to conference the call with a third party using another cell phone. The two cell phone parties should sound significantly robotic by now, due to all the reencoding.

Unfortunately, most of the control over this issue resides with the service providers. While it's possible for a SIP client to negotiate GSM codec directly with a cell phone network for a call that needs no transcoding or reencoding, most telephone network operators don't facilitate this kind of advanced signaling yet. This is because they use SS7, which is out of the reach of most enterprise VoIP users.

The good news is, many more small voice carriers and large enterprise NOCs (network operations centers) will be able to use SS7 signaling in the near future. Verisign and Transaction Network Services, Inc., both offer SS7 connectivity that allows VoIP switches to directly signal and negotiate capabilities for each call, including codec selection, in order to increase quality.

Calls across a wide area call path have dropouts in the audio

As a rule, this means that the latency or jitter on the call path's route is a problem. If the wide area link in question is used for voice and data, use routers that support IP precedence to promote VoIP over non-VoIP traffic. If the link is running at capacity, consider increasing its capacity or using a bandwidth-reservation QoS measure across the link. Also, make sure that, for call paths that have limited bandwidth, calls are selecting a thrifty codec (not G.711).

You and your caller find yourself interrupting each other a lot

This problem is either caused by latency on the call path or poor conversational manners. Most often, latency is the culprit. Do what you can to decrease latency on this call path. Are you using a small packet interval? What about packet-loss conceal-ment? Do you need to be using it? What about jitter buffering? How fast is the link? Is the call path a VPN? All of these things contribute to latency.

When a caller begins to speak, you can't hear the first word or two of his sentences

Normally, this problem is caused by poor silence suppression techniques. Silence suppression conserves bandwidth by not transmitting sound frames during periods of silence. But since a sound must occur in order to resume transmission, some of the initial frames may not actually get sampled. Of course, since sampling frames isn't what eats bandwidth (*transmitting* them does), any cessation of sampling during silence-suppressed periods is just poor design. The lesson here is this: if you're going to use silence suppression, try before you buy. Listen to phones that support silence suppression with, and without, that feature enabled. You'll be able to tell very easily if the silence suppression support is any good.

The power went out and so did all the phones

Power provisioning is a common challenge with IP telephony. If IP phones are replacing TDM phones throughout the office, then you've probably decided to use 802.3af to provide power to the IP phones. If not, you'll need to place UPS (uninter-ruptible power supplies) and AC/DC power packs at each phone. With a dozen phones or more, the cost of doing so more than justifies the use of 802.3af switches or power injection panels. This way, power backup can be centralized, connected to a heavy-duty UPS and/or power generator, so that when the power goes out, the phones don't! The bottom line is: don't forget to properly provision electrical power during your VoIP build-out.

And, while we're on the subject of power…

I love the Cisco IP phones but they don't accept 802.3af inline power. What do I do?

There are a couple of things you could do. First, if your budget permits, the obvious (though proprietary) answer is to use Cisco PoE switches to power the phones. Some other switch makers, like Foundry Networks, also support Cisco's proprietary PoE standard. If you can't afford to forklift your switches and instead want to power your Cisco phones by way of power injectors, then you should consider Cisco PoE-compatible injectors like those made by PowerdSine (*http://www.powerdsine.com*). But if you can't do that either, do the next best thing: *hack*.

 Hacking inline power will almost certainly void your IP phone's warranty, and probably your switch's or power injector's too. A short circuit could fry your switch *and* phone if you're not careful. Proceed with caution!

By flipping wires 4 and 7 and wires 5 and 8 on a standard UTP Ethernet patch cable, you've basically made a "compatibility cable" that lets you plug Cisco IP phones into an 802.3af source, as in Figure 15-2.

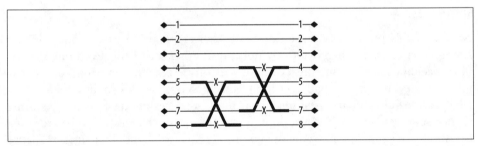

Figure 15-2. The wiring diagram for a hacked PoE cable

Make sure your switch lets you program, port by port, which ports get power and which ones don't. This is necessary because, in the Cisco PoE solution, Cisco IP phone power requirements are autodetected, so power can turn itself on and off as necessary on each port. There's no such provision when using a hacked cable to supply 802.3af power to a Cisco PoE-using phone. If this is a problem, and 802.3af won't work with the hacked cable, then try using a device that does the two-pair flip but also works with autodetection, such as 3Com's 48-volt Intellijack switch converter, part number 3CNJVOIP-CPOD.

 The Cisco 7970G IP phone *does* support 802.3af power sources, unlike the more popular (and less expensive) 7960 and 7940 phones.

When my PBX routes a call to IAXTel or another Internet voice destination, the sound quality is awful

Chances are, you're trying to use a bandwidth-absorbent codec like G.711 over the Internet VoIP trunk that connects to IAXTel. Switch to a more miserly codec like GSM or, if the destination supports it, Speex.

A clumsy keystroke took the softPBX down during peak business hours

This is a rare occurrence, but every seasoned system administrator knows the danger of a rashly typed command or a misplaced mouse click. One second, everything's humming along fine, and the next, a critical system is, well, *gone*. Aside from careful consideration of every administrative move you make, your best defense against this scenario is, of course, a good backup of your PBX configuration. This will also protect you against another possibility: a hard drive crash or data corruption on a PBX.

Some commercial PBXs offer built-in backup capabilities and even an undo feature to roll back a botched config. But if you're using a noncommercial solution like Asterisk or Open H.323, you'll need to brew your own backup recipe. Here's a shell script that you could trigger from cron on Linux to back up the Asterisk configuration:

```
mkdir /var/backup/$(date)
cp -R /etc/asterisk /var/backup/$(date) > /var/backup/backup.log
tar -cf /var/backup/$(date)/$(date).tar /var/backup/$(date)/*
gzip -c /var/backup/$(date)/$(date).tar
mv /var/backup/$(date)/$(date).tar.gz /var/backup
rm -Rf /var/backup/$(date)
```

This script copies the Asterisk configuration directory, */etc/asterisk*, to a temporary directory in /var/backup, compresses it into a *gzip* file, and drops it in */var/backup* before deleting the temporary directory. You would want to run this daily, and use Veritas, *cpio*, *tar*, or a similar cassette backup software to copy it onto a cassette daily, too. That way, your rollback window, the time between the most recent backup and a critical failure, is never longer than a single day.

Make a script that quickly rolls back your active configuration, so that, if you've hosed the config and you're not sure exactly *how* you hosed it, you can back the hosed config up to figure out your mistake later, and roll it back to its prior, working condition.

The old-timers are complaining about the new phones, the new voice mail greeting, or the new _____

As nice as new IP phones are, if your rollout is big enough, somebody is going to complain about *something*. It could be the angle of the new handsets causing (probably exaggerated) neck strains, or it could be that the buttons on the keypads aren't dimpled, making it harder to dial without looking at the keypad. There may be slight differences in the sound of the phone receivers that aggravate certain users (MOS scoring can help you evaluate demonstration equipment *before* you make a big purchase).

The key to minimizing complaints after the implementation is to evaluate end user input before it. This means engaging end users, and allowing their requirements to drive the project. It also means defining your service-level agreement in terms that are as specific and practical as possible. The mean opinion score metric is a good guarantor of SLA attainment (MOS is covered in Chapter 9).

When the legacy equipment is gone, though, attrition will deal with the matters of opinion and nuance that you *just can't do anything* about. If complainers are the usual suspects, those people who complain no matter what, then you've probably done a good job. Eventually, the dust will settle, and the ROI process will begin. Once the rewards of the VoIP switch are in hand (increased efficiency and decreased cost), the noisy complaints of Ebenezer Evans in the Accounting Department about the size of his keypad buttons will suddenly seem frivolous.

My IP telephony salesperson said I would be able to do overhead or zone paging using the IP phones as broadcast speakers, but I can't!

Well, you can. But you may have to do a bit of hacking. The problem is, IP telephony isn't an ideal technology for broadcast applications such as overhead paging. Some proprietary solutions support IP phones for overhead paging using IP multi-cast, but there really isn't an authoritative standard for dealing with this need yet.

This simple hack will let you do intercom paging on a Cisco SIP 7960. It uses autoanswer to pick up incoming calls on a second line, which we'll call a paging line, so that the paging party can broadcast his voice. A 7960 phone configured in that way can act as a paging speaker because, as soon as a call comes in on the paging line, it's automatically answered. In principle, a similar approach should work on any phone that supports autoanswer. So the hack is—configure a line appearance on the 7960 to autoanswer, and you've got a basic intercom paging system.

With a little work, you can make a group of 7960s with paging lines act as "overhead" speakers for a paging zone. To do this, your PBX server will have to support some form of ad hoc conferencing. Create, in the dial-plan, an extension that adds all of the 7960s to the conference room, using their paging lines so they autoanswer, and then gives the dialing party a beep to let her know she can begin her announcement. If you can, limit the absolute timeout of the conference to 10 or 15 seconds to ensure all of the paging lines are "hung up" at the conclusion of the page. Now, instead of merely having a single phone acting as an intercom, you've got a group of phones acting as a paging zone.

Keep in mind that this technique is unicast in nature—it doesn't preserve wide area bandwidth the way a multicast paging setup would. And, with 96 kbps per simultaneous paging endpoint on the LAN, you could clobber your local area bandwidth if you have more than a few hundred phones on each zone. Yet, if your building has lots of obstructions (like office walls), you may need to use a lot of paging endpoints so that everybody can hear zone pages. Since there tends to be a phone on each desk, they make great paging speakers in this situation. If the building where you're placing the paging zone is open and has no interior walls to block the sound—like a warehouse or factory—then you'd probably be better off using a few large, analog speakers for zone paging.

The VoIP budget was, well, too small

The old adage "You get what you pay for" applies aptly to VoIP. But if your ROI payoff is going to be real, all of your project's potential costs must be recognized fully and truthfully. This means adding the cost of consulting, training, and migration to the sticker price on that shiny new VoIP hardware. Depending on your particular implementation, the hardware itself may represent only a third of the overall cost. When you create your RFP, ask responders to include a cost per user for training and implementation and then tack that cost onto the project's bottom line.

The phone company missed a critical circuit switchover deadline

You spent months doing MOS testing, interviewing end users, looking over the shoulder of the corporate receptionist, taking detailed notes about call flow, and maybe pulling your hair out doing return-on-investment analysis.

Now, the big day for the ambitious change has come. Today, you'll be turning on 500 new IP phones, two new fully survivable softPBX servers, and a brand-new 48-trunk PSTN connect point with two PRI gateways attached. But at the moment of truth, as everybody's breath is held and fingers are crossed, the phone company makes a provisioning mistake and one of those T1 smart jacks has bright red alarm lights. Your VoIP network has no way to reach the outside world, and now that you've flipped the proverbial switch, nobody from the outside world can reach that group of users—users who, yesterday, had a flawlessly functioning legacy phone system. Ouch.

This isn't a situation you want to end up in. If the phone company misses your switchover date, it might only mean that your service will remain on your old phone system. But if the trunks used for the new system have the same phone numbers as those used with the old system, you could be in a mess. This is because they can't be connected to both systems (old and new) at the same time. If they could, switching PSTN connectivity from one trunk group to another wouldn't be such a critical task. Here's how you can avoid a botched trunk move turning into phone downtime:

1. Don't switch over all of your inbound phone numbers to a new trunk circuit. Bring over only the "main" numbers—the numbers the outside world knows. That way, if the switchover doesn't happen, you'll still have usable (secondary) outbound trunks on the old system. You'll have to worry about inbound traffic only to those "main" numbers.

2. In the event of a failed switchover, first determine whether you still have dial-tone on your old trunks. If you do, you can continue to use them with your old system until the switchover can be rescheduled.

3. If the old trunks don't have dial-tone any longer and the new trunks aren't working, have the phone company forward all calls destined for those "main" inbound numbers at the CO to the secondary trunks that you haven't switched over. These trunks are, hopefully, still connected to your old system. (See item 1.)

4. If possible, coordinate the switchover with the phone company so that any failures can be picked up on immediately and the CO reverted back to its prior state until the switchover can be attempted again. (Third-party implementers of PBX systems are experienced at managing interaction with the phone company.)

5. Plan to do your switchover during off-hours. (And not an hour before business on a Monday morning.) Give yourself plenty of time for testing of the internal dial-plan, too. Even if your shiny new PRIs are working right, there may be flaws in your call-flow logic that need to be worked out before users can start making and receiving calls.)

I've read all about QoS and proper converged network design, but I'm still paranoid about quality issues

If IP precedence and Ethernet CoS have you totally convinced that you can replace a thousand TDM phones with a single switched Ethernet segment and a palette full of SIP phones, then don't sweat it. You're not alone. A perception of poor quality, whether deserved or undeserved, was the top reason for the industry's slowness to adopt VoIP during the late 90s and early 00s.

If you have money to burn and you want to do absolutely everything in your power to make sure your VoIP links are tip-top—if they're Ethernet or WAN links—then you can dedicate them only to VoIP. By eliminating all other traffic from these network segments, you can take advantage of IP telephony without having to share the transport with non-telephony apps. In essence, this means building and running a second, independent IP network alongside your existing one. More than just a VLAN, this network is an operationally independent network that has no bridges to the other network, and therefore no way of comingling traffic with it.

Why would anybody go to such lengths for VoIP, when the IETF and IEEE have invested so much time and effort into QoS? That's a philosophical question that I'll leave to you.

If the skeptical decision-maker in your company has the mistaken impression that all IP telephony sounds about as good as the share-ware copy of Internet Phone that he used to use with Windows 95, just give him a call from your Packet8 or BroadVoice service, and ask how he thinks it sounds. The point is, most people can't tell the difference between the sound of an IP phone call and the sound of a land-line or cell phone call.

Key Issues: What Can Go Wrong?

- Echo is a problem that is exacerbated by latency and the use of hybrid inter-faces—those that provide VoIP servers with a way of using legacy phone lines. Avoid latency and legacy phone lines, and you solve the echo problem. If you can't avoid them, then use high-quality hybrid interfaces like those available for Digium's TDM400P or Intel Dialogic cards. Asterisk users can also enable aggressive echo suppression, as in Chapter 9.

- SIP registrations and RTP media streams will fail through a NAT firewall unless a DMZ-based SIP proxy is used. SIP servers need non-NATted access to both the Internet and the private IP network, so that they can serve as an application proxy for SIP endpoints. DMZ provides this non-NATted access.

- If an ATA for a TSP such as Vonage or Packet8 works fine for a few days and then suddenly fails, requiring a reboot to remedy the failure, it's likely that your broadband router has obtained a new IP address and the ATA doesn't know what it is. Rebooting the ATA allows it to "sense" the new address, and reestab-lish its path to the TSP's server.

- Callers sound robotic when one too many codecs or sampling conversions occurs. Going from analog to G.729A to analog to G.711 ALaw will cause quite a robotic-sounding call, so avoid excessive transcoding and re-sampling.

- Your private-premises VoIP system must draw power from your own facilities, not from the phone company's. Make sure there's adequate backup power, not just for your servers, but for your endpoints. PoE is a good solution to this prob-lem (see Chapter 8 for a review of PoE).

- Use a regular backup routine to keep several of the most recent revisions of your dial-plan and voice peering configurations on hand, in case of a natural or administrative disaster. Keep copies of the backups off-site if possible.

CHAPTER 16

VoIP Vendors and Services

Telephony software is everywhere, and a majority of it is free. Much of it is open source. It's a good idea to experiment with as much of this stuff as you can, because the VoIP family of technologies is evolving continuously. The capabilities and efficiency of these tools rise daily.

Whether you use Linux, Solaris, Windows, or Macintosh, plenty of fully developed softphones, servers, troubleshooting tools, APIs, and sample code are available. What follows is a partial list of some excellent VoIP software and service providers. Many of them have been used in projects throughout this book.

We'll also briefly cover the growing list of vendors that manufacture modular and turnkey telephony hardware.

Softphones and Instant Messaging Software

OhPhone

OhPhone and OhPhoneX are H.323 softphones that support Windows, Linux, and Macintosh. Unlike NetMeeting, OhPhone is open source, but it is not as polished. OhPhone is a part of the Open H.323 distribution (*http://www.openh323.org*).

X-Lite, X-Pro, and Eyebeam

Xten Networks produces several VoIP software products. No VoIP test lab is complete without X-Lite, the company's free SIP softphone for Windows, Mac, and thin-client (web-based) setups. Chapter 11 contains a few projects that make use of X-Lite.

Xten also produces X-Pro, a fully featured commercial SIP softphone for Windows, Mac, and MS PocketPC. Using a PocketPC with built-in WLAN and X-Pro, you could create a handy cordless SIP phone. Xten also makes Eyebeam, a SIP video phone (*http://www.xten.com*).

LIPZ4

Zultys, a provider of embedded VoIP gateway devices, makes LIPZ4, a graphical SIP softphone for Linux and X11. The phone can be downloaded from *http://www. zultys.com/LIPZ4.htm*.

Firefly (SIP and IAX)

Virbiage produces a softphone that supports SIP and IAX signaling. This makes it a good choice for Asterisk-based networks or for using services like IAXTel, which support the IAX protocol (*http://www.virbiage.com/firefly/faq.php*).

IAXComm

IAXComm is an open source softphone for Windows, Linux, and Mac OS X. It supports IAX only (*http://iaxclient.sourceforge.net/iaxcomm/*).

NetMeeting

Microsoft's NetMeeting softphone has been shipping with Windows since 98. It's a multipurpose H.323 client, offering voice, video, text chat, and white-boarding applications. It has a built-in directory services client that can make use of an H.323 gatekeeper or an ILS (Internet location service) server. NetMeeting is a worthwhile tool for any H.323 administrator (*http://www.microsoft.com/windows/netmeeting/*).

MSN Messenger

A potential replacement for NetMeeting, Microsoft's MSN Messenger uses SIP to signal telephony and messaging applications that are tied to the proprietary MSN network service. Unlike NetMeeting, MSN Messenger supports the Macintosh, too (*http://www.msn.com*).

iChat

Apple's instant messaging software includes voice and video messaging for users of Mac OS X. The instant messaging framework used is AOL Instant Messenger (AIM), a leading IM standard. iChat also makes use of SIP for call signaling, but can't be used independently of the AIM network, so it really isn't a good VoIP hacker's tool.

iChat supports videoconferencing in resolutions up to full screen, if you have enough bandwidth. The version that ships with Mac OS X 10.4 lets up to three participants videoconference simultaneously using a very slick, futuristic OpenGL interface. Perhaps the only drawback of iChat is that it doesn't support non-Mac users. Find out more about iChat at *http://www.apple.com/ichat*.

Skype

Skype is a voice telephony application that uses a unique peer-to-peer network model to enable a global VoIP network. While proprietary, Skype does provide an API that allows Windows developers to link their applications to the Skype client software. The Skype network allows inbound and outbound PSTN calling, too. Skype has minimized sound latency exceptionally well—even over the Internet, and all calls use encrypted media channels, so nobody can listen in on you.

Skype has been downloaded millions of times, and its network is regularly used by a million to two million users simultaneously. Features of its UI are similar to instant messaging apps: a buddy list, an availability indicator, and a searchable user directory. Aside from Windows, Skype supports Mac OS X, Linux, and PocketPC. Skype also offers a hardphone manufactured by Siemens. You can obtain Skype from *http://www.skype.net*.

Skype Web Links

When you install Skype, the `callto:` URI prefix is enabled on your web browser. This means that, whenever you click a `callto:` link on a web site, your Skype client will attempt to place a voice call to the user referenced in the link. In the same way, you could add a link to your web site that allows Skype users to call your Skype client from their web browsers:

```
<a href="callto://SkypeUserName">Skype Me</a>
```

Conference Calls

Like other IM clients with voice calling features, Skype permits basic conference calling. Using the Conference button on Skype's toolbar, you can start a conference with up to five participants, who can use a mixture of Windows, Linux, and Mac OS X versions of the program.

 Skype conference calls are no fun if participants can hear echo! The more participants in a call, the more echo can be a problem. So all participants should use a headset to cancel any microphone feedback.

Skype Handset

Aside from a basic headset, Skypers can use a USB handset that resembles a standard telephone. It has a dial-pad, microphone, and speaker and is powered by the USB bus. You can use it to answer incoming calls without having to bring the Skype interface up on your screen. If you do a lot of Skype calling, this handset can be a time-saver.

Also available are corded handsets, like the VoIPVoice Cyberphone K, available at *http://www.skype.voipvoice.com*. There are also cordless USB handsets, like the VTech USB7100, available at *http://www.vtech.com/phones*.

Other Softphones to Check Out

Try Googling these other softphones and messaging apps: GnoPhone, EyeP Phone Lite, KPhone, LinPhone, and BonePhone.

Other Desktop Telephony Software

Call Soft Pro

This application provides basic voice mail/answering machine functionality on a Windows host. It also allows its users some remote control of the PC from the telephone when they call in. Call Soft Pro is made by TOSC and is available from *http://www.mycallsoft.com*.

CallerID

Using Soft-Haus Solutions CallerID, users can use named party ID (caller ID) to trigger a custom WAV file playback for each caller or display a GIF or JPEG avatar for each caller on a Windows PC (*http://www.soft-haus.com*).

GeckoPhone

GeckoPhone is a combination softphone and voice mail server for Windows. It is skinnable and has a base of supportive users (*http://www.geckophone.com*).

FaxMachine

Nico Cuppen's Windows software, FaxMachine, turns your PC with scanner, printer, and modem into a fax machine with queuing and on-screen previewing. This no-frills software has a very simple interface (*http://www.nicocuppen.com/*).

FaxStatus (Mac)

This application shows a menu bar icon that indicates the current status of the OS X Panther fax service—i.e., whether it's sending or receiving. You can download it from *http://www.macupdate.com/info.php/id/15377*.

PhoneValet (Mac)

Parliant Software's commercial PhoneValet application gives Mac OS X users voice mail, call logging, call recording, and limited remote control of the Mac using a telephone. PhoneValet requires an included USB adapter that allows connection of a POTS line to the Mac (*http://www.parliant.com/phonevalet*).

Voicent Gateway

Voicent's Gateway application for Windows allows you to create IVR applications that can be tied to web sites. Gateway has limited voice recognition in addition to DTMF recognition (*http://www.voicent.com/gateway.php*).

SpeechSoft

This commercial package includes scriptable IVR, fax, and basic call-routing abilities. It requires Intel Dialogic interface hardware running on a Windows PC (*http://www.speechsoft.com/prodsw_cm.htm*).

Developer Tools and SoftPBX Systems

Asterisk

Asterisk has earned a reputation as the most feature-complete and ready-to-use soft-PBX available for free. Since it is covered significantly elsewhere in the book (chapters 3 and 17, specifically), we'll look at some of the worthwhile add-ons for Asterisk.

AstLinux

AstLinux is a distribution of Gentoo Linux with Asterisk v1 installed, specifically aimed at small-footprint applications. It can fit on 256 MB flash media, though you won't have excessive voice mail recording capacity. AstLinux is available from *http://www.krisk.org/astlinux*.

Asterisk Management Portal

AMP is a PHP- and Perl-based suite that implements a web-based administrator interface for the Asterisk PBX. It makes use of MySQL as well. AMP is available from Coalescent Systems via their web site at *http://amp.voxbox.ca*.

LDAPGet

LDAPGet is a module for Asterisk that allows programmatic use of LDAP inquiries from within the Asterisk dial-plan. This could be useful for multiserver setups with complex dial-plans or for find-me-follow-me lookups and presence applications. LDAPGet is available from *http://www.mezzo.net/asterisk.*

TAFM

The Asterisk Fax Manager adds fax-receiving capability to Asterisk extensions via a set of Python scripts. TAFM users can retrieve their faxes by web interface or, with a little programming, by email (*http://tafm.sourceforge.net/*).

Another fax option for Asterisk is SpanDSP, which was covered in Chapter 14.

Asterisk Perl modules

The Perl scripting language has excellent support for Asterisk, thanks to the work of the open source community. A group of modules are available at *http://asterisk. gnuinter.net.*

JAsterisk

Where you find Perl, you're likely to bump into Java. JAsterisk is a thread-based implementation of Java interfaces for Asterisk. It's available from *http://sourceforge. net/projects/jasterisk/.*

AstWind

AstWind is a Win32 implementation of Asterisk. It isn't intended for production, but rather to allow Windows users to experiment with Asterisk. Digium's web site states that production environments should run Digium on Linux. It was created by N2Net and is available at *ftp://ftp.nacs.net/asterisk/astwind.*

Non-Asterisk

Open H.323

Open H.323 is a complete H.323 stack implemented for Linux, though a majority of it has been ported to Mac OS X and Solaris. Using H.323 IP phones, Open H.323, and QuickNet voice interface cards, it's possible to build feature-complete softPBX networks with Open H.323. Download it from *http://www.openh323.org.*

VOCAL

The Vovida Open Communication Application Library is a set of open source components that form a very capable framework for building IP telephony systems. Many telephone network operators regard VOCAL as the standard-bearer of VoIP interoperability.

VOCAL is sponsored by Cisco Systems. As a result, many of the projects described at the project's home page are beholden to Cisco's VoIP hardware. This doesn't detract from VOCAL, though: Cisco's switches and routers are in use on a majority of enterprise VoIP networks. So Cisco's ubiquity in the server closet certainly works in VOCAL's favor.

Developers have used VOCAL code to create IP phones, gateways, RADIUS authentication servers, SIP-to-H.323 translators, conferencing applications, and STUN servers. VOCAL developers offer free and open implementations of G.729A, MPEG-4, and other compression and streaming schemes, as well as a massive developer support archive and very active mailing list. Its web site is *http://www.vovida.org*, and the authoritative book about VOCAL is O'Reilly's *Practical VoIP with VOCAL*.

Intel Dialogic products

Intel's line of NetStructure gateways and Dialogic interface cards for telephony applications includes an array of device drivers, open source application examples, and lots of documentation. Intel also offers commercial licensing. More information is available at *http://www.intel.com/design/network/products/telecom*.

SIP Express Router

SER from IPTel is an open source SIP server that implements a registrar and proxy. It is designed to be RFC 3261 compliant (*http://www.iptel.org*).

Envox

This commercial software toolset implements a softPBX that uses the Intel Dialogic and NetStructure voice interface products (*http://www.envox.com*).

Bayonne

A product of the GNU project, Bayonne offers a scriptable telephony service for building voice response systems on Linux. Bayonne allows you to add telephony features to Perl for home automation, automated attendant, voice messaging systems, and so forth (*http://www.tuxmobil.org/*).

VOCP

This tool adds Perl-based fax, messaging, and IVR capabilities to Linux (*http://www.vocpsystem.com*).

Telos ISDN H.323 Gateway

This software allows your Linux PC to act as a PSTN-to-H.323 gateway, connecting your H.323-based VoIP network to the public telephone system. *http://www.telos.de/linux/H323/default_e.htm*

TFTP Servers and Tools

TFTP (Trivial File Transfer Protocol) is the means by which IP phones obtain firmware loads and, sometimes, their operating configurations. This basic, insecure file-transfer system allows you to store a repository of VoIP-related files—like SIP phone config files, firmware patches, XML user directory files, and ring tones—in a central place. Your DHCP server can tell IP phones the IP address of the TFTP server so they can download the files they need.

Linux, FreeBSD, and Mac OS X have the common TFTP server *tftpd*. A fantastic GUI add-on for the OS X version of *tftpd* is TFTPServer, which you can download from *http://www.versiontracker.com/dyn/moreinfo/macosx/18561*.

Windows users with Cisco CCO accounts can download a good TFTP server from *http://www.cisco.com/cgi-bin/tablebuild.pl/tftp*, while other Windows users can get a shareware one from *http://www.tftp-server.com*.

VoIP Service Providers

Service providers sell dial-tone or call-origination applications, bulk voice trunking, or wholesale interexchange signaling. Some provide all three.

Dial-Tone and Call Origination

Companies that offer dial-tone service or call origination are often called telephone service providers. Next-generation TSPs are using broadband Internet links to replace the services ordinarily provided by the regional Bell operating companies like SBC, BellSouth, and Verizon—services like simple local and long-distance calling.

The essential broadband TSP offers an IP-based trunk, usually using SIP, to connect a single line appearance to its network operations center (NOC), where calls are originated to, and received from, the PSTN. The NOC acts as a CO for the customer, who usually has a VoIP gateway device to connect his analog phones to the calling service. A broadband Internet links the customer's network to the TSP's.

This VoIP gateway device is either an ATA or a more elaborate VoIP server. ATAs are used to support analog phones, while, less frequently, embedded VoIP servers may be used to support IP phones. Overwhelmingly, SIP is the endpoint signaling protocol used between the VoIP gateway and the NOC. Some smaller TSPs have begun using the IAX protocol, due to its immunity to problems traditionally imposed by NAT firewalls.

The customer provisioning model for each TSP varies slightly. Some TSPs charge a fee for the gateway device and rebate it after a few months' paid use of the service. Other TSPs give the ATAs away free, but demand them back if you cancel prematurely. Still others allow you to bring your own VoIP gateway or ATA. Some even let you use a full-blown softPBX as your SIP client, which, in essence, transforms the SIP link to the TSP into a PBX dial-tone trunk.

TSPs in this vein include those listed in Table 16-1.

Table 16-1. Independent VoIP telephone service providers (TSPs)

Company	URL	Supplied subscriber equipment	Bring your own device	PSTN phone number
BroadVoice	http://www.broadvoice.com	1 FXS ATA	Yes	Yes
BroadVox Direct	http://www.broadvoxdirect.com	1 FXS ATA	No	Yes
Free World Dialup	http://www.pulver.com/fwd	X-Lite softphone	Yes	No
Net2Phone	http://www.net2phone.com	1 FXS ATA	No	Yes
Nikotel	http://www.nikotel.com	1 FXS ATA	No	Yes
Packet8	http://www.packet8.net	1 FXS ATA	No	Yes
SOYO	http://phone.soyo.com	SOYO broadband router with ATA	No	Yes
VoicePulse	http://www.voicepulse.com	1 FXS ATA	No	Yes
Vonage	http://www.vonage.com	1 FXS ATA	No	Yes

A variation on the broadband TSP formula is to use PC-based call origination instead of hardphones and VoIP gateways. Skype uses such a model, allowing its users to place all calls, even those bound for the PSTN, from its proprietary desktop messaging application. Most TSPs that support hardphone calling offer some kind of softphone solution for PC-based calling, though Skype is unique in requiring it.

Bulk trunk providers

Bulk trunk providers offer business-grade subscribers with dial-tone service en masse. This type of service is intended to replace high-density legacy technologies such as T1 and PRI with a high-density packet-based interface into the PSTN that serves a large group of endpoints in a single office. The use of these VoIP trunks were described in Chapter 13.

Bulk trunk providers may or may not place a VoIP gateway on the customer premises. This depends largely on what technology is used to provide the link to the provider's NOC—ATM, IP, and so on. Whether the customer premises equipment is a VoIP gateway or just a group of IP phones connected by an Ethernet segment, the trunk provider dictates the trunk signaling standards used. Most are presently using SCCP, MGCP, or SIP, A VoIP-compatible bulk trunk service is available from most large network carriers, including Covad, AT&T, BellSouth, MCI, SBC, ICG, and Verizon, and by metropolitan carriers like N2Net (*http://www.n2net.net*).

Wholesale IXCs

Traditionally, interexchange carriers have formed a backbone that connects long-distance networks around the world. IXCs dealt in voice bandwidth at the wholesale level, and phone companies tapped into IXC networks using SS7. This kept carrier-grade connectivity relegated to the elite, large operators who had industrial-class SS7 switches.

But more recently, a few IXC networks have opened their doors to next-generation VoIP operators. In fact, a majority of VoIP traffic on the Internet today is being signaled and hauled by a few IXCs that specialize in servicing next-generation VoIP TSPs like Vonage.

Level3 Communications, Verisign, Global Crossing, and Transaction Network Services have launched services that allow SIP-based last-mile providers and other VoIP TSPs to access facilities on the big IXC, ILEC, and LD networks. This could, for example, allow a startup TSP to gain DID phone numbers and PRI circuits in very faraway places, allowing them to better compete with the already-established incumbent LECs. More information is available from *http://www.level3.net*, *http://www.verisign.com*, *http://www.globalcrossing.com*, and *http://www.tnsi.com*.

Hosted PBX

A number of TSPs can host a softPBX at their data center, allowing many small organizations to share a single VoIP server. The subscriber's PSTN connect point is at the TSP rather than its own premises. This makes sense for those who want to avoid the expense of hosting VoIP services in-house. But it doesn't make sense for everybody. If you've got fewer than a dozen users, it might be better to stick with VoIP trunks and a cheap PBX, which costs only a few thousand dollars. Hosted PBX services cost $30–$100 per month per user, so the benefit of outsourcing your call routing is greater as your user count grows.

> Most hosted PBX offerings use a SIP-based set of standards called IP Centrex. Coincidentally, IP Centrex has nothing to do with traditional, analog Centrex from the phone company.

ICG Communications' VoicePipe service connects a network of VoIP phones to a managed telephony applications suite using a single IP WAN link or VoATM (Voice over ATM) connection. In this fashion, the hosted PBX provider is responsible for maintaining the telephony switching servers and dial-plan, and it often means that the subscriber *cannot* integrate its own PBX with the service. Check with your prospective provider to see what it supports.

Web-based call center services

A relatively new entrant into the TSP arena is the web-based call center provider. A step beyond the hosted PBX model, this type of service provides ACD, IVR, call queues, and other features commonly required in call centers and hosts them like an application service provider (ASP) at an off-site data center. VoIP is used to connect the call center endpoints with the ASP's backend. Companies offering this service include:

- ICT Group, Inc. (*http://www.ictgroup.com*)
- Echopass, Inc. (*http://www.echopass.com*)
- Cosmocom (*http://www.cosmocom.com*)
- Aspect (*http://www.aspect.com*)

Some smaller-scale call center services geared toward the small/medium business sector are available from Vonage and Packet8, too.

Telephony Hardware Vendors

Behind all the software and services that enable VoIP, there's hardware. Whether it's a dedicated SIP call-routing server, an old-fashioned key system, or a PSTN POTS gateway, we've covered a great deal of ground when it comes to silicon and iron in the telephony business. Table 16-2 provides a partial list of key manufacturers of hardware devices.

Table 16-2. Telephony hardware vendors

Manufacturer	URL	Devices
3com	*http://www.3com.com*	Phones, switches, and power products
Aculab	*http://www.aculab.com*	TDM interface cards
Alcatel	*http://www.alcatel.com*	PBX systems, phones, TDM equipment
Altigen	*http://www.altigen.com*	PBX systems, phones, TDM equipment
AP Connections	*http://www.netequalizer.com*	QoS enforcement devices
Audiocodes	*http://www.audiocodes.com*	VoIP gateway equipment
Avaya	*http://www.avaya.com*	PBX systems, phones, TDM equipment
Brix	*http://www.brixnet.com*	QoS monitoring devices
Cisco Systems	*http://www.cisco.com*	VoIP servers, gateways, and phones
Convedia	*http://www.convedia.com*	Carrier-grade VoIP servers
Digium	*http://www.digium.com*	TDM interface cards
DSP Group	*http://www.dspg.com*	Phones
Edgewater Networks	*http://www.edgewaternetworks.com*	VoIP gateway equipment
Elma Electronic	*http://www.elma.com*	Voice server chassis and TDM interface equipment
Epygi	*http://www.epygi.com*	Voice conferencing server

Table 16-2. Telephony hardware vendors (continued)

Manufacturer	URL	Devices
Ericsson	http://www.ericsson.com	VoIP gateway equipment
Eutectics	http://www.eutecticsinc.com	USB handsets
Hughes Software Systems	http://www.hssworld.com	VoIP, TDM, and SS7 server components
i3	http://www.i3net.it	IVR systems, VoIP system-building components
InGate	http://www.ingate.com	VoIP-friendly firewalls
Intel	http://www.intel.com	Microchips and TDM interface hardware
Inter-tel	http://www.inter-tel.com	PBX systems, phones, TDM equipment, VoIP servers
Mediatrix	http://www.mediatrix.com	VoIP gateway equipment
Microsoft	http://www.microsoft.com	VoIP hardware drivers for Windows
Minacom	http://www.minacom.com	VoIP and TDM service tools
Mitel	http://www.mitel.com	PBX systems, phones, TDM equipment
Motorola Embedded Communications	http://www.motorola.com/computers	Microchips, interface devices, and embedded software components
Multi-Tech Systems	http://www.multitech.com	VoIP gateway equipment
NEC	http://www.necunifiedsolutions.com/ip	VoIP-enabled PBX systems and TDM phones
Netfabric	http://www.netfabric.com	VoIP gateway equipment
Netrake	http://www.netrake.com	VoIP-friendly firewalls
Network Instruments	http://www.networkinstruments.com	Troubleshooting and management tools
Nortel Networks	http://www.nortel.com	PBX systems, phones, TDM equipment
Octasic	http://www.octasic.com	Microchips for audio encoding
Pannaway	http://www.pannaway.com	VoIP gateway equipment
Pika Technologies	http://www.pikatechnologies.com	TDM interface equipment
Power Dsine	http://www.powerdsine.com	Inline power (PoE) devices
SEI Power	http://www.seipower.com	Inline power (PoE) devices
Shoretel	http://www.shoretel.com	PBX systems, phones, TDM equipment
Siemens	http://www.siemens.com	PBX systems, phones, TDM equipment
SNOM	http://www.snom.de	IP phones and softPBX servers
Teltronics	http://www.teltronics.com	PBX systems
Texas Instruments	http://www.ti.com/broadband	Voice-over-broadband interface equipment
Toshiba	http://www.telecom.toshiba.com	PBX systems, phones, TDM equipment
VegaStream	http://www.vegastream.com	VoIP gateway equipment
Verilink	http://www.verilink.com	VoIP gateway equipment
VOCAL Inc.	http://www.vocal.com	ATAs
VocalData	http://www.vocaldata.com	Carrier-grade VoIP servers
VocalTec	http://www.vocaltec.com	Voice-over-broadband interface equipment
Zoom	http://www.zoom.com	VoIP gateway equipment, routers

## CHAPTER 17

Asterisk Reference

Earlier in the book, we covered compiling and installing Asterisk, installing a legacy interface card and drivers, and setting up some simple PBX applications. Later on, we used Asterisk configurations to illustrate some common enterprise telephony concepts. This chapter is geared toward the person who's comfortable with the earlier material in the book and wants a deeper understanding of Asterisk.

Asterisk is a deep subject that touches disciplines of networking, code-writing, protocols, and standards. This chapter won't make you an Asterisk expert, but it should help you go a step or two beyond the essentials we've already covered. We'll cover channel configuration—with an emphasis on Zaptel and SIP channels, dial-plan syntax elements like variables and string processing, and the commands you can use in building an Asterisk dial-plan.

How Asterisk Is Supported

Asterisk's principle sponsor is a firm called Digium, based in the United States. The company provides development leadership and commercial support for the open source system. It also manufactures and distributes the Wildcard interface devices and iAXY ATA used in some of the projects in this book. A number of independent consultants provide commercial support for Asterisk. Susbcribing to the Asterisk Business Users' email list is a great way to solicit commercial assistance from telephony experts.

To use Digium's support lists, visit *http://lists.digium.com*, where you'll find several lists covering commercial announcements, business-oriented discussion, running Asterisk on BSD operating systems, Asterisk's documentation, DUNDi, SS7, and more. Before posting a question to Digium's (very busy) support lists, be sure to search for the answer first using the online web archive. Some of the lists generate dozens to hundreds of messages a day, so it's better for everybody if you post a new issue only if you absolutely have to.

Asterisk's Configuration Files

Asterisk's operations are governed by a set of text configuration files, like Apache and other common Unix software. Everything from the assignment of extension numbers to call queuing to low-level hardware interfacing is established by these files, which are located in */etc/asterisk*. Now would be a good time to peruse a few of the most important ones:

asterisk.conf

> Contains the locations of Asterisk software components, sound files used for music-on-hold, automation scripts (AGIs), and other files used by Asterisk.

extensions.conf

> Contains the all-encompassing *dial-plan*, a comprehensive configuration of phone users, voice mailboxes, call-processing instructions, and default behaviors.

features.conf

> Tells Asterisk how to handle features such as call park.

h323.conf

> Contains instructions on how Asterisk should interact with devices using the popular H.323 VoIP signaling protocol, such as Microsoft's NetMeeting softphone, if you've compiled H.323 support.

iax.conf

> Tells Asterisk how to handle Inter-Asterisk Exchange protocol peers.

manager.conf

> Configures security restrictions for Asterisk's Manager Socket API.

mgcp.conf

> Configures Asterisk's support of the Media Gateway Control Protocol.

modules.conf

> Tells Asterisk which modules, or telephony applications, to load when it starts up.

sip.conf

> Contains instructions on how Asterisk should interact with VoIP devices using the SIP signaling protocol, such as Xten's X-Lite softphone and Cisco's 7960 SIP hardphone.

skinny.conf

> Allows SCCP phones like the Cisco 7970 to be used with Asterisk in a manner similar to SIP phones.

logger.conf

> Tells Asterisk where to store its logfiles and how detailed they should be.

voicemail.conf

> Tells Asterisk how to operate its built-in voice mail server, called Comedian Mail.

zapata.conf and zaptel.conf

> Tell Digium's signaling kernel modules (*wcfxs*, etc.) and the Asterisk PBX what type of legacy interface hardware is installed and how it is configured (*zaptel. conf* is in */etc*; it's the only config file not in */etc/asterisk*).

There are many other configuration files. At a bare minimum, you'll need to familiarize yourself with *extensions.conf* and *SIP.conf*, *h323.conf*, or *skinny.conf* in order to build a bare-bones VoIP server that can handle simple calls. If you're using legacy interface hardware such as the X100P card, you'll also need to learn about *zaptel.conf* and *zapata.conf*.

The files in */etc/asterisk* follow a similar convention, mostly. Many have organizational sections that begin with a string of text in square brackets ([and]). Following each bracketed heading is a list of configuration settings and directives. A setting is a string followed by = and a user-specifiable value, like this:

```
writeprotect=yes
```

A directive has a slightly different syntax than a setting. Directives tend to have a more complex definition, and convention dictates that they use the => operator rather than simply = (though either will work). This definition's structure will vary depending on what kind of directive is instantiating it. For example, variable assignments are directives that have a simple syntax in *extensions.conf*:

```
LINE1 => Zap/1
```

But an extension directive appears more complex:

```
exten => 2101,1,Dial(Zap/3/5552101)
```

A directive could be an extension assignment or a priority of an extension assignment (more on this later). Many directives and settings only work in one of the configuration files and are invalid in the others. For example, you can't use voicemail settings in the *extensions.conf* file; you can only use them in *voicemail.conf*.

Including Configuration Files

You can't assign extensions in any files other than *extensions.conf*, unless you specifically include those other configuration files. All of the configuration files support an inclusion directive, which allows them to pull settings and commands from another file that you specify, like this:

```
#include myCustomExtensions.conf
```

If you put this line in *extensions.conf*, Asterisk would read the file called *myCustomExtensions.conf* and treat it as though it were a part of *extensions.conf*. The file you included would need to have a structure that is valid for *extensions.conf*, too.

Included configuration files are expected to have a structure that is also valid for the file they are included *into*.

Asterisk Dial-Plan

All calls placed to, from, and through the Asterisk PBX are handled on logical voice pathways. A pathway may consist of a single hard line, like an analog telephone line interfaced with Asterisk via an X100P card, or there may be hundreds of pathways sharing a single physical connection, like a bunch of SIP phones connected to the Asterisk server via an Ethernet interface. In any scenario, the pathways are called *channels*, and the Asterisk PBX's purpose is to handle their voice traffic according to a set of rules known as a *dial-plan*. The effect the dial-plan has on a call is called a *call flow*.

Many conventional PBX systems use dial-plans to deal only with calls that can be placed and answered when a human is present and require an offboard voice mail server with an auto-attendant to answer calls when nobody's in the office. But the Asterisk PBX uses dial-plan for a much larger purpose: complete call processing for both scenarios. The Asterisk dial-plan includes rules that specify what to do when:

- A call comes in on a particular channel or from a certain caller—i.e., if the call is from *this* line, handle it a certain way.
- A call comes in at a certain time of day or on a certain day of the week—i.e., if the call comes in after hours, play an out-of-office message.
- A certain callee does not answer in a timely manner—i.e., if Bob doesn't answer within 25 seconds, try Sue instead.
- A caller dials certain keys in response to recorded prompts—i.e., the caller is prompted to press 1 for sales, 2 for support, and so on, and the proper action is taken.
- A caller is placed on hold or needs to enter a hold queue—i.e., during hold, the user may hear music or a commercial; she may be on hold indefinitely or for a limited time before something else happens.
- A caller performs a multiparty conference or transfers a telephone call to another extension—i.e., several people are participating in a single call, and the dial-plan describes the parameters for that call.
- A subscriber wants to manage his voice mailbox—i.e., Bob presses his voice mail button on his IP phone because his message-waiting indicator is lit.
- …and lots of other scenarios.

The dial-plan is the command center of the Asterisk system. Almost every dial-plan function you can access requires a visit to the dial-plan's main configuration file: *extensions.conf*. This file defines all of the possible call destinations (extensions), which may include users' telephones, hold queues, voice mailboxes, or call processors.

extensions.conf describes call-routing scenarios, called *extensions*. Other configuration files, like *SIP.conf* and *zaptel.conf*, describe endpoint channels and trunk channels—the actual originators and receivers of calls.

In Asterisk, *extensions* aren't defined as physical PBX telephones in the traditional sense, but as what-to-do scenarios that are referenced by numbers. Put another way, each extension answers the question, "What should the PBX do when this extension number is invoked?" The answer isn't always to ring a phone—not all extension numbers are actually related to phones as they tend to be on a traditional PBX. Indeed, on Asterisk, some extensions may be related to voice mailboxes or custom call processors. Extension number 501 might ring a phone, but extension number 105 might play a voice mail announcement and then record the caller's message.

The numbers—phone numbers—provide legacy compatibility so that Asterisk extensions can be used with the old 12-key dial-pad on legacy phones. This conforms to the conventions of the ITU's numbering standard, E.164. But, as VoIP becomes more prevalent, you'll discover that Asterisk can be used with alphanumeric addresses, which are common with SIP.

When the extension asks "What should the PBX do when this extension is invoked," the answer comes in the form of an Asterisk application command—a task-specific call-handling function. There are many application commands for different tasks, and each answers that question with its own functionality. Here's a (very) partial list:

Dial
> The Dial command is one of the most commonly used in Asterisk. Its only purpose is to open a second channel, place a call, and connect that call with the caller on the current channel.

Voicemail
> This command plays back a voice mail greeting and optionally attempts to record a message on behalf of the caller on the current channel.

Background
> The user hears a sound file, and the digits she dials are captured and handled.

Monitor
> This command records the conversation on the current channel and saves it into a file.

Zapateller
> This command plays a special tone designed to block telemarketers on the current channel, much like the popular TeleZapper device marketed by Royal Appliance Manufacturing Co.

System
> A command line is relayed to the Linux shell for immediate execution.

Variables

Asterisk's *extensions.conf* and *voicemail.conf* files can use variables in a manner similar to shell scripts. Variables can be contained within a particular voice channel or be made "global," available to all contexts of the dial-plan. (More on contexts later.) Variables are referenced using curly brackets ({ and }) around the variable name, preceded by a dollar sign ($). Here's an example of a variable called announceBox being assigned, then recalled:

```
exten => 1,1,SetGlobalVar(announceBox=15)
exten => 1,2,NoOp(${announceBox})
```

Built-in variables

Asterisk has a number of built-in variables to handle specific tasks. They behave like functions: ${DATETIME} provides a reference to the current system time, while ${CALLERID} provides caller ID information. Built-in variables are almost always channel-specific—that is, they don't have much meaning outside the scope of a particular call.

The PBX can behave differently depending on the time of day, the caller's identification, and even the channel used to handle a call. The more you use configuration settings, commands, and variables with Asterisk's configuration files, the more the syntax of the configuration files will make sense. For now, let's talk about the place where commands and variables are most often employed: the dial-plan.

The layout of extensions.conf

There are three section categories in the *extensions.conf* file, each headed by a square-bracketed ([and]) string that contains each section's name. The first section category, called [general], allows you to change two settings—static and writeprotect. Neither of these settings has much consequence in learning Asterisk, so you can ignore both of them or even leave them out of the file altogether for now.

The second section category, called [globals], is for defining global variables, which are administrator-assignable values that can be recalled throughout the dial-plan. A good example would be assigning a vague channel name (Zap/1) to a meaningful variable (TELEPHONECOMP), like this:

```
[globals]
TELEPHONECOMP => Zap/1
```

You could also use the globals section to store a list of hard-to-remember channel names for a group of SIP or analog phones:

```
[globals]
CAFEPHONE => SIP/105@oreilly.com:5060
OFFICEPHONE => SIP/106@oreilly.com:5060
CORDLESSPHONE => Zap/2
```

Notice how the structure of a channel name varies—a Zaptel legacy channel has a very simple name structure (Zap/2), but a SIP channel has a complex name structure (SIP/106@oreilly.com:5060). SIP has a URI naming, while legacy channels have no such convenience. Asterisk therefore names legacy POTS and PRI channels by referencing their hardware interfaces (Zap/1, Zap/2, and so forth).

 A channel called SIP/106@oreilly.com:5060 is the same as a channel called SIP/106, as long as 106 is defined as a peer in *sip.conf* and *oreilly.com* is the default SIP domain name. See "SIP Channels," later in this chapter, for more details..

The variables assigned in the globals section don't actually cause Asterisk to do anything. They merely store information that can be used elsewhere in the dial-plan.

Contexts

The third section of *extensions.conf* consists of multiple sections known as contexts. While there can only be one general and one globals section in *extensions.conf*, there can be many context sections. In fact, there can be as many contexts as you like. But, just what *is* a context?

With a cutomizable softPBX like Asterisk, you have the ability to set up limitless dial-plan scenarios. So, you may still have day and night modes, but you may also have lunchtime mode, happy hour mode, etc., depending on your users' needs. Asterisk refers to these modes of operation *contexts*. Contexts organize related extension numbers in the dial-plan so that they create call flows that are explicitly defined for any combination of:

- What channel the call is coming in on.
- What time of day it is.
- What type of endpoint originated the call.
- Specifically, which endpoint originated the call—i.e., was it Bob's phone or Sue's?
- What extensions should be reachable by the current caller.

For example, an initial catchall greeting might say, "Press 1 for Marketing, 2 for Sales, and 3 for Service." Let's say the user presses 1. He may then hear another greeting that says, "Press 1 for the Head of Marketing, 2 for the Secretary of Marketing, or 3 for the Ad Buyer."

So, in one context, option 1 is for the Marketing Department, but in a subsequent context, option 1 is for the head of marketing. Contexts give you, the administrator, programmatic control over call flows, but they also give you access control abilities at the application layer.

Here's the configuration of three contexts to re-create the call flow described in Figure 17-1:

```
[default]    ; (main context)
exten => s,Playback(main-greeting)
exten => s,2,WaitExten( )
exten => 1,1,Goto(Sales,1)
exten => 2,1,Goto(Marketing,1)

[Sales]
exten => s,1,Playback(sales-dept-greeting)
exten => s,2,WaitExten( )
exten => 1,1,Goto(NewCustomers,1)
exten => 2,1,Goto(ReturningCustomers,1)

[Marketing]
exten => s,1,Playback(marketing-dept-greeting)
exten => s,2,WaitExten( )
exten => 1,Dial(${headOfMarketing})
exten => 2,Dial(${secyOfMarketing})
```

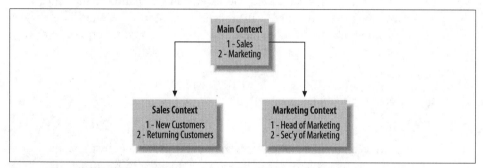

Figure 17-1. Contexts define extensions that are localized and grouped logically

So, extensions 1 and 2 are actually defined several places in the dial-plan, but they have different meaning in different contexts. In the main context, extension 2 transfers the caller to the Marketing Department, but in the Sales context, extension 2 transfers the call to the secretary of marketing's phone.

Including contexts

In the Asterisk dial-plan configuration, it is possible to include one context within another, so that all of the extensions defined in a context are inherited by the context that includes it. Suppose you have this context:

```
[voipco]
include => sales
exten => 102,1,Dial(${TEDSPHONE})
exten => 103,1,Dial(${KELLYSPHONE})
exten => 104,1,Dial(${MADDIESPHONE})
```

Note the include directive just below the section heading voipco that names this context. The include directive causes this context to implicitly inherit all of the directives of the sales context. So, from the voipco context, I can dial extensions 1 or 2 from the sales context in the previous example.

> The default context in Asterisk is called, amazingly enough, *default*.
> This is where calls exist if no other context is specified.

The syntax of extensions

Notice the exten directives that begin each extension definition in the prior examples. In order to establish an extension in *extensions.conf*, you'll need this directive. The second half, the definition, contains three elements, separated by commas:

```
exten =>extension number, priority, application command
```

The first element of the definition is the extension number—generally, this means that a caller or a process has requested the extension by dialing it into their telephone set, or by signaling it during a call transfer. The second element is the priority, or the order in which the third element, the application command, is executed. Consider the following:

```
[sales-after-hours]
exten => 110,1,Answer
exten => 110,2,Playback(sales-greeting)
exten => 110,3,Voicemail(110)
exten => 110,4,Hangup
```

Even though there are four lines of configuration, this is actually just a single extension (110) that makes use of several application commands. In the order defined by the priorities, extension 110 is answered by Asterisk, a sound file called *sales-greeting* is played to the caller, a voice mail message is recorded, and then the channel is hung up. In this way, the question, "What should the PBX do when someone dials 110?" is answered.

Besides user-defined extensions, Asterisk has a very important built-in extension, called the *start extension*, or simply *S*. This is the de facto extension for calls not associated with any other extension—like calls coming in to the Asterisk PBX from a single-line X100P interface. Using the start extension in place of the extension number in a definition tells Asterisk how to handle these calls:

```
[incoming]
exten => s,1,Dial(SIP/105)
exten => s,2,VoiceMail(105)
```

In this case, all unqualified calls are sent first to SIP phone 105 and, if that fails, to voice mailbox 105.

Extension pattern matching

Not all extensions are defined as static strings; some are defined as pattern-matched strings. If you're familiar with regexp or some other pattern-matching concept, this idea is nothing new. Consider the following:

```
914403651964
914403652642
```

Two different strings of digits, right? Not always. When the following pattern mask is applied, the second string and the first string are both the same extension definition, even though they are actually different phone numbers. Take a look at the pattern that makes these two numbers become a single extension definition:

```
_91440XXXXXXX
```

Like all Asterisk pattern masks, this one begins with the underscore character. It's purpose is to match any 12-digit phone number beginning with 91440. When this pattern mask is used in the dial-plan, it causes all numbers that match its pattern to be treated as though they are all the same extension. The following extension definition matches all numbers dialed that begin with 91 and connects them to the public telephone company using the Zaptel channel provided by the X100P card, assuming the variable ${TELCO} contains the name of the Zaptel channel:

```
exten => _91.,1,Dial(${TELCO}/${EXTEN})
exten => _91.,2,Congestion( )
```

The first priority of this extension is to use the Dial( ) application command to connect the caller's channel to the telephone company's line at the phone number contained in ${EXTEN}, a variable which always contains the most-recently dialed sequence of digits on the current channel. If this priority fails, the caller will hear a busy tone, as indicated by the Congestion( ) application command.

The wildcard characters that are used to match patterns consist of:

- X matches any digit from 0 to 9.
- Z matches any digit from 1 to 9.
- N matches any digit from 2 to 9.
- [126-9] matches any digit or letter in the brackets (in this example, 1, 2, 6, 7, 8, and 9).
- . matches one or more alphanumeric characters.
- 1, 2, 3, 4, 5, 6, 7, 8, 9, #, and * are always taken literally.

Matched patterns can be any length. You might wonder why you would need extremely long patterns, but if you're using IVR applications, you may be collecting long data entries like credit card numbers (16 digits or more). While you can use other techniques to collect this data, *extensions.conf* is the only place where you can use powerful pattern matching for extensions. So routing calls based on credit card numbers, social security numbers and so on, is easy to do here. Pattern matching can be applied to both human-dialed and machine-signaled strings.

Asterisk pattern matching is relatively tame compared to regular expressions, which is supported by other soft-switch systems like SIP Express Router. But if you really need complex pattern matching with Asterisk, you can pass data as arguments to the Asterisk Gateway Interface (AGI), where you can use regexp in Perl or Python programs, and then pass it back into the dial-plan.

With the flexibility of Asterisk's application commands, variables, priorities, and contexts, there are nearly infinite ways of combining contexts and extension definitions in the dial-plan.

Special extensions

Special extensions are alphabetic characters that have a logical meaning in the dial-plan. They are used to describe scenarios in which an extension *number* doesn't exist, as when a call first enters the context from the outside world.

A—Asterisk
> This is where calls go if callers press the star key during a voice mail greeting.

i—Invalid
> This is where calls end up if the user has dialed an invalid DTMF or if the dial-plan has a bug that transfers a call to a non-existent extension.

s—*Start*
> The de facto extension that handles all incoming calls on channels that don't specify any other extension when first connecting.

h—Hangup
> This is the extension you can program to handle what to do after the call has been disconnected.

t—Timeout
> This is the extension that handles programmed timeouts—those time limits specified by dial-plan commands.

T—Absolute Timeout
> The extension that handles programmed timeouts triggered by the `AbsolouteTimeout` dial-plan command.

o—Operator
> The extension that handles users who dial "0" in the current context.

talk—Talk Detect
> The extension that handles callers who've been detected talking using talk detection in the dial-plan.

fax
> The special extension that receives calls when Asterisk has detected a fax signal. (This extension is able to receive faxes only when the SpanDSP fax module is

installed; otherwise, it can be used to connect the call with a fax machine on an analog channel.)

String Processing in the Dial-Plan

Since the dial-plan defines call flows, it infers programmatic logic. So, when you set up a dial-plan, you're really programming the PBX. For that reason, Asterisk includes some syntactic capabilities that make your life as a PBX programmer simpler. One of the most important of these capabilities is string processing.

In most dial-plan commands, arguments are supplied as textual expressions. They may contain a mixture of literal values and variables. These expressions may be used in whole or in part—through Asterisk's extensive chopping capabilities. Consider the following:

```
exten => 1,1,SetVar(wholename=DestinysEnd)
exten => 1,2,SetVar(firstword=${wholename:1:8})
```

In this example, ${wholename} is assigned the value DestinysEnd. Then, in the next priority, ${firstword} is assigned the value of the first eight characters of ${wholename}. The syntax for string chopping and substringing is as follows:

```
${substring} = ${string:offset:length}
```

If the offset is less than 0, then it will be an offset from the end of the string, not the beginning. If length isn't supplied, the remainder of the string, from the offset on, will be included in the yielded substring. The following two examples produce identical results:

```
exten => 1,1,SetVar(wholename=DestinysEnd)
exten => 1,2,SetVar(lastword=${wholename:9})

exten => 2,1,SetVar(wholename=DestinysEnd)
exten => 2,1,SetVar(lastword=${wholename:-1:4})
```

In either extension 1 or 2, ${lastword} is *End*.

Getting string lengths

The ${LEN} function measures the length of strings. Use it with SetVar to assign the length of an expression to a variable:

```
exten => 1,1,DBGet(creditCard=${CALLERIDNUM}/creditcard)
exten => 1,2,SetVar(length=${LEN(${creditCard})})
exten => 1,3,GotoIf(${length}!=16?100:200)
exten => 1,100,Playback(thank-you)
exten => 1,101,Hangup( )
exten => 1,200,Playback(invalid-credit-card-number)
exten => 1,201,Goto(default,1,1)
```

In this example, a credit card number is pulled from the Asterisk database. If its length is 16, the system plays a thank you message. Otherwise, it plays an invalid message and changes contexts.

Concatenating strings

As in Perl and PHP quoted string expressions, multiple strings can be concatenated simply by putting them inline:

```
exten => 1,1,SetVar(wholename=DestinysEnd)
exten => 1,2,SetVar(firstword=${wholename:1:8})
exten => 1,3,SetCIDName("Gilneas ${firstword}")
; Caller ID Name is now "Gilneas Destiny"
exten => 1,4,SetCIDName("Gilneas${firstword}")
; Caller ID Name is now "GilneasDestiny"
```

Dial-Plan Command Reference

Each extension you define in the dial-plan can have one of many commands. These commands are, as pointed out earlier, executed in order of their priority. Some of the commands are dependent upon one another. Here is a list of the most commonly used commands.

AbsoluteTimeout

```
AbsoluteTimeout(seconds)
```

Sets absolute maximum time of call, in seconds. Once this time elapses, the call will be disconnected or handled by the t special extension, if one is defined in *extensions. conf*. Could be useful for calling card applications.

AddQueueMember

```
AddQueueMember(queue|[channel])
```

Adds the current channel as a member to a call queue as configured in */etc/asterisk/queues.conf*. If *channel* is specified, uses that channel instead of the current one.

ADSIProg

```
ADSIProg(script)
```

Programs an ADSI (Analog Display Services Interface) script into an ADSI-capable phone. ADSI is a standard developed by the telecom industry for creating text-based interactive applications that augment traditional calling services.

AGI

```
AGI(program|arguments)
```

Executes an Asterisk Gateway Interface program on the current channel. See "The Asterisk Gateway Interface," later in the chapter, for more information on AGI.

Answer

```
Answer( )
```

Answers the incoming call on the current channel. Should be followed up with other commands or the calling party will just hear silence.

AppendCDRUserField

```
AppendCDRUserField(value)
```

Appends the specified string value to the Call Detail Record user field.

 Asterisk's Call Detail Record field map was spelled out in Project 5.1.

BackGround

```
BackGround(filename)
```

Plays the specified GSM format sound file on the current channel and ends, giving control of the context to the next priority, if any, or goes back to the s extension if not. This can be used for IVR prompts like "Dial 1 for sales, 2 for support," and so on. If these numbers are valid within the context, then the context will handle them appropriately.

BackgroundDetect

```
BackgroundDetect(filename)
```

Plays a sound file on the current channel while waiting for the caller to dial some digits or speak. Upon speaking, the user will be transferred to the special talk extension if it exists.

Busy

```
Busy( )
```

Plays a busy signal on the current channel and waits for the caller/callee to disconnect.

ChangeMonitor

```
ChangeMonitor(filename)
```

Changes the filename to which the audio dump of the current channel is being or will be sent, when used with Monitor.

ChanIsAvail

```
ChanIsAvail(channel)
```

Checks to see if the specified channel is available, and store its name in ${AVAILCHAN}. Generally, this is used with the Cut command to reformat the channel name for use later in the dial-plan.

```
; Check if line 3 is available. If not, try line 2, then line 1.
exten => s,1,ChanIsAvail(Zap/3&Zap/2&Zap/1)

; ${AVAILCHAN} may now contain the value: Zap/2-1
; Strip off the session ID, putting just the channel number in ${MyChannel}
exten => s,2,Cut(MyChannel=AVAILCHAN,,1)

; ${MyChannel} would now contain the value: Zap/2
; Dial the intended number on this channel, which we'll assume is in ${PSTTNumber}
exten => s,3,Dial(${MyChannel}/${PSTNNumber})
exten => s,4,Hangup

; If neither line is available, play a busy message.
exten => s,102,Playback(all-circuits-busy-now)
exten => s,103,Hangup
```

CheckGroup

```
CheckGroup(count)
```

Determines whether the current channel's channel group is handling more than *count* calls. If so, redirects current call to priority equal to the current priority plus 101.

Congestion

```
Congestion( )
```

Plays a congestion message to the caller/callee and waits for disconnect.

ControlPlayback

```
ControlPlayback(filename,skip,forward,rewind,stop,pause)
```

Plays the specified sound file, giving fast forward, rewinds and exit controls to the user on the current channel. Dialed digits are used to control playback. The arguments are the name of the GSM sound file, the number of milliseconds to use as an interval for rewinding and forwarding, and the dial-pad digits (including # and *) that will be used to control the playback. Control of the call proceeds to the next priority when the end of the sound file is reached.

```
exten => 2112,1,ControlPlayback(syrinx-song.gsm,4000,*,#,1,0)
```

Cut

```
Cut(expression,delimiter,fieldspec)
```

Function that splits delimited strings. Could be used to split area codes from phone numbers.

```
; OldString is 'A&B&C&D'
exten => s,1,Cut(NewString=OldString,&,2)
; NewString is 'B'
```

DBdel

```
DBdel(keytree/key)
```

Drops the named key from the Asterisk database.

DBdeltree

```
DBdeltree(keytree)
```

Drops a key tree from the Asterisk database.

DBget

```
DBget(varname=keytree/key)
```

Gets a value from the Asterisk database. The following example gets a customer who is calling in to enter some digits and then uses those digits to look up the extension of that customer's service agent in the Asterisk database. Then, it transfers the call to that extension.

```
[ServiceIVR]
exten => s,1,Answer( )
exten => s,2,Background(enter-your-8-digit-account-number.gsm)

exten => _NXXXXXXX,1,dbget(repExten=${EXTEN}/rep)
exten => _NXXXXXXX,2,Goto(${repExten})
```

DBput

```
Dbput(keytree/key=value)
```

Sets a value in the Asterisk database. value can be a variable or a string literal.

DeadAGI

```
DeadAGI(program)
```

Runs an AGI program on a channel after the call has ended.

Dial

```
Dial(channel/identifier[,timeout][,options][,url])
```

Makes an outbound call on the specified channel and bridges it with the current channel. Dial is one of the most capable, and most frequently used commands in Asterisk. Here are its arguments:

channel/identifier

In the form *channel type/channel number/phone number*. So, a using Zap/1/ 5551212 would dial 5551212 on the first Zaptel channel. Using SIP/105 would

```
```

attempt to call the SIP peer defined as 105 on the local server. Using `SIP/`
`ted@sip.oreilly.com:5060` would attempt to open contact with the user ted at
the *oreilly.com* SIP domain on the SIP-standard port of 5060.

timeout

Specifies the number of seconds to wait for an answer before giving up. As a
rule, the current channel hears a ringback tone during this time.

options

A string of characters that represent different possibilities:

t   Allows the called user to transfer the call.

T   Allows the calling user to transfer the call.

r   Generates a ringback tone for the calling party immediately, rather than
    waiting until the remote system signals ringback. Avoid this option, as it
    stops real call progress indicators from being heard by the caller. Cannot be
    used with m.

R   Generates a ringback tone for the called party upon connection.

m   Plays music-on-hold to the caller until the call is answered. Cannot be used
    with r.

M(macro)

Runs the macro called `macro` upon connection of the call.

P   Forces the caller to provide caller ID by way of Privacy Manager before con-
    necting the call.

h   Permits the called party to hang up by pressing the star key (*).

H   Permits the caller to hang up by pressing the star key (*).

C   Resets the CDR for this call. This is similar to using the `NoCDR` command.

g   When the call is disconnected, continues to execute more commands in the
    current context.

A(filename)

Plays a GSM sound file to the called party before connecting the call.

S(seconds)

Automatically hangs up the call the specified number of seconds after
connection.

D(digits)

Once the callee answers, sends *digits* as a DTMF stream, then connects
the call to the current channel. Should be used only on Zaptel channels or
G.711 calls.

L(x[:y])

Time-limits the call to *x* ms, give an audible warning when *y* ms are left,
repeated every *z* ms) Only *x* is required, *y* and *z* are optional. The voice
heard for the warnings is that of a sound macro that's built in to Asterisk.

> *f*   Forces caller ID to match the extension number of the outgoing channel. For example, some phone companies don't allow caller ID that doesn't match the trunk that it's being used with.

url

Finally, the optional *url* parameter will be sent to the called party's endpoint upon connection, if the endpoint supports URLs (some IP phones support this).

DigitTimeout

```
DigitTimeout(seconds)
```

Sets the maximum time out, in seconds, allowed between dialed digits on the current channel only.

Directory

```
Directory( )
```

Presents the user on the current channel with an IVR directory of voice mailboxes.

DISA

```
DISA,password|context
```

Gives the user direct inward system access capability, or the ability to obtain dial-tone and the ability to originate calls from the current channel in the specified context. User must enter a password using the dial-pad that matches the specified password.

Echo

```
Echo( )
```

Immediately plays back the incoming audio signal back on the same channel it's coming from. Used to give testers an idea of how significant echo is on any given connection. Users can exit to the next priority by pressing the pound key (#).

EnumLookup

```
EnumLookup(extension)
```

Looks up the specified extension number in ENUM and returns the SIP or TEL URI to the variable called ENUM. If a SIP entry exists, the next priority will be reached. If a TEL entry exists, the next priority plus 50 will be reached. If no entry is found, the next priority plus 100 will be reached. This example least-cost-routes all calls that start with a 9 to SIP via ENUM first before routing them to the PSTN.

```
; Any calls starting with 9 will undergo ENUM lookup.
exten=> _9[1-9]XXX.,1,BackGround(enum-doing)
exten=> _9[1-9]XXX.,2,EnumLookup(431${EXTEN:1})
; ${EXTEN:1} is the dialed number with the leading 9 removed.
```

```
;EnumLookupsets ${ENUM} on success. On failure jumps to priority+101.
exten=> _9[1-9]XXX.,3,BackGround(enum-successful)
exten=> _9[1-9]XXX.,4,Dial(${ENUM},30)
exten=> _9[1-9]XXX.,5,Goto(104)
; No answer on SIP after 30 seconds, so dial the number on the PSTN
exten=> _9[1-9]XXX.,103,BackGround(enum-failed)
exten=> _9[1-9]XXX.,104,Dial,${TRUNK}/${EXTEN:1}
```

Festival

```
Festival(speech text)
```

Says the specified text using the Festival voice synthesizer, if it's installed. The audio is sent to the current channel, and the user has the option of interrupting by dialing digits.

```
exten => 200,1,Answer
```

Flash

```
Flash()
```

Sends a flash signal on the current Zaptel channel. Works only on Zaptel channels. Not useful unless the connected trunk has flash signaling features like call-waiting or three-way calling.

GetGroupCount

```
GetGroupCount()
```

Obtains the number of channels in the group the current channel belongs to and stores it in ${GROUPCOUNT}.

Goto

```
Goto([priority],[extension],[context])
```

Immediately changes the current channel to the specified context, priority, or extension.

```
Goto(101) ; go to priority 101 in current extension
Goto(incoming,101) ; go to extension 101 in context 'incoming'
Goto(incoming,101,2) ; go to priority 2 in extension 101 in context 'incoming'
```

GotoIf

```
GotoIf(condition?target1[:target2])
```

Go to *target1* if the condition is true or to *target2* if the condition is false.

When *target1* and *target2* are Goto-compatible targets, see "Goto" for target syntax.

condition is a string of the same expression syntax used for normal string processing within the dial-plan. This example does some simple ACD when an incoming call dials the 210 extension:

```
; See if area code is 216 and go to priority 3 or 5, depending.
Exten => 210,1,SetGlobalVar(myAreaCode=${CALLERIDNUM}:0:3)
exten => 210,2,GotoIf($[${myAreaCode}=216]?3:5)
exten => 210,3,Dial(${SPHONE1},15,rt)
exten => 210,4,Hangup
exten => 210,5,Dial(${PHONE2},15,rt)
exten => 210,6,Hangup
```

GotoIfTime

```
GotoIfTime(timecondition,[priority],[extension],[context])
```

Immediately changes the current channel to the specified context, priority, or extension, if the supplied time condition is true. The time condition is provided in the format of time-based context includes, which are described in "Time-based context includes" in Chapter 12.

Hangup

```
Hangup( )
```

Hangs up the current channel, ending the call, but going to the next priority for this extension, if one exists.

LookupBlacklist

```
LookupBlacklist( )
```

If there's a caller ID for the caller on the current channel, looks up her name and number in the Asterisk database key tree called blacklist. If it exists, the command changes the priority to the next priority plus 100.

MailboxExist

```
MailboxExist(mailbox)
```

Checks to see if the specified voice mailbox exists. Useful for intercepting misdialed numbers.

Math

```
Math(result,operation)
```

Performs a simple mathematical operation and assigns its value to the variable named by *result*. Here are some examples:

```
exten => 1,1,Math(var,3*2) ; multiplication
exten => 1,2,Math(var,10-2) ; subtraction
exten => 1,3,Math(var,${EXTEN}>1000) ; greater-than
exten => 1,4,Math(var,${EXTEN}>=1000) ; greater-than or equal-to
```

The Math command is available only in Asterisk 1.01 and higher.

MeetMe

MeetMe(*confno*[|*options*][|*password*])

Creates an Asterisk conference bridge. Requires Zaptel hardware to be installed, even if all bridge participants aren't on Zaptel channels. This is because the Zaptel hardware provides the conference mixing and timing functions needed for multiparty bridging. The configuration file */etc/meetme.conf* is used to configure the Asterisk conference mixer. (Users without Zaptel hardware may consider compiling the *ztdummy* timing module instead.) *options* may be:

m   Sets monitor mode. Listen to conference only; current channel will not be heard by other participants.

t   Sets talk mode. Current channel will be heard by other participants, but will not be able to hear other participants.

p   Allows user on current channel to leave the conference by pressing the pound key (#).

X   Allows user on current channel to exit the conference by dialing a valid single-digit extension within the current context.

v   Uses video mode if the endpoint supports video.

q   Stops the conference mixer from playing alert sounds when users enter and leave the conference.

d   Creates a conference.

D   Creates a conference, prompting for PIN number. When prompted to establish the PIN, if the user wants no PIN created for the conference, he should press the pound key (#).

M   Plays music-on-hold when the conference has only a single participant.

e   Selects any existing empty conference (i.e., party line).

E   Selects any existing empty conference that doesn't require a PIN (i.e., party line).

b   Runs the AGI script specified in ${MEETME_AGI_BACKGROUND}. The script affects only Zaptel channels and is useless for SIP channels.

a   Sets admin mode, allowing the current channel to lock or unlock the conference. This has the effect of freezing the conference so that nobody can join it.

s   Enables the conference menu, which prompts the participant or conference admin appropriately. The conference menu is called by pressing the star key (*) during the conference. For participants, the menu allows the user to toggle mute on her channel only by dialing 1. For administrator participants, the menu adds the ability to toggle locking/unlocking of the conference by dialing 2.

A   Marks a user. Marked users can be used to trigger automatic teardown of the conference bridge when they leave and to prevent anybody from using the conference bridge until they connect.

x   Terminates the conference and disconnects all users when last marked user exits.

w   Waits until a marked user enters the conference before allowing any participants to send audio. While waiting, such participants will hear music-on-hold.

MeetMeAdmin

```
MeetMeAdmin(conference,command[,user])
```

Issues conference administration commands to Asterisk conferences. The options are as follows:

K   Kicks all users from the conference. Conference will be destroyed if it were dynamically created.

k   Kicks the specified user from the conference.

L   Locks the conference.

l   Unlocks the conference.

M   Mutes the conference.

m   Unmutes the conference.

MeetMeCount

```
MeetMeCount(conference,[variable])
```

Plays back the conference participant count on the current channel. If a variable is specified, records the count in that variable and skips playing back the count.

Milliwatt

```
Millwatt( )
```

Generates a 1,000 Hz signal at 0 dBm on the current channel. This tone can be used by a lineman to measure transmissions on a phone line from a remote site, such as a residence or office for which the Asterisk server provides an FXO port.

Monitor

```
Monitor(format,name[,M])
```

Records the telephone conversation from the current channel to a sound file or caller and callee parts of the conversation to separate sound files. *format* is the desired sound file format. The default is wav. *name* is the base name of the output file(s), and the M flag can be added if you want SoxMix to automatically mix the two parts of the conversation into a single file upon termination of the call. The Sox package must be compiled and installed in order for this to work.

MP3Player

```
MP3Player(filename)
```

Plays an MP3 sound file on the current channel. Playback can be stopped and the priority incremented by dialing any digit.

MusicOnHold

```
MusicOnHold(class)
```

Plays hold music indefinitely from the specified *class*. Music-on-hold classes are established with simple directives in */etc/asterisk/musiconhold.conf*. Normally these directives just point a certain class to a certain directory that contains MP3 files, but as this example shows, you can also use a streaming source:

```
; musiconhold.conf
default => mp3:/var/lib/asterisk/onhold,http://www.streamvoip.us/royalty-free:8000
```

The config needs a blank directory called onhold in */var/lib/asterisk* using:

```
# mkdir /var/lib/asterisk/onhold
```

NoCDR

```
NoCDR( )
```

Insures Asterisk will not save any CDR data for the current call.

NoOp

```
NoOp(value)
```

Echos values to the console for debugging purposes.

ParkedCall

```
ParkedCall(extension)
```

Answers the parked call at the specified parked call extension.

Playback

```
Playback(filename)
```

Answers the call on the current channel and plays the specified sound file.

Playtones

```
Playtones(tone)
```

Play a tone from among those specified in *letc/asterisk/indications.conf* while executing other commands. Tone will continue to play indefinitely until StopPlaytones is called. Can also accept a tone list that consists of frequencies and durations, rather than a simple tone name. The tone list must be in the format specified in *tones.conf*.

PrivacyManager

```
PrivacyManager( )
```

Requires a phone number to be entered on incoming calls when no caller ID signals are received. If the user doesn't enter the phone number, the priority advances to next plus 100. Currently supports only 10-digit phone numbers. Configured by *letc/asterisk/privacy.conf*.

Queue

```
Queue(queue|options)
```

Puts the current call into a call queue specified by *queue*. *options* is one or more of the following:

t   Permits the callee to transfer the call.

T   Permits the caller to transfer the call.

d   Marks the call as minimum delay, for use with modem calls, presumably on Zaptel channels only.

H   Permits the caller to hang up by pressing the star key (*).

r   Rings while queued instead of playing music-on-hold.

The hold queue configuration is stored in *letc/asterisk/queues.conf*. Consult its comments for configuration instructions.

Random

```
Random(likelihood,priority)
```

Goes to a random priority based on the supplied likelihood, which is expressed as a percent. Useful for controlling traffic splitting.

Read

```
Read(variable[|filename][|maxdigits])
```

Reads a string of digits from the user on the current channel, storing them in *variable*. Optionally, *filename*, a sound file, can be played prior to receiving the digits. *maxdigits* indicates the maximum number of digits that you'll allow, which can be no greater than 255.

```
exten => s,1,Answer( )
exten => s,2,Read(credCardNum|enter-cred-num|16)
```

```
exten => s,3,Read(credExpDate|enter-exp-date|4)
exten => s,4,Playback(thank-you)
exten => s,5,Noop(${credCardNum} ${credExpDate})
```

ResetCDR

```
ResetCDR( )
```

Resets CDR data for the current call, as if it were beginning now.

ResponseTimeout

```
ResponseTimeout(seconds)
```

Sets maximum timeout awaiting a dialing activity from the user. Default is 15 seconds, after which time the call will be transferred to t or hung up if t doesn't exist.

Ringing

```
Ringing( )
```

Plays a ringback tone on the current channel.

Rpt

```
Rpt( )
```

Allows ham radio systems and two-way repeaters to be interfaced with Asterisk through special hardware, so that radio calls can be signaled from the PBX. For more information, visit *http://zapatatelephony.org/app_rpt.html*.

SayAlpha

```
SayAlpha(string)
```

Audibly says the *string*, one character at a time, on the current channel. Besides letters and number, SayAlpha has sounds for #, *,-,+, !, @, and $.

SayDigits

```
SayDigits(string)
```

Audibly says the string, ignoring any non-numeric characters.

SayNumber

```
SayNumber(string)
```

Uses full-word pronunciation to say the number, from 0 to 99 million. If the string is 405120, the caller will hear "Four hundred five thousand, one hundred twenty."

SayPhonetic

```
SayPhonetic(string)
```

Says the *string* using NATO phonetic pronunciation. Recommend using Festival instead for most applications.

SayUnixTime

```
SayUnixTime([unixtime],timezone,format)
```

Says the date and time on the current channel. *unixtime* is a Unix timestamp, the number of seconds since January 1, 1970. If *unixtime* is omitted, the current time is used. *timezone* is a timezone consistent with the configuration of */usr/share/zoneinfo*. *format* is a string that uses the strftime(3) function's convention. For more information, see the manpage for strftime(3).

SendDTMF

```
SendDTMF(digits)
```

Sends the DTMF *digits* on the current channel.

SendText

```
SendText(text)
```

Sends a text message to the endpoint on the current channel. Works only with devices that support text messaging, such as SIP clients.

SendURL

```
SendURL(url[,wait])
```

Sends the device on the current channel a web page URL to display, if it can. Devices that don't support HTML transport will result in going to the next priority plus 100.

SetAccount

```
SetAccount(account)
```

Sets the CDR account code to be used for this call. Also, stores the account code in ${ACCOUNTCODE} for further reference within the dial-plan.

SetCallerID

```
SetCallerID(newcallerid)
```

Overrides the existing caller ID string for the current call. SetCallerID is overridden if SIP peer has a fromuser configuration.

SetCDRUserField

```
SetCDRUserField(user)
```

Sets the name of the user to appear in the CDR record for this call.

SetCIDName

```
SetCIDName(name)
```

Sets the caller ID name for originating calls from the current channel, using, for example, the Dial command.

SetCIDNum

```
SetCIDNum(number)
```

Sets the caller ID number for originating calls from the current channel.

SetGlobalVar

```
SetGlobalVar(variable=value)
```

Sets a global dial-plan variable to the value supplied. Value can be another variable's value. Here are some examples:

```
; set the name of a PSTN trunk to a Zaptel channel name
exten => _N*,1,SetGlobalVar(PSTNtrunk=Zap/1)
; store the extension in a variable
exten => _N*,2,SetGlobalVar(CallInProgress=${EXTEN})
exten => _N*,3,Noop(${CallInProgress})
```

SetLanguage

```
SetLanguage(language)
```

Changes localization setting for sound playback of digits, characters, and fully pronounced numbers.

SetMusicOnHold

```
SetMusicOnHold(class)
```

Sets default hold music class for the current channel.

SetVar

```
SetVar(variable=value)
```

Sets a local variable (local to this call) to the value supplied.

SIPdtmfMode

```
SIPdtmfMode(option)
```

Changes DTMF mode during a SIP call in progress on the current channel. *option* is one of these three: info, inband, or rfc2833. inband should be used only with G.711.

SMS

```
SMS(options)
```

Sends and receives SMS (short messaging service) messages from within the dial-plan. For more information, see *http://www.smsclient.org/*.

SoftHangup

```
SoftHangup(channel)
```

Ends the call on the specified channel. To hang up the current channel, use Hangup( ) instead.

StopMonitor

```
StopMonitor( )
```

Stops monitoring a channel that is currently being monitored using Monitor( ).

StopPlaytones

```
StopPlaytones( )
```

Stops playing a tone list or a tone. Returns audio on the current channel to the call in progress.

StripLSD

```
StripLSD(number)
```

Strips the least-significant digits—i.e., from the right of the current extension. The number of digits to strip is specified by *number*. So, if extension 5241900 stripped down to 1900 at the first priority, processing would continue at the second priority of extension 1900.

StripMSD

```
StripMSD(number)
```

Strips most significant digits—i.e., from the left of the current extension. The number of digits to strip is specified by *number*. Continues processing at the next priority in the new extension. So, if extension 5241900 stripped down to 524 at the first priority, processing would continue at the second priority of extension 524.

System

```
System(command)
```

Executes a system command in a shell as the user that runs Asterisk. Can substitute variables within the command line, as in this example, which creates a new directory for every call before recording it into a sound file in that directory using Monitor( ):

```
exten => s,1,Answer( )
exten => s,2,System('mkdir /var/spool/calls/${exten}')
exten => s,3, Monitor(WAV,/var/spool/calls/${exten}/call,M)
```

Transfer

```
Transfer(extension)
```

Transfers the call on the current channel to the specified extension.

VoiceMail

```
VoiceMail([options]box)
```

Connects the current channel to a local voice mailbox specified by *box*. *options* is one or more of the following (or none of them):

s   Skips the voice mail instructions ("Please leave your message after the tone. When done, hang up, or press the pound key.") and just plays the subscriber's greeting.

u   Causes the unavailable message to be played (instead of the busy message). By default, the message says, "The person at extension 2112 is unavailable," but the voice mail subscriber may record his own unavailable message if he wants.

b   Causes the busy message to be played (instead of the unavailable message). By default, the message says, "The person at extension 2112 is busy," but can be overridden with a recorded message by the subscriber.

VoiceMailMain

```
VoiceMailMain(box)
```

Connects the current call to the generic voice mail subscriber greeting. If *box* is given, the voice mail server assumes the specified subscriber is trying to log in to retrieve messages or manage her voice mailbox.

 Chapter 14 describes how to configure Asterisk's voice mail system using the */etc/asterisk/voicemail.conf* file.

Wait

```
Wait(seconds)
```

Waits for the number of seconds specified before proceeding to the next priority.

WaitExten

```
WaitExten(duration)
```

Waits for the number of seconds specified for the user on the current channel to enter a valid extension; otherwise, proceeds to next priority. Also, can accept the duration in decimal numbers for tenths of a second (1.2, 10.5, 0.5, etc.).

WaitMusicOnHold

```
WaitMusicOnHold(seconds)
```

Identical to `Wait()`, but plays the music-on-hold class associated with this call or context while the caller is waiting.

Zapateller

```
Zapateller()
```

Sends the do-not-call signal to incoming callers.

ZapBarge

```
ZapBarge(channel)
```

Allows caller to clandestinely listen in on another channel's call in progress, as long as it's a Zaptel channel. Neither party in the conversation will know that the call is being monitored by the "barging" caller. If you don't provide a channel name, the caller will be prompted for one. Entering 1# would listen in on Zap/1, while entering 3# would listen in on Zap/3.

ZapScan

```
ZapScan()
```

Scans Zaptel channels to monitor calls. Pressing # moves to the next channel. Pressing the star key (*) exits to the next priority.

Asterisk Channels

While *extensions.conf* is the main place Asterisk's dial-plan is configured, other files are needed to set up the VoIP and TDM interfacing necessary to allow the Asterisk server to communicate with the outside world. These files include *zapata.conf*, *zaptel.conf*, and *sip.conf*.

Zaptel Channels

Digium's Wildcard interface cards enable legacy channels so Asterisk can communicate using T1s and analog phones or POTS lines. Chapter 3 describe how to obtain and install Asterisk with the Linux drivers for a kernel interface to the Wildcard product line, an interface known as Zaptel.

Once installed, the *ztcfg* program, included with the drivers, can be used to query the *zapata.conf* file to validate your Zaptel channel configuration. These files determine:

- The ID number for each legacy interface channel that will be referenced in *extensions.conf*. Each card must be assigned a unique number.

- The kind of signaling that each legacy interface card will use (T1, FXO, FXS, etc.).
- What localized ringback tones will be used with each interface card. English ringback tones sound different than American ones. Consult the latest Zaptel build to find out which localities are currently supported.
- What telephony features are enabled on each channel (caller ID, autocallback, etc.).
- How distinctive ringing is handled on trunks and how it is generated for analog phones connected to the server.

Executing */usr/sbin/ztcfg* will tell you if the *zaptel.conf* file is valid for the hardware you've got installed. If *ztcfg* exits without giving you any output, your configuration is valid. If you'd like to see more verbose output, add a -v or -vv.

Using analog interfaces with Asterisk

The *zaptel.conf* file contains information used by Asterisk to determine what interface modules you're using with your Wildcard hardware. Each section in the *zaptel. conf* file describes a single interface's configuration. The following example *zaptel. conf* is for a system with two X100P cards and one TDM400P card with four FXS interfaces on it. It can support two POTS lines and four analog telephones:

```
; zaptel.conf example
loadzone = us
defaultzone=us

; X100P cards
fxsks=1,2

; TDM400P card
fxoks=3-6
```

Depending upon the analog interface cards you've selected, you'll need configuration entries similar to those in Table 17-1.

Table 17-1. zaptel.conf analog interface configuration combinations

Card	Module configuration	zaptel.conf entry
1 X100P card	1 FXO interface	fxsks=1
2 X100P cards	1 FXO interface on each card	fxsks=1-2
1 S100U card	1 FXS interface	fxoks=1
2 S100U cards	1 FXS interface on each card	fxoks=1-2
1 TDM400P card	4 FXO interfaces	fxsks=1-4
1 TDM400P card	4 FXS interfaces	fxoks=1-4
1 TDM400P card	2 FXO interface, 2 FXS interfaces	fxsks=1-2

Card	Module configuration	zaptel.conf entry
2 TDM400P cards	4 FXO interfaces on one card; 4 FXS interfaces on the other card	fxsks=1-4
2 TDM400P cards	2 FXO interfaces on the first card; 2 FXS interfaces on the first card; 4 FXO interfaces on the second card	fxsks=1,2,5,6,7,8

Besides FXO and FXS, Zaptel channels can be configured to use other types of signaling, including the T1 family, E&M, and in-band single frequency (SF). It's unlikely you'll need to use SF unless you're interfacing with some exotic legacy hardware.

> *zaptel.conf* is the only Asterisk configuration located in */etc* instead of */etc/asterisk*.

While *zaptel.conf* establishes the choice of signaling for each piece of interface hardware, another file, *zapata.conf*, defines the telephony configuration of each channel. It establishes what telephony features the channel can use (caller ID, call-waiting, and so on). It describes how the channel behaves when used with distinctive ringing (see Project 12.5 for an example of a distinctive ring configuration) and whether the channel fits into a T1 interface configuration (T1s carry 24 channels).

In *zapata.conf*, the configuration parameters of each channel occur *before* the channel is designated with a number. (This is different from *extensions.conf*, where the extension number occurs at the top of its own section.) Consider the following *zapata.conf* file, which defines channels to correspond with the preceding *zaptel.conf* file:

```
[channels]
language=en
context=default
signalling=fxs_ks
usecallerid=yes
hidecallerid=no
callwaiting=yes
callwaitingcallerid=yes
threewaycalling=yes
transfer=yes
callreturn=yes
channel => 1
channel => 2

signalling=fxo_ks
channel => 3
channel => 4
channel => 5
channel => 6
```

For channels 3 through 6, all but the signalling setting are inherited from the previous definitions used first for channels 1 and 2. Table 17-2 lists commonly used *zapata.conf* settings and describes what they do.

Table 17-2. zapata.conf frequently used channel configuration options (default values are in boldface)

Setting	Description	Common values
accountcode	Free-form account code to be associated with this channel's calls in the CDR	—
adsi	Whether or not to use analog display services interface signals with this channel	yes; **no**
cadence	Establishes a dinstinctive ring pattern for this channel	length on, length silent, {length on},{length silent}
callerid	For trunk interfaces, overrides the caller ID number for this channel; to retain the supplied caller ID number, use asreceived	asreceived
callreturn	Whether or not to allow auto call return via *69	yes; **no**
callwaiting	For FXO channels, whether or not to allow call-waiting	yes; **no**
callwaitingcallerid	Whether to allow caller ID signals for call-waiting calls	yes; **no**
context	The context calls originated from this channel are directed to	default
dring*n*	Recognizes the ring pattern for incoming distinctive ring of type *n*	
dring*n*context	Sets the context in which to place incoming calls that match ring pattern *n*	
echocancel	Enables echo cancellation on this channel to reduce unwanted echo between IP destinations and legacy endpoints or the PSTN	yes; **no**
echocancelwhenbridged	Enables echo cancellation when the call path is all-TDM; recommend disabling this option on all channels	yes; **no**
echotraining	Enables automatic calibration of echo cancellation on Zaptel channels	—
flash	The length of time in ms to use for flash-hook signaling	750
group	The number of the outgoing rollover group this channel is a member of	—

Table 17-2. zapata.conf frequently used channel configuration options (default values are in boldface) (continued)

Setting	Description	Common values
hidecallerid	Whether or not to send caller ID signals on this channel	yes; no
jitterbuffers	Sets up jitter buffering for all calls on this channel, each buffer is 20 ms in length	4
language	The local language to be used with this channel	—
mailbox	The voice mailbox associated with this channel; produces stutter dial-tone when message waiting	—
musiconhold	Which class of music-on-hold will be played on this channel	default
overlapdial	Allows dialed DTMF digits to overlap	yes; **no**
rxwink	The length of time in ms to wait before receiving DID signals, if applicable	0
signalling	The signaling protocol to use with this channel	em; em_w; featd; featdmf; fxs_ls; fxs_gs; fxs_ks; fxo_ls; fxo_gs; fxo_ks; pri_cpe; pri_net
switchtype	The type of PRI switching to use on this channel, if applicable	national; dms100; 4ess; 5ess; euroisdn; ni1
threewaycalling	Whether to permit three-way calling	yes; no
transfer	Allow flash-hook call transfer during three-way calling	yes; no
usecallerid	Whether to use caller ID signals from this channel	yes; no
usedistinctiveringdetection	Whether or not to try to pick up distinctive rings on this channel	yes; no

Using T1 interfaces with Asterisk

Zaptel channels can also be used on T1 or E1 interfaces. Digium's T100P, T400P, T405P, T410P, TE110P, TE405P, and TE410P cards offer support for as many as four T1 or E1 circuits per card. These cards come equipped for different bus voltages, so double-check yours and indicate whether you need 3.3 volts or 5 volts when ordering. Cards with model numbers that begin with *T* work with T1s, while *TE* cards work with T1s or E1s.

In order to use any Zaptel interfaces with Asterisk, the right kernel modules must be loaded at boot time. For T1 and E1 interfaces, the module to load is *wct1xxp* for single-span T1/E1 cards and *wc4xxp* for quad-span T1/E1 cards. TE110P and TE4xxP cards use *wcte11xp* and *wct4xxp*, respectively.

Most T1 setups will use PRI signaling for trunking of PSTN dial-tone. But Zaptel T1 channels could be used for other purposes as well. Connected to a channel bank, a T1 channel can support up to 24 legacy phones. A T1 channel can also connect directly to another PBX that also supports T1/E1 connections.

The most common T1 channel, for a PRI trunk circuit, is straightforward to set up. First, a span definition in *zaptel.conf*. A span is a single leg of a point-to-point T1 connection link, defined at the data link layer. Just as DSU/CSUs deal with only a particular span of a T1 circuit, Zaptel deals with only a single span. Spans are defined in the following format:

```
span=span number,timing,LBO,framing,coding
```

`span number` is a value you assign to this span to identify it. `timing` can be 0, 1, or 2. 1 and 2 means that Zaptel will use this span as a primary or secondary synchronization source, respectively. 0 means Zaptel won't use it as a synchronization source. `LBO`, or line build out, is a value that tells Zaptel the distance to the T1/E1 smart jack and roughly what gain level ought to be used to accommodate for attenuation. Most setups will use a value of 0, which is good for up to 133 feet or roughly 40 meters. `framing` is an option that describes what kind of T1 framing method to use—superframe, extended superframe, etc. Most T1 channels will use `esf` for extended superframe. Finally, `coding` is used to describe the data coding technique, which will always be `b8zs` for T1s and `hdb3` for E1s.

B8ZS, also called clear channel, is the most popular coding scheme used on T1 circuits. B8ZS uses an electrical signaling trick called bipolar violations to mark time syncs into the stream of data. The receiving device recognizes these bipolar violations and uses it to keep synchronized with the sending device.

Following the `span` declaration, `dchan` and `bchan` options describe which channels are the bearer channels and which the signaling channel:

```
; zaptel.conf
span=1,1,0,esf,b8zs
bchan=1-23
dchan=24
```

Corresponding to this configuration are entries in the *zapata.conf* file, that describe how the telephony features will work on the channels served by this T1:

```
; zapata.conf
switchtype=national
context=default
signalling=pri_cpe
group=1
callerid=1-800-534-1550
channel => 1-23
```

zaptel.conf sets up a T1 span, configures the 24 DS0 channels on the T1, and sets the framing and coding. Then *zapata.conf* configures the telephony apparatus for those channels: national PRI signaling, a custom caller ID string, and the default dial-plan context. The channels are grouped into a single rollover group (1). Finally, on the last line, these settings are applied to Zaptel channels 1 through 23. Within the dial-plan, these channels will be referenced as Zap/1, Zap/2, and so on.

 E1 circuits will have 31 bearer channels and typically use HDB3 coding instead of B8ZS. Keep that in mind if you're working outside North America.

SIP Channels

Asterisk implements SIP only partially. Though SIP itself describes a total PBX replacement approach using VoIP, Asterisk employs SIP mainly for connecting IP phones and for trunking to other systems that also use SIP—like Zaptel. Asterisk deals with its SIP subsystem in terms of channels: legs of a call. Two channels are needed to complete a call between two SIP phones. A call between an analog phone and a SIP phone also requires two channels: one Zaptel channel and one SIP channel.

Asterisk refers to devices that communicate with it by SIP *peers*. A SIP channel is established when a call is received from, or directed to, a SIP peer. SIP phones are peers, and so are remote SIP servers and SIP ATAs. Anything with a SIP UA and SA is considered a peer, and all SIP peers are configured in the same file: *sip.conf*.

The *sip.conf* file has been used throughout the book to set up various projects. In Project 3.1, a basic SIP peer was set up to allow an IP phone to interact with Asterisk's default voice menus and prompts. In Project 4.2, a group of SIP phones was configured to behave as a ring group.

There's almost nothing you can't do with a SIP channel that you can with a traditional TDM channel. Those few things you can't do—creating your own ring cadences, for example—have their own SIP equivalents anyway. Instead of cadences, most SIP phones just let you use WAV or GSM sound files with distinctive ringing.

SIP peer configuration

The */etc/asterisk/sip.conf* file is organized into a general section that's followed by individual, peer-specific sections. The general section establishes parameters that apply globally to Asterisk's SIP module, while the peer-specific sections deal only with individual SIP endpoint configurations.

In the general section, you can establish what codecs are allowed and restricted for all SIP peers, the default context for incoming SIP calls, which type-of-service tags will be used with SIP media channels, and whether or not peers will be authenticated.Table 17-3 shows the settings that can be used in the general section of *sip. conf*:

Table 17-3. Asterisk general SIP configuration settings (default values are in boldface)

Setting	Description	Possible values
allow	Used with `disallow`; describes which codecs can be used with SIP peers on this server.	mlaw; Alaw; gsm; g729; g723; g726; speex; **all**
autocreatepeer	Causes Asterisk to forgo all SIP authentication; probably useful only in a carrier network where there are substantial controls on network access.	yes; **no**
bindaddr	The IP address you want Asterisk's SIP SA to listen on, if other than one that's already bound.	—
canreinvite	Allows SIP peers to establish media channels directly to each other using the REINVITE method instead of using the server as a call path.	yes; **no**
context	The default context in which to place incoming SIP calls.	—
defaultexpirey	The interval, in seconds, between registration attempts when Asterisk is used as a SIP client.	**120**
disallow	Specifies codecs that won't be permitted with any SIP channels unless specifically allowed for at the channel level.	—
externip	The IP address to use in SIP headers if the Asterisk server is behind a static NAT firewall. The address you use would actually be bound on the firewall, not on the Asterisk server. See also `nat`.	—
fromdomain	Sets the domain name to use for outgoing SIP calls when Asterisk is used as a SIP client.	—
maxexpirey	The longest time, in seconds, a registered SIP client can remain so before reregistering.	**3600**
nat	Tells the Asterisk server whether or not it is behind a NAT firewall.	—
port	The port number to bind the SIP SA to, if other than the default.	**5060**
srvlookup	Allows Asterisk to call SIP users by their SIP URI via a DNS lookup. Only works when DNS provides an SRV record for the SIP domain in question.	yes; **no**
tos	Establishes the ToS bits sent with SIP channels' RTP media packets on this server. To specify DiffServ Expedited Forwarding, use `0xB8`.	—
useragent	Specifies a custom SIP user agent header. Normally, this header describes the type of IP phone being used to place the call.	—

Once you've established the Asterisk server's global SIP peer functionality in the general section, you can configure individual peers—the actual IP phones and trunk connections. A few settings, like `allow` and `disallow`, are applicable to the individual peers as they are to the general section, as shown in Table 17-4.

Table 17-4. Asterisk SIP peer settings (default values are bolded)

Setting	Description	Possible values
accountcode	Associates all calls to and from this peer with an account code in the CDR. Can be used for billing purposes.	—
Amaflags	Standardized billing flags for use in CDR records.	**default**; omit; billing; documentation
callgroup	The number of the call group that this SIP channel is a member of; default is none.	—
Canreinvite	Allows the SIP peer to be used in calls with independent call paths via the SIP REINVITE method.	yes; **no**
Context	The dial-plan context for calls from this peer.	**default**
Defaultip	The IP address to contact a SIP peer for calls in the event it isn't registered. Requires that the host setting be dynamic.	—
dtmfmode	Instructs Asterisk on how to send and receive DTMF signals to/from this peer. The inband method only works well with G.711 codec.	inband; info; rfc2833
Fromuser	Overrides the caller ID username for calls to SIP SAs only, not for calls that ultimately reach a legacy endpoint.	—
Fromdomain	Establishes the SIP domain to associate with calls placed from this peer.	—
Host	The method by which this peer can be reached. Peers using dynamic must be registered locally or you must provide a defaultip setting too. Otherwise, provide a static IP address for host.	**dynamic** or an IP address
incominglimit	Maximum number of simultaneous calls allowed from this peer.	—
insecure	Forgoes matching the IP address and port of this peer with its configuration in *sip.conf*. Probably only useful in the lab to toggle this check on and off. Peers that need a constantly changing IP address can use host=dynamic.	yes; **no**
language	A setting that allows IVR prompts to be localized when this peer is interacting with the voice menus on the server.	—
Outgoinglimit	The maximum number of simultaneous calls allowed to this peer.	—
Mailbox	The mailbox number to associate with this peer's message-waiting indicator, if applicable.	—

Setting	Description	Possible values
Md5secret	The MD5 hash of the peer's SIP username and password; optional measure designed to make password transmittal more secure.	—
Musiconhold	The default class of music-on-hold to use with calls to and from this peer.	**default**
Nat	Allows this peer to be behind a NAT firewall and still work with Asterisk.	yes; **no**
Pickupgroup	The number of the pickup group that this SIP channel is a member of. Default is none.	—
Port	The SIP port on this peer's UA.	5060
Restrictcid	Allows the caller ID signals for calls from this peer to be hidden from the receiver.	yes; **no**
Rtptimeout	The number of seconds before Asterisk will disconnect calls to/from this peer due to no RTP activity (unless the call is on hold).	—
rtpholdtimeout	The number of seconds before Asterisk will disconnect calls to/from this peer due to no RTP activity while on hold. Must be greater than rtptimeout.	—
Secret	The SIP password (shared secret) used for all INVITE methods to or from this peer.	—
Type	The kind of peer this is. Friends can call and receive. Peers can receive only. Users can call only.	**friend**; peer; user
Username	The SIP username used to authenticate INVITEs to and from this peer.	—

The Asterisk CLI

Asterisk's CLI is where you, the administrator, control and monitor the Asterisk PBX. The Asterisk CLI provides you with real-time information about voice channels, extensions, contexts, and more. Using the CLI, you can start and stop the Asterisk server, as described earlier in the chapter. You can do status queries about the voice channels, place and hang up calls, and add or remove extensions and contexts.

If the Asterisk server is already running, you can launch the Asterisk CLI by starting another instance of Asterisk in client mode:

```
# asterisk -r
```

The first command you should probably learn is *help*, which displays a list of valid commands or, when used as follows, gives command-specific assistance:

```
pbx*CLI> help show channels
```

Notice the pbx*CLI> prompt. This prompt will vary depending on the Unix hostname of your machine—in this case, it's pbx.

Inspecting Channels

There are several types of voice channels in Asterisk. There can be analog FXO (foreign exchange office) channels, such as those facilitated by a POTS telephone company line and an X100P card. There can also be voice channels provided by H.323, SIP, IAX, SCCP, and MGCP (all different VoIP signaling protocols), as well as by FXS (foreign exchange station) devices like QuickNet's Phone Jack interface card and PRI interfaces, too.

You can use the Asterisk CLI to inspect each type of channel, but the details provided are different for each:

- zap show channels
- iax2 show channels
- sip show channels

Each of these commands provides a list of channel numbers that can correspond to a numeric reference in the channel name, as described earlier (example: Zap/1). Take a look at a zap show channels command:

```
pbx*CLI> zap show channels
Chan. Num. Extension  Context    Language   MusicOnH
   1                  incoming              default
```

The command output shows that there is a single Zaptel channel available for use by the Asterisk PBX. In our experimental server, this logical channel represents the physical X100P interface with a telco line connected to it. The command output tells that the context for calls on this channel is incoming because the start extension in extensions.conf is established within that context. This channel is referred do in the dial-plan as Zap/1.

You can dig deeper using the CLI, to the configuration of a specific channel, by running the zap show channel <channel#> command, like this:

```
pbx*CLI> zap show channel 1
Channel: 1
File Descriptor: 17
Span: 1
Extension:
Context: incoming
Caller ID string:
Destroy: 0
Signalling Type: FXS Kewlstart
Owner: <None>
Real: <None>
Callwait: <None>
Threeway: <None>
Confno: -1
Propagated Conference: -1
Real in conference: 0
DSP: no
Relax DTMF: yes
```

```
Dialing/CallwaitCAS: 0/0
Default law: ulaw
Fax Handled: no
Pulse phone: no
Echo Cancellation: 128 taps, currently OFF
Actual Confinfo: Num/0, Mode/0x0000
```

There are many channel-specific settings, most of which are user-definable. You can see from this list that, even within the scope of a single Zaptel channel, Asterisk is extremely flexible. In our test server, this channel is an FXO channel—that is, a channel that connects to the telephone company. In order to interact correctly with the phone company's switch, the X100P (or other compatible) interface has to use FXS signaling, so that, to the phone company, it appears to be a regular telephone answering the line and not a newfangled Asterisk server (FXO/FXS signaling was introduced in Chapter 4). Other types of channels (SIP, H.323, etc.) have their own channel-specific settings.

Inspecting the Dial-Plan

The show dialplan CLI command can be used two ways: showing the entire system-wide dial-plan, or only showing a specific context or extension. Here's the output for a small, but complete, dial-plan:

```
pbx*CLI> show dialplan
[ Context 'default' created by 'pbx_config' ]
  '3101' =>          1. Dial(SIP/3101@10.1.1.101)        [pbx_config]
                     2. Voicemail(u4101)                 [pbx_config]
  '3102' =>          1. Dial(SIP/3102@10.1.1.102)        [pbx_config]
                     2. Voicemail(u4101)                 [pbx_config]
  '3103' =>          1. Dial(SIP/3103@10.1.1.103)        [pbx_config]
                     2. Voicemail(u4101)                 [pbx_config]
  '8500' =>          1. VoiceMailMain( )                 [pbx_config]
  '_9NXXXXXX' =>     1. Dial(${PSTN}/{EXTEN:1})          [pbx_config]
                     2. Congestion( )                    [pbx_config]

  Ignore pattern => '9'                                  [pbx_config]

[ Context 'incoming' created by 'pbx_config' ]
  's' =>             1. Answer( )                        [pbx_config]
                     2. Dial(SIP/3103@10.1.1.103|30)     [pbx_config]
                     3. Voicemail(u4101)                 [pbx_config]
                     4. Hangup( )                        [pbx_config]
```

In the default context, several extensions are configured to ring calls on specific SIP telephones with static IP addresses and an extension pattern that matches outbound calls. The 8500 extension is one that allows private users of the Asterisk system to access the voice mail management greeting.

There's also an incoming context that attempts to ring a certain phone with the s extension and then transfers the call on the current channel to a voice mailbox if nobody answers the phone.

To get a specific extension's dial-plan only, run the show dialplan CLI command with an argument of extension@context:

```
pbx*CLI> show dialplan 3102@default
[ Context 'default' created by 'pbx_config' ]
  '3102' =>          1. Dial(SIP/3102@10.1.1.102)          [pbx_config]
                     2. Voicemail(u4101)                   [pbx_config]
```

In this case, only the extension definition for 3102 in the default context is shown.

Administering the Dial-Plan Using the CLI

Earlier, we covered the primary configuration file for the dial-plan: *extensions.conf*. You can also alter the dial-plan using CLI commands. The difference between using the CLI and using the configuration file is this: CLI dial-plan changes, such as additions and removals of extensions, occur immediately, whereas changes to *extensions. conf* require a restart of Asterisk (or at least a config-reload), which introduces downtime.

Consider the following CLI command:

```
pbx*CLI> add extension 120,1,Dial(SIP/110@oreilly.com:5060) into local
```

The command add extension is followed by the same definition string you would see in an exten directive in *extensions.conf*—that is, the extension number, priority, and application command for the extension. The final part, into local, tells Asterisk which context to add the extension into—in this case, local.

The CLI can also be used to drop extensions, using the remove extension command, and to re-create the *extensions.conf* file based on the current "live" dial-plan, using the save dialplan command.

If you prefer to use the configuration file to maintain your dial-plan, you can avoid total Asterisk restarts by issuing a reload extensions command, which re-reads the *extensions.conf* file and affects the dial-plan accordingly. That said, using the CLI may be the preferred method for programming the dial-plan, because it gives you immediate feedback in case of syntax errors or other common mistakes. In order to save changes made at the CLI, your *extensions.conf* must contain the writeprotect=no setting in its general section, and you must issue a save dialplan command at the CLI. This will cause the current dial-plan configuration to overwrite anything in your *extensions.conf* file (including comments).

CLI Command Reference

Administrator commands

!   Execute the following shell command. Example:

```
pbx*CLI> !ls AM*
AMP-1.10.002.tar
AMP-1.10.002
```

abort halt
> Cancel a running halt.

add extension
> Add new extension into context. The spec of the extension is the same as it would be in *extensions.conf*.

add ignorepat
> Add a new ignore pattern to the specified context.

```
Pbx*CLI> add ignorepat _25XX into default
```

add indication
> Create the given indication to the country specified. The indication is a tonelist similar to those found in *indications.conf*.

```
Pbx*CLI> add indication us busy 480+620/500.0/500
```

debug channel
> Enable debugging on a specific channel.

dont include
> Remove a specified include from the named context.

```
Pbx*CLI> dont include longdistance in localonly
```

help
> Display help list or specific help on a command—i.e., help dont include.

include context
> Include context in other context.

```
Pbx*CLI> include tollfree in localonly
```

load
> Load the specified Asterisk module.

logger reload
> Reopen logfiles. For use after rotating the logfiles.

meetme
> Shows information about currently ongoing conference calls, if any.

no debug channel
> Disable debugging on the specified channel.

pri debug span
> Enable debugging on a span that uses PRI.

```
Pbx*CLI> pri debug span 1
```

`pri intense debug span`

    Enable detailed PRI debugging on a span.

`pri no debug span`

    Disables PRI debugging on a span.

`remove extension`

    Remove the specified extension.

`remove ignorepat`

    Remove the specified ignore pattern from the specified context.

        `Pbx*CLI>` **`remove ignorepat _25XX from default`**

`remove indication`

    Remove the given indication from the country (see `add indication`).

`save dialplan`

    Overwrite your current *extensions.conf* file from the current state of the dialplan based on the commands you've issued at the CLI.

`set verbose`

    Set level of verbosity for messages displayed on the console: 0 means display as few as possible; 9 means display a lot of detail. This has the same effect as launching `asterisk -vvvvvvvvv` at the shell prompt.

`show agents`

    Show status of agents. Agents are named users of the system.

`show applications`

    Show all registered applications—i.e., available dial-plan commands.

`show application`

    Show a synopsis of the specified application.

`show channel`

    Display information about the specified channel.

`show channels`

    Show information about all channels.

`show codecs`

    Display information about codecs.

    The `show codecs` command doesn't indicate which codecs your system is using or can use. It just provides information about binary and decimal tokens used to refer to codecs within Asterisk.

`show dialplan`

    Show the entire dial-plan or, if a context is specified, show only the dial-plan for that context.

`show indications`

    Show a list of all countries and indication tones, per *indications.conf*.

show manager commands
: Show a list of Asterisk Manager API commands.

show manager command
: Show a synopsis of the specified Manager API command.

show manager connected
: Show a list of connected Manager API users.

show parkedcalls
: Display a list of currently parked calls.

show queues
: Show a report of queue activity, including the current status of the queue and the utilization of each agent that receives calls from the queue.

show translation
: Display codec transcoding matrix, a grid of latency times incurred by transcoding one channel's codec into another's.

show voicemail users
: List active voice mailboxes, according to /etc/asterisk/voicemail.conf.

show voicemail zones
: List active voice mail zones, according to /etc/asterisk/voicemail.conf.

soft hangup
: Request a hangup on the specified channel, similar to SoftHangup( ).

Process control commands

extensions reload
: Reload extensions and only extensions (not channels/peers).

reload
: Reload configuration, including channels/peers.

restart now
: Restart Asterisk immediately, disconnecting all calls in progress.

restart when convenient
: Restart Asterisk when there are no active calls, but not before.

show modules
: List running modules and information about them.

show uptime
: Show the Asterisk PBX's uptime, which may vary from the OS uptime.

show version
: Display the Asterisk version.

stop now
: Shut down Asterisk immediately, disconnecting all current calls in progress.

stop when convenient
> Shut down Asterisk when there are no calls, but not before.

unload
> Unload an Asterisk module by name.

AGI commands

dump agihtml
> Print a list of AGI commands in HTML format.

show agi
> Show AGI commands or specific help on a particular AGI command.

Database commands

database del
> Remove an Asterisk database key/value.

database deltree
> Remove an Asterisk database key tree/value.

database get
> Get an Asterisk database value, like DBGet( ).

database put
> Add/update an Asterisk database value, like DBPut( ).

database show
> Show the Asterisk database contents.

IAX v2 commands

iax2 debug
> Enable IAX2 debugging, dumping all IAX protocol debug info to the console.

iax2 no debug
> Disable IAX2 debugging.

iax2 set jitter
> Set IAX2 jitter buffer to a certain maximum length. Can be optionally supplied a call ID so that it only applies to a certain call.

iax2 show cache
> Display IAX2 cached dial-plan—i.e., IAX2 destinations for which recent lookups have occurred, so calls can be connected without resolving them again.

iax2 show channels
> Show the currently active IAX2 channels.

iax2 show peers
> Show all defined IAX2 peers, per */etc/asterisk/iax.conf*.

`iax2 show registry`
>   Show the status of IAX2 peers for which this server is a client—i.e., show a list of servers to whom this server is registered as an IAX2 client.

`iax2 show stats`
>   Display real-time statistics about IAX2 packets.

`iax2 show users`
>   Show all defined IAX2 users, per *iax.conf*.

`iax2 trunk debug`
>   Request an IAX2 trunk debug.

IAX v1 commands

 IAX Version 1 is deprecated. Use IAX Version 2 instead.

`iax debug`
>   Enables IAX (Version 1) debugging. Debug output goes to the console.

`iax no debug`
>   Disables IAX debugging.

`iax set jitter`
>   Sets IAX jitter buffer like `IAX2 set jitter`.

`iax show cache`
>   Displays IAX cached dial-plan like `IAX2 show cache`.

`iax show channels`
>   Shows active IAX channels like `IAX2 show channels`.

`iax show peers`
>   Shows defined IAX peers.

`iax show registry`
>   Shows the status of IAX peers for which this server is a client.

`iax show stats`
>   Displays real-time IAX statistics.

`iax show users`
>   Shows defined IAX users.

Crypto commands

`init keys`
>   Initializes RSA key pass codes.

show keys
> Displays RSA key information for peers with whom RSA keys are used in authentication.

H.323 commands

H.323 media channels are supported by Asterisk but does not compile by default. Installation instructions for H.323 support are found in */usr/src/asterisk/channels/ h323/README*. Asterisk cannot act as an H.323 gatekeeper, only as a gateway. A fully H.323 VoIP network would also need an H.323 gatekeeper, such as the excellent GnuGK, which is covered in Chapter 7.

h.323 debug
> Enable H.323 debugging if the H.323 module is installed.

h.323 gk cycle
> Manually register with an external gatekeeper.

h.323 hangup
> Manually hang up the specified call.

h.323 no debug
> Disable H.323 debugging.

h.323 no trace
> Disable H.323 stack tracing.

h.323 show codecs
> Show available codecs.

h.323 trace
> Enable H.323 stack tracing.

SIP commands

SIP support is built in to Asterisk via the *chan_sip* module, which comes with it. Asterisk can act as a SIP proxy, a SIP registrar, and a SIP client. Named SIP call routes, which are created in */etc/asterisk/sip.conf*, are called SIP peers, and their destinations can be SIP phones or other SIP servers. (Refer to "SIP Channels," earlier in this chapter.)

sip debug
> Enables SIP debugging. Debug output goes to the console.

sip no debug
> Disables SIP debugging.

sip reload
> Reloads SIP configs, including peer changes, from *sip.conf*.

sip show channels
> Shows active SIP channels—i.e., current ongoing calls.

`sip show channel`
> Shows information about a SIP channel.

`sip show inuse`
> Shows real-time call volume of each SIP peer.

`sip show peers`
> Shows defined SIP peers that can register with the Asterisk server's SIP registrar.

`sip show registry`
> Shows SIP registration status (when Asterisk registers as a client to a SIP registrar).

`sip show users`
> Shows defined SIP users per *sip.conf*.

MGCP channel commands

The Media Gateway Control Protocol is supported by Asterisk out of the box. The MGCP module is included in the standard distribution. MGCP channels are configured in */etc/asterisk/mgcp.conf*.

`mgcp audit endpoint`
> Displays the capabilities of the specified MGCP endpoint.

`mgcp debug`
> Enables MGCP debugging. Debug output goes to the console.

`mgcp no debug`
> Disables MGCP debugging.

`mgcp show endpoints`
> Shows defined MGCP endpoints/peers.

SCCP channel commands

SCCP (Skinny Client Control Protocol) is a leg-only signaling protocol for VoIP endpoints. It is supported most prominently by all Cisco IP phones, among others. The *chan_skinny* Asterisk module provides SCCP compatibility.

`skinny debug`
> Enables SCCP (skinny) debugging.

`skinny no debug`
> Disables SCCP (skinny) debugging.

`skinny show lines`
> Shows SCCP lines on each registered client.

CAPI channel commands

CAPI channels are used to support BRI channels with Asterisk. The CAPI module can be obtained from *http://www.junghanns.net/asterisk*.

```
capi debug
```
Enables CAPI debugging.

```
capi no debug
```
Disables CAPI debugging.

```
capi info
```
Shows CAPI information.

Zaptel channel commands

Zaptel interface hardware for POTS, FXO/FXS, T1, and E1 links is manufactured by Digium, and it is supported natively by Asterisk using the *chan_zap* module, which is loaded by default when you install it.

```
zap destroy channel
```
Destroys the specified channel.

```
zap show channel
```
Shows information about a particular Zaptel channel.

```
zap show channels
```
Shows active Zaptel channels.

Integrating Asterisk with Other Software

While the *extensions.conf* file and the CLI are great places to command and control your Asterisk PBX, certain administrative applications may demand more capabilities than these tools offer—like integration with other applications or with a web server. For instance, if you were running a paid-use service using Asterisk to provide the dial-tone and call routing, then you might need a way to automatically set up and tear down extensions and voice mailboxes; this might even mean integrating with a billing application or self-service web application.

Asterisk CLI Wrapper

At such times, it's good to know that Asterisk's CLI functions are available to the outside world via two mechanisms: the Asterisk application in client mode, and the Asterisk Manager API. Running Asterisk in client mode lets you pass any Asterisk CLI command to the Asterisk executable in a shell as an argument. To pass an Asterisk command, run it with the x option. Have a look:

```
asterisk -r -x add extension 120,1,Dial,SIP/110@oreilly.com:5060 into local
```

Using the shell command functions of scripting languages such as Perl and PHP, it is possible to integrate almost every CLI command into a custom-made application. For example, you could build a web-based administration tool for Asterisk (some have attempted this, but using the Manager API), or you could use a Perl script to automatically remove extensions when customers' bills are overdue in an external billing system.

Consider the following line of Perl:

```
system("asterisk -r -x add extension $newExt,$newPri,$appl,$args into $context");
```

This Perl command launches Asterisk in client mode and runs the add extension CLI command with an entirely dynamic extension definition—one that another part of the Perl script establishes according to the programmer's design.

 In order to use the CLI wrapper from a CGI script, the web server running the script must have permission to execute Asterisk. Usually, this means putting Asterisk and your web server in the same user group. This could pose a potential security risk, so use this trick with care.

The Asterisk Gateway Interface

One of the most compelling ingredients in the Asterisk recipe, AGI gives developers the ability to create custom interfaces between the dial-plan and outside systems. The AGI is to Asterisk what CGI is to a web server. In a nutshell, it allows you to use your favorite programming language—Perl, Python, Pascal, RealBASIC, whatever—to affect or control interaction with a telephone user who's attached to the PBX. Here a few examples of things you can do with the AGI:

- Show the caller ID of an incoming phone call on your desktop computer.
- Pull the database records of an incoming caller so an answering call center operator can view them on the computer screen as he's answering the call, and during the call.
- Feed distributed call-billing information to a non-Asterisk CDR billing system.
- Set up a telephone backdoor for you, the administrator, to check the server load, memory consumption, and so on. You could even have AGI cycle Apache or reboot the server for you if necessary.
- Tell you the current weather report when you call in.
- Create elaborate find-me-follow-me applications.
- Communicate with software systems that don't have built-in support for Asterisk but can already be customized with your favorite programming language.

Though AGI bears some similarity to CGI, the output from an AGI program is completely different from that of a CGI program. Generally, a web CGI produces HTML or XML output, but AGI programs produce output in the form of AGI commands. These commands are sent to the Asterisk server for processing, and then Asterisk sends data back to the AGI program in response. A list of the commands can be found in Appendix C. A Perl library for writing AGI programs is available from *http://asterisk.gnuinter.net/*.

Asterisk Manager Socket API

Like the AGI, the Asterisk Manager socket API provides a way for external software to communicate with Asterisk. Unlike AGI, the purpose of the Manager API isn't building telephony applications. Instead, the Manager API is used for controlling administrative and process control functions of Asterisk from external software. Here are some things you could do with the API:

- Build a program that lets you monitor call activity. (The *astman* program uses the API for this purpose.)
- Add and remove extensions and channels from a Perl, VBScript, or AppleScript program. Of course, as long as you can create a socket, you can use any language you like.
- Do basic call control, like hanging up channels and transferring calls. This way, you could create a web-based or fat-client attendant console application.

All API sessions, which are TCP connections to the Asterisk server on port 5038, are text-based exchanges. They must begin with a login exchange to authenticate the connecting party. The security permissions for each Manager API user are stored in */etc/asterisk/manager.conf*, as discussed in "Monitoring Asterisk" in Chapter 3.

Sending a TCP packet with the following payload would log a user named Jake into the Asterisk server. Follow each line with a single Retrun and the entire section with two Returns:

```
Action: login
Username: Jake
Secret: monsterjam
```

You can telnet to port 5038 on your Asterisk server in order to try this. Just make sure your *manager.conf* is set up to allow somebody to log in.

Once you're authenticated, you can issue other text-based commands to the Manager API and parse the output as needed. Appendix D describes the Manager API syntax.

Key Issues: Asterisk Reference

- Asterisk uses text-based configuration files. Almost all of them (*zaptel.conf* is in */etc*) are found in */etc/asterisk*.
- Variables, pattern matching, and string chopping can be used to control call flows.
- Contexts define related groups of extensions based on time of day, the identity of the caller, or the channel used to originate the call.
- SIP and Zaptel channels are the most common, and best supported, types of channels supported by Asterisk.

- There are dozens and dozens of dial-plan commands, some with very sophisticated syntax, that you can use in *extensions.conf*. These commands define the telephony experience of the system users.

- The Asterisk CLI is used to monitor and maintain the system.

- Asterisk CLI commands can be issued with shell and Perl scripts by running the Asterisk program in client mode and passing the CLI commands with the -r -x options.

- Asterisk can be integrated with other systems using AGI (Asterisk Gateway Interface) and the Asterisk Manager API.

SIP Methods and Responses

Table A-1 shows SIP methods.

Table A-1. Methods

INVITE	A SIP device is being invited to participate in a call.
ACK	Confirms that the client has received a final response to an INVITE request.
BYE	Terminates a SIP call. Can be sent by any party involved.
CANCEL	Cancels any pending call but does not terminate a call that has already been connected.
OPTIONS	Queries the capabilities of servers without requesting to establish a call.
REGISTER	Registers an IP with a SIP registrar.
PRACK	Insures reliability of provisional 1xx responses if a UAS offers them.
UPDATE	Updates a previously made offer for a not-yet-established session.
REFER	Initiates a call transfer by telling the recipient (specified by URI) to contact a third party using the contact information provided in the request.
SUBSCRIBE	Subscribes to be notified of an event occurrence; for example a user presence update.
NOTIFY	Used to notify that an event has occurred.
MESSAGE	A method signifying the payload is an instant message.

Table A-2 shows SIP responses.

Table A-2. Responses

100	Trying
180	Ringing
181	Call Is Being Forwarded
182	Queued
183	Session Progress
200	OK
202	Accepted

Table A-2. Responses (continued)

300	Multiple Choices
301	Moved Permanently
302	Moved Temporarily
305	Use Proxy
380	Alternative Service
400	Bad Request
401	Unauthorized to Use This Registrar
402	Payment Required
403	Forbidden
404	User Not Found
405	Method Not Allowed
406	Not Acceptable
407	Proxy Authentication Required
408	Request Timeout
410	This User Is Gone From Here
413	Request Entity Too Large
414	Request URI Too Long
415	Unsupported Media Type
416	Unsupported URI Scheme
420	Bad Protocol Extension
421	Extension Required
423	Interval Too Brief
480	Temporarily Unavailable
481	Call/Transaction Does Not Exist
482	Loop Detected
483	Too Many Hops
484	Address Incomplete
485	Ambiguous
486	Busy Here
487	Request Terminated
488	Not Acceptable Here
491	Request Pending
493	Undecipherable s/MIME Part
500	Server Internal Error
501	Method Not Implemented on This Server
502	Bad Gateway
503	Service Unavailable

Table A-2. Responses (continued)

504	Server Timeout
505	Version Not Supported
513	Message Too Large
600	Busy Everywhere
603	Decline
604	Does Not Exist Anywhere
606	Not Acceptable

AGI Commands

ANSWER

Answers the current channel if not already answered.

Returns -1 upon failure, or 0 upon success.

CHANNEL STATUS [*channel*]

Returns the status of the specified *channel*. If none is given, returns the status of the current channel. Here is what the status codes mean:

0: Channel is on hook and available.
1: Channel is on hook, but reserved.
2: Channel is off the hook but no digits have been dialed.
3: Digits have been dialed.
4: The line connected to this channel is ringing.
5: A called line connected to this channel is ringing.
6: A called line connected to this channel has a call in progress.
7: A called line connected to this channel is busy.

DATABASE DEL *family key*

Deletes a value in the Asterisk database for the specified family and key.

Returns 1 if successful, 0 if not.

DATABASE DELTREE *family* [*keytree*]

Deletes a family or and/or keytree within a family in the Asterisk database.

Returns 1 if successful, 0 if not.

DATABASE GET *family key*

Retrieves a value in the Asterisk database for the specified family and key.

Returns 0 if *key* does not exist. Returns 1 and the requested value in parentheses if *key* exists, as in this example:

```
200 result=1 (value)
```

DATABASE PUT *family key value*

Creates or updates a value in the Asterisk database for the specified family, key.

Returns 1 if successful or 0 if not.

EXEC *command options*

Executes the specified dial-plan `command` with given `options`. The options are to be given in the syntax appropriate for the command. Commands are listed in Chapter 17.

Returns the same value as the called command would normally return, or -2 if the command is invalid.

GET DATA *file* [*timeout*] [*maxdigits*]

Streams the specified sound file and receives DTMF digits on the current channel.

Returns the digits received from the channel, if any

GET VARIABLE *variable*

Gets a variable from the current context and extension.

Returns 0 if `variable` is null or 1 if `variable` has a value. Also returns the variable in parentheses, as in the following return code:

```
200 result=1 (value)
```

HANGUP [*channel*]

Hangs up the specified `channel` or the current channel if none is given.

Returns 1.

RECEIVE CHAR *timeout*

Waits for a character of text on the current channel. Specify a timeout of 0 to wait indefinitely.

Returns the decimal (8-bit) ASCII value of the character if one is received, 0 if the channel does not support text, or -1 if the channel is hung up.

RECORD FILE *filename format* '*digits*' *timeout* [*beep*]

Records the sound on the current channel to a file until the specified digit sequence is received. Format can be wav, WAV, or gsm. `timeout` specifies the maximum recording length. `beep` will cause a beep to be played before recording begins.

Returns -1 only on error.

SAY NUMBER *number* '*digits*'

Say the specified number, stopping if the specified *digits* are received on the channel.

Returns 0 if playback completes, the ASCII decimal value of the digit if one was received, or -1 on error.

SEND TEXT "*text*"

Sends *text* on the current channel, if it's supported by the connected device. Don't forget to use quotes around *text*.

Returns -1 only on hangup, otherwise always returns 0, even if the device doesn't support text.

SET CALLERID *number*

Changes the caller ID of the current channel to *number*.

Returns 1.

SET CONTEXT *context*

Sets the context in which to return when the AGI program is complete.

Returns nothing.

SET EXTENSION *extension*

Sets the extension to which to return when the AGI program is complete.

Returns nothing.

SET PRIORITY *priority*

Sets the priority in which to return when the AGI program is complete.

Returns nothing.

SET VARIABLE *variable value*

Sets the value of the specified *variable* to *value*.

Returns 1.

STREAM FILE *file 'digits'*

Streams the specified sound file, allowing playback to be interrupted by the given *digits*. Don't include the file extension in the filename.

Returns 0 if playback completes without receiving a digit, the ASCII decimal value of the digit if one was received, or -1 on error.

TDD MODE *on|off*

Enables or disables TDD transmission/reception on a channel. TDD is a signaling protocol used to assist deaf phone users in communicating via text messages.

Returns 1 if successful, 0 if not.

VERBOSE *message level*

Sends the specified message to the console via verbose logging. *level* is the verbosity level, from 1 to 4.

Returns 1.

WAIT FOR DIGIT *timeout*

Waits a maximum of *timeout* milliseconds for a dialed digit on the current channel. If *timeout* is -1, the AGI program will wait indefinitely for a dialed digit.

Returns -1 on failure, 0 if no digit is received, or the decimal (8-bit) ASCII value of the digit if one is received.

Asterisk Manager Socket API Syntax

AbsoluteTimeout *Channel Timeout*

Sets an absolute timeout in seconds for the specified channel. The call will be ended after the time has elapsed. The following example limits the call on the current channel to 10 minutes:

```
Action: AbsoluteTimeout
Channel: SIP/201
Timeout: 600
```

ChangeMonitor *Channel File*

Equivalent of ChangeMonitor( ).

```
Action: ChangeMonitor
Channel: Zap/1-1
File: Zap1-1-incsound
```

Command *command*

Execute the specified dial-plan command. The command must include all arguments necessary for it to work.

GetVar *Channel Variable*

Gets a variable from the specified channel.

Hangup *Channel*

Hangs up specified channel. Equivalent to SoftHangup( ).

IAXpeers

Lists IAX peers. Equivalent of IAX2 show peers CLI command.

ListCommands

Lists available Manager API commands.

Logoff

Closes the connection to the Manager.

MailboxCount *Mailbox*

Gets the message count for the specified mailbox.

MailboxStatus *Mailbox*

Gets the message-waiting indication for the specified mailbox.

Monitor *Channel*

Begins monitoring the specified channel. Equivalent of Monitor( ).

Originate *Channel Context Exten Priority Callerid*

Places a call on the specified channel to the specified context, extension, and priority using the specified caller ID.

```
Channel: SIP/201
Context: default
Exten: 18005044556
Priority: 1
Callerid: 3125551212
```

ParkedCalls

Lists currently parked calls.

Ping

Sends a keep-alive message to the Manager.

Redirect *Channel ExtraChannel Exten Context Priority*

Connects the call on the specified channel to the specified context, extension, and priority using the specified channel (ExtraChannel) and caller ID.

SetVar *Channel Variable Value*

Sets the specified channel's named variable to the value supplied.

Status *Channel*

Provides a status report on the specified channel. Similar to AGI command CHANNEL STATUS.

StopMonitor *Channel*

Stops monitoring the specified channel. Equivalent of StopMonitor( ).

Glossary

adaptive differential pulse code modulation (ADPCM)
An advanced, compressed version of plain PCM digitizing. This algorithm is used by the G.726 codec.

additive noise
any unwanted signal that distorts a phone call by increasing the strength of the sound signal.

Alaw
The scale used by PCM digitizing algorithms in many countries outside North America.

asynchronous transfer mode (ATM)
A buffered switching system that transports packets called cells across packet-based, often optical, data links.

autoattendant
A telephony application that provides interactive prompts and greetings for incoming calls to a PBX, so that calls can be routed without intervention from a human reception operator.

automated location identifier (ALI)
The signaling using to notify emergency dispatchers of a caller's postal address on the 911 system.

automatic call distribution (ACD)
Also called automatic call *delivery*, a telephony application that routes inbound calls to the appropriate answering party based on a logic-based algorithm, such as the originating area code or the language the caller speaks—i.e., Spanish or English, etc.

baby bells
Nickname for incumbent local exchange carriers (ILECs), the companies that sprung from the AT&T antitrust breakup of the early 1980s.

basic rate interface (BRI)
An ITU signaling standard that allows two simultaneous phone calls on a digital phone line. The slang term for BRI is *ISDN line*.

bit stream
A continual linear stream of binary data, such as those in a T1 or a TDM bus.

call center
An area of high concentration of telephone connections and operators, usually using queuing and hunting applications.

CallManager
Cisco Systems' Windows-based softPBX.

central office (CO)
The building where an exchange switch resides, supporting telephone service for a geographically narrow group of PSTN subscribers, usually from a few hundred to tens of thousands.

centrex
A type of business phone service available from RBOCs that offers enhanced calling features.

channel bank
A device that splits T1 circuits into digital channels or analog signals for use by TDM or analog phones, respectively.

Cisco Discovery Protocol (CDP)

A hardware-locating technique that enables Cisco's proprietary E911 ALI solution; also, the method by which Cisco PoE is requested from a Cisco switch.

codec

An algorithm that quantizes and packages an analog signal for transport across a packet network. Common telephony codecs include G.711 and G.729A.

code-excited linear prediction (CELP)

An algorithm used by some voice codecs to reduce the amount of bandwidth required to reliably transmit spoken sounds.

comfort noise generation

The simulation of white noise by an endpoint so that its listener is less aware of silence suppression during times in the call when nobody is speaking.

class of service (CoS)

Refers to the group of QoS standards that use precedence and buffering to improve network availability for voice applications.

demarc (or D-mark)

The point at which telephone company–owned facilities end and subscriber-owned facilities begin. Usually a cross-connect panel, DS1 mounting, or a similar wiring terminal.

dial-peer

In Cisco media gateways, a configuration for switching phone calls to an H.323, MGCP, SCCP, or SIP endpoint based on the digits signaled by the caller. Equivalent to Asterisk's Dial command, though much simpler.

dial-tone trunk

A circuit connecting a softPBX to the PSTN that provides inbound and/or outbound calling service.

DiffServ

A QoS mechanism that provides class of service (CoS), precedence-based treatment of packets for voice applications, but no guarantee against capacity overages.

DS0

A single channel of a T Carrier circuit. T1s have 24 DS0s.

DS1

A T1 circuit.

DSU/CSU

A device that terminates one end of a T1 circuit, allowing its channels to be assigned for voice and data use. DSU/CSUs provide signaling and error checking for the T1's bitstream, too.

dual-tone multifrequency (DTMF)

A standard for in-band signaling of dialed digits on legacy telephones.

E.164

An ITU convention for phone numbers on the PSTN and in H.323 VoIP networks. E.164 is commonly supported in MGCP and SIP networks, too.

E1

The European equivalent of T1. E1 circuits provide eight additional DS0s per circuit than their North American cousins.

E911

Enhanced 911 service; the signaling and human element protocols for handling of public safety dispatch calls in the United States. Predated by (nonenhanced) 911 service. The most significant enhancement of E911 is automatic location identification signaling that allows dispatchers to know the postal address of an emergency caller.

Erlang

A unit of measure for voice traffic capacity in a PBX environment. Erlang ratings help system builders decide how many trunks are needed between two voice networks, like the PSTN and a PBX.

Ethereal

A freely available packet analysis software tool for Windows, Linux, and Macintosh.

Ethernet

The most common standard for local area networking connectivity. Modern Ethernet offers transmission speeds from 10 mbps to 10,000 mbps.

extension

An alphanumeric code that has a programmed purpose on a PBX, such as calling a specific phone or voice mailbox. Legacy PBXs may support only numeric E.164 codes.

Federal Communications Commission (FCC)

The national government agency in the United States that is charged with regulating communications systems that cross state lines, such as the Internet and long-distance voice networks.

find-me-follow-me (FMFM)

A telephony application that can attempt to locate a mobile user on her desk phone, cell phone, satellite phone, or Internet phone, when the PBX receives a call for her.

five nines reliability

A service level commitment that dictates that, when a caller tries to reach somebody on the PSTN, his call will be reliably connected 99.999% of the time.

frame-relay

A wide area networking technology that permits many subscribers to use the infrastructure of a long-distance carrier's network for transport of packet-based data.

full-duplex

Refers to the characteristic of data traveling simultaneously between two participants in both directions on the same pathway. A full-duplex speakerphone allows a caller to speak while listening to his calling partner. See also *half-duplex*.

G.711

The most common voice codec in telephony applications; it uses PCM to encode sound signals on an 8-bit quantization scale at the PSTN-standard sampling rate of 8 kHz. G.711 uses 64 kbps of bandwidth for a single one-way sound signal.

gateway

A device or software program that provides a proxy or intermediary between two systems with incompatible technologies. Example: a media gateway for connecting legacy telephony circuits like POTS lines to a VoIP server.

H.323

An ITU recommendation for the use of Ethernet and TCP/IP for voice connectivity.

half-duplex

Refers to the characteristic of data traveling in only one direction at a time between two participants in a conversation. A half-duplex speakerphone allows a caller to hear his partner's voice only while he himself is silent. See also *full-duplex*.

hunt group

A group of extensions or phone company trunks that ring simultaneously or in sequence when one of them is called. (Simultaneous hunt groups are sometimes called ring groups.)

hybrids

The interfaces used to connect legacy phone lines to a TDM bus, for instance to connect a POTS trunk to a PBX.

IAXTel

A TSP that links Asterisk users and IAX-compatible VoIP networks free of charge.

in-band signaling

Call control signals that occur within the audio spectrum or electrical frequencies also used to transmit the sounds of the phone call. In the context of a T1, *in-band* means substitution of bits normally used for audio for bits used to indicate call control signals. See also *robbed-bit signaling*.

incumbent local exchange carrier (ILEC)

See *regional Bell operating company (RBOC)*.

instant messaging (IM)

A type of desktop computer application that allows users to send text messages, voice, and video back and forth using the Internet.

lintegrated access device (IAD)

Pronounced "eye-add," a router with a built-in data link interface, such as a DSL modem or cable modem. An IAD is designed to replace two or more access devices on the customer premise. Some IADs may contain a router, DSL modem, firewall, Ethernet switch, VoIP ATA, and/or SIP proxy.

integrated services digital network (ISDN)
A family of ITU standards that dictate how transport and signaling work on digital telephone lines, including T1s.

Inter-Asterisk Exchange Protocol (IAX)
A signaling and media packaging protocol designed to connect Asterisk systems. Unlike SIP, IAX is intended for use in telephony applications only and uses only one socket, so it's compatible with NAT firewalls. IAX is pronounced "eeks." The current version of IAX is 2.

International Telecommunications Union (ITU)
The body that developed most of the standards in use on the PSTN and many cell phone networks. Formerly known as the CCITT, a French abbreviation that, roughly translated, means Consulting Committee for International Telephones and Telegraphs.

Internet Engineering Task Force (IETF)
The advisory body responsible for the specification of SIP and many other Internet protocols.

IntServ
The family of resource QoS reservation recommendations recommended by the IETF. Chief among them is RSVP.

key system
A legacy device that allows several business phones to share several telephone company POTS trunks. Also known as a KSU (key system unit) or KTS (key telephony system).

KSU
See *key system.*

line
In traditional telephony, a two- or four-wire circuit that connects a telephone to a switch. See also *trunk.*

local area network (LAN)
A communications network that links devices on a geographically narrow basis—i.e., computers or phones in a single office. See also *wide area network (WAN).*

local exchange carrier (LEC)
A regional company that operates a local telephone network. Examples include SBC, Verizon, and ATX.

Ma Bell
Nickname for the old AT&T national telephone network. Also, a pet name for regional Bell operating companies (RBOCs) that came from the antitrust breakup of AT&T.

mean opinion scoring (MOS)
The predominant technique for measuring users' opinions about the quality of their phone service.

media gateway
A device that allows traditional telephony technologies, like T1 circuits, POTS lines, and analog phones, to be used with a VoIP network.

method
In SIP terms, a request from a SIP user agent to a SIP server agent.

mouth-to-ear
An adjective that refers to sound latency—i.e., one-way latency from speaker to listener.

μlaw
The scale used by PCM digitizing algorithms in North America and a few other parts of the world.

multiprocotol label switching (MPLS)
A stackable switching standard intended to augment and replace ATM.

NetMeeting
An H.323 softphone with video and white-board capabilities that comes with Microsoft Windows operating systems.

off-hook
The state a phone is said to be in when it is being dialed, when it has a call in progress, or when it hasn't been hung up.

OhPhone
An H.323 softphone available for Windows, Linux, and Mac OS X.

out-of-band signaling

Call control signaling that occurs outside the audio spectrum and electrical frequency range used to carry the sound of the call. Out-of-band signaling may occur on a separate network altogether, like SS7, or just in a separate logical connection, like SIP. See also *in-band signaling*.

overmodulation

The noise that results when a sound signal is distorted because it was too strong for the transducer that played it back or the microphone that recorded it. See also *additive noise*.

packetcable

A privately developed group of standards for enabling cable television network operators to deliver voice services using IP using the cable network as the carrier.

packet interval

The amount of time that passes between the transmission of each frame in a VoIP sound stream. Inversely expressed as the packet rate.

pathping

A Windows software utility that, among other things, allows the user to see which routers have QoS enabled.

Peer

In Asterisk terms, a defined channel that can be used to route calls. SIP and IAX trunks are called peers. See also *dial-peer*.

permanent virtual circuit (PVC)

In frame-relay networks, the point-to-point logical pathway established for use by the subscriber through the provider's network cloud.

plain old telephone service (POTS)

The simplest kind of telephone service available from the phone company

power over Ethernet (PoE)

A way of centrally powering IP phones over an Ethernet Cat5 cable plant using DC voltage. The common POE standard is 802.3af. Also known as inline power.

predictive dialing

A CTI application that makes outbound calls from a database of phone numbers on behalf of an operator. Frequently used by telemarketers.

presence

The notification of a user's status, often including her availability or desire to receive calls. Presence is common in instant messaging apps.

primary rate interface (PRI)

Sometimes called *Prime* in marketing materials, a signaling standard that permits 23 simultaneous voice calls on a T1 circuit using PCM encoding.

private branch exchange (PBX)

A device that provides centralized call-routing and telephony applications for a group of business telephones. PBXs use a TDM bus to support analog or digital endpoints and often have a built-in autoattendant and voice mail capability.

public safety answering point (PSAP)

An office with a dispatcher who can receive calls from the PSTN on behalf of a local public safety jurisdiction. See also *automated location identifier (ALI)*.

Public Switched Telephone Network (PSTN)

The privately operated, publicly regulated global network that facilitates telecommunications.

Public Utilities Commission (PUC or PUCO)

The state-run agencies that regulate local telephone system operators' services.

pulse code modulation (PCM)

The simplest digitizing algorithm for analog signals of audio or video, PCM is employed by the G.711 voice codecs in their North American and non–North American varieties, μlaw and Alaw.

Q.931

A ubiquitous family of PSTN signaling recommendations that includes ISDN.

quality of service (QoS)

The general group of techniques and standards for improving performance in a VoIP network.

Quality of Service (QoS)

The specific group of standards that provide end-to-end bandwidth reservation for media channels on a converged network. One such standard is RSVP.

regional Bell operating company (RBOC)
An American local telephone company that was born from the breakup of AT&T in 1983, also called ILEC.

Resource Reservation Protocol (RSVP)
An IETF QoS standard for bandwidth control on wide area routed networks.

response
In SIP terms, a message from a SIP server agent to a SIP user agent, usually in response to a method. See also *method*.

robbed-bit signaling
The nickname for in-band signaling on T1 circuits, which is rarely used for new voice applications today.

rollover group
A trunk group that can receive many simultaneous calls destined for the same phone number. See also *trunk group*, *hunt group*.

sampling
The process wherein sound waveform amplitudes are quantified by digital or analog means; for example, graphing points on a sound wave. Sampling is the first step in pulse code modulation (PCM).

secure RTP (SRTP)
A protocol for encrypting RTP media payloads.

Session Description Protocol (SDP)
A simple mechanism used by SIP endpoints to communicate their capabilities during call setup.

Session Initialization Protocol (SIP)
An IETF protocol for enabling telephony, media delivery, and instant messaging applications on IP networks.

side-tone
The sound of your own voice that you hear in your analog telephone.

Signaling System 7 (SS7)
The network that provides billing and calling signaling for the PSTN. Also known as Common Signaling System 7.

silence suppression
A method of conserving bandwidth by not transmitting any data during periods of silence in a VoIP phone call. Often accompanied by comfort noise generation.

Skype
A desktop softphone application for Windows and Macintosh that uses an Internet peer-to-peer network to facilitate calls between its subscribers.

Snort
An open-source security software designed to assist system administrators with intrusion prevention and detection on IP networks.

softPBX
A server that performs call-routing functions, replacing the traditional legacy PBX or key system.

softswitch
See softPBX.

subtractive noise
Any distortion of a sound signal that causes loss of the intended signal's amplitude or strength.

Success Delta
The amount of cost savings reaped through a switch from legacy telephony to new VoIP-based systems. (Also, the cost savings reaped through any technology adoption project.)

switch
A device that directs voice or data traffic at OSI layer 2. Ethernet can be switched technology, as can ATM. Also, a device that performs telephony call routing based on a dial-plan, such as a PBX.

synchronous optical network (SONET)
The physical layer specification for optical carrier circuits that are used to connect CO switches and in very high-density data applications.

T1
An ITU specification for carrying multiplexed data on a digital, four-wire circuit over medium distances (under about 5 miles). Repeaters can be used to extend the range of the circuit. See also *E1*.

T carrier

The hierarchy that describes digital circuits with DS0 channels, including T1s and T3s.

telephony service provider (TSP)

An IP network operator that provides voice calling services to its subscribers. Examples include Vonage and BroadVox Direct.

time division multiplexing (TDM)

The method by which multiple voice signals are combined into a single bit stream for transport over a digital bus, such as the one inside a PBX or a T1 circuit.

Trivial File Transfer Protocol (TFTP)

A simple protocol commonly used by IP phones to obtain their operating config and firmware updates, which are files stored on a TFTP server.

trunk

Any physical or logical pathway between two VoIP networks or switches. An analog two-wire POTS line between a PBX and a CO switch is considered a trunk.

trunk group

A group of trunks that reach a common logical destination, such as an incoming rollover group.

Uniform Resource Indicator (URI)

A spec for the format of SIP user aliases. Example: *sip:ted@sip.oreilly.com.*

universal serial bus (USB)

A broad-purpose physical/data link interface technology that allows PC softphones to be used with special handsets to re-create the look and feel of a hard phone. USB is also commonly used for connecting printers, mice, keyboards, etc.

virtual private network (VPN)

A way of securely transporting IP and non-IP packets between two private networks across the Internet.

VLAN

Also known as Virtual LAN, a specification for partitioning broadcast domains on Ethernet segments and across WANs.

voice mail

A telephony application that allows callers to record messages for recipients who aren't available to answer the call at the moment.

voice over ATM (VoATM)

Technique for transporting encoded voice data on an ATM network without the overhead of TCP/IP encapsulated in the ATM cell.

VoIP trunk

A switch-to-switch trunk comprised of VoIP rather than a direct copper analog loop or other physical/data link layer trunk.

wide area network (WAN)

A network that uses routers or switches to connect a geographically diverse campus. The Internet is the world's largest WAN, of course.

Zapateller

A feature of Zaptel FXO interfaces that allows signaling of automatic removal instructions to telemarketers during incoming call attempts.

Zaptel

The TDM channel type Asterisk uses to interface with POTS lines and T1 circuits. In order to use these legacy channel types, Asterisk requires Zaptel driver software and PCI interface cards made by Digium.

Index

Symbols

= (assignment), in configuration file, 73, 384
=> (assignment), in configuration file, 73, 384
{...} (curly brackets), referencing variables, 387
$ (dollar sign), preceding variables, 387
! (exclamation point), command, Asterisk CLI, 424
_ (underscore), beginning pattern mask, 391

Numbers

100BaseT Ethernet, 22
10Base2 Ethernet, 22
3com 3102 IP Phone, 123
802.11 radio Ethernet, 317
802.11 Wireless Networks: The Definitive Guide, 79, 318
802.11b standard, 175
802.11g standard, 175
802.1p standard, 192, 196
802.1q VLAN standard, 192, 203–206
802.3af standard, 176, 364
911 emergency dispatch, 268, 349–352

A

A extension, Asterisk, 392
abort halt command, Asterisk CLI, 424
AbsoluteTimeout command, Asterisk, 394
AbsoluteTimeout command, Asterisk Manager API, 443

access control
    for PSTN, 221
    for VoIP, 224–228
accounting (see call accounting)
ACCPAC faxServe, 339
ACD (automatic call distribution), 284, 445
ACF (ARQ Confirm), 139
ACK method, SIP, 151, 435
ACR (automatic call return), 104
actions (see applications)
actual consumption cost model, 169
adaptive differential pulse code modulation (ADPCM), 445
add extension command, Asterisk CLI, 423, 424
add ignorepat command, Asterisk CLI, 424
add indication command, Asterisk CLI, 424
additive distortion, 190
additive noise, 189, 445
AddQueueMember command, Asterisk, 394
addressing scheme, 16
administrator tools, 353–355
ADPCM (adaptive differential pulse code modulation), 445
ADSIProg command, Asterisk, 394
ADSPEC header, 208
advertising, for directory services, 323, 324
AF (Assured Forwarding) class, 199
aggressive suppressor algorithm, Zaptel, 196
AGI (Asterisk Gateway Interface), 432
    Asterisk CLI commands for, 427
    commands, list of, 439–441

We'd like to hear your suggestions for improving our indexes. Send email to *index@oreilly.com*.

bit stream, 445
blind call transfer, 96
books
    802.11 Wireless Networks: The Definitive
        Guide, 318
    Building Secure Servers with Linux, 203,
        235
    Cisco IOS in a Nutshell, 298
    Cisco Routers for the Desperate, 298
    Ethernet: The Definitive Guide, 206
    Linux iptables Pocket Reference, 203
    Managing Security with Snort and IDS
        Tools, 244
    Practical VoIP with VOCAL, 246, 376
    Virtual Private Networks, 297
Border Gateway Protocol (BGP), 295
BRI (Basic Rate Interface), 76, 445
BRI trunks (see ISDN BRI trunks)
bridging, 107, 274
broadcast speakers, using IP phones as, 366
BroadFax service provider, 339
BroadVoice, 378
BroadVox Direct, 378
budget for VoIP rollout, 171
Budgetone 101 phone, 85
bugs, as security risks, 228
Building Secure Servers with Linux, 203, 235
bulk trunk providers, 378
burglary alarm systems, 340
business environment, compatibility with
        VoIP, 166–174
business use of VoIP, 9
Busy command, Asterisk, 395
BYE method, SIP, 151, 435

C

C7 (Common Signaling System 7), 65
cable plant, 69
cabling, LAN, 175
call accounting, 97–100, 222
call centers, 108, 445
call flow, 385
call forwarding, 101, 272
call logs, 101
call origination service provides, 377–380
call parking, 103
call paths, 125, 127
call quality, 10
    (see also QoS)

call routing
    at PSTN connect points, 282–287
    based on distinctive ring, 287
    software vendors for, 374
call setup, tracing with log
        comparison, 258–261
Call Soft Pro, 373
call transfer feature, 96
caller ID, 97, 222
    PSTN trunks and, 281
    software vendors for, 373
caller ID blocking, 222
caller identification (see caller ID)
CallerID software, 373
CALLERID variable, Asterisk, 387
call-handling applications, 95–97, 103
calling party identification (see caller ID)
CallManager, 31, 161, 445
    hererogeneous signaling and, 163
    softPBX not on call path, 125
    Version 4.1 modem support, 332
callto: prefix, Skype supporting, 372
camera surveillance, 340
CANCEL method, SIP, 151, 435
capabilities negotiation, 253–257
capacity
    inadequate, 360
    Moore's Law and, 314
    of IP networks, 6
    of PSTN, 5, 6, 59
    (see also bandwidth)
CAPI channels, Asterisk CLI commands
        for, 430
capi debug command, Asterisk CLI, 431
capi info command, Asterisk CLI, 431
capi no debug command, Asterisk CLI, 431
carrier, xviii
carrier channel, 66
CAS (channel-associated signaling), 70
CAT-5 cable, not using to connect T1
        circuits, 278
CCS (common channel signaling), 65
CDP (Cisco Discovery Protocol), 446
CDR default fields, Asterisk, 99
cdr-csv file, Asterisk, 54
cell phone, phreaking on, 222
cells (see datagram)
CELP (code-excited linear prediction), 331,
        446
Central Office (CO), 62, 331, 445

Centrex, 68, 445
Centrex groups, 278
Centrex line, failure of, 314
Centrex service, 4
Centrex trunks, 266–270
CER (Cisco Emergency Responder), 351
ChangeMonitor command, Asterisk, 395
ChangeMonitor command, Asterisk Manager
    API, 443
ChanIsAvail command, Asterisk, 353, 395
channel bank, 75, 292, 445
CHANNEL STATUS command, AGI, 439
channel-associated signaling (CAS), 70
channelized T1 circuit, 277
channels, 111, 385
    Asterisk, 55, 411–420
    decoding and playback, 122–126
    digitizing, 111
    encoding, 112–118
    inspecting, with Asterisk, 421
    line speed of, 112
    media channels, 111
    sampling, 111
    SIP, Asterisk and, 417–420
    transport, 119–122
    Zaptel, Asterisk and, 411–417
CheckGroup command, Asterisk, 396
circuit noise, 190
Cisco 7960 SIP firmware, 307
Cisco ATA-186 ATA, 40, 123
Cisco AVVID IP/VC, videoconferencing
    using, 341
Cisco CallManager, 31, 161, 445
    heterogeneous signaling and, 163
    softPBX not on call path, 125
    Version 4.1, modem support, 332
Cisco CallManager Express, 184
Cisco Discovery Protocol (CDP), 446
Cisco Emergency Responder (CER), 351
Cisco IOS in a Nutshell, 298
Cisco IP Firewall IOS firmware, 297
Cisco IP phones
    7960 IP phone, 83, 84, 123
    loading configuration files with
        TFTP, 326
    not accepting 802.3af inline power, 364
    SIP display phone, 325
    web-based applications used with, 328
Cisco media gateway, 127, 142
Cisco media gateway routers, 338
Cisco PoE solution, 176
Cisco PoE-compatible injectors, 364

Cisco Routers for the Desperate, 298
Cisco SCCP (see Skinny Client Control
    Protocol)
Cisco SRST (Survivable Remote Site
    Telephony), 317
city codes, 65
CLAS (custom local area calling
    services), 101
Class of Service (see CoS)
CLEC (competitive local exchange
    carrier), 61, 267, 268
CLI, Asterisk (see Asterisk CLI)
client-server model, 306
CO (Central Office), 62, 331, 445
codec algorithm, 446
codecs, 114, 114–116
    bandwidth requirements for, 118
    custom selection of, setting up, 127
    how many to support, 125
    IP phones and ATAs supported by, 123
    packet rates, 116–118
    peer-by-peer selection of, 321–325
    recommended, 125
code-excited linear prediction (CELP), 446
coder/decoders (see codecs)
Comedian Mail, Asterisk, 342
comfort noise generation, 126, 446
Command command, Asterisk Manager
    API, 443
common channel signaling (CCS), 65
Common Open Policy Service (COPS), 198
Common Signaling System 7 (C7), 65
competitive local exchange carrier
    (CLEC), 61, 267, 268
compression, 114
computer telephony integration (CTI)
    applications, 57, 107
conference bridge server (see MCU)
conference calling, 96, 372
configuration files, Asterisk, 383, 384
Congestion command, Asterisk, 391, 396
connect state, H.323 protocol, 137
connectionless networking, 17
connection-oriented networking, 18
console, 101
console key word, Asterisk, 243
consultative transfer, 96
contact information for this book, xix
Contact One, 351
contexts, Asterisk, 388–390
Controlled Load service level, RSVP, 209
ControlPlayback command, Asterisk, 396

ILEC (incumbent local exchange
carrier), 267, 447
emergency dispatch requirements, 268,
349
RBOCs and, 61
IM (see instant messaging)
implementation scenario, compatibility with
VoIP, 166, 181–186
in-band faxing, 331
in-band signaling, 63, 362, 447
include context command, Asterisk CLI, 424
including configuration files, 384
incumbent local exchange carrier (see ILEC)
independent call path, 125, 129
INFO method, SIP, 151
init keys command, Asterisk CLI, 428
init script, Red Hat, starting Asterisk
using, 47
inline power device vendors, 380–381
inline power (see PoE)
InnoPort service provider, 340
in-out function, 101
inside wiring, 69
instant messaging (IM), 447
SIP supporting, 133
vendors for, 370–371
integrated access device (IAD), 213, 447
integrated services digital network
(ISDN), 448
Intel Dialogic products, 376
interactive directory, creating, 325–328
interactive voice response (see IVR)
Inter-Asterisk Exchange protocol (see IAX
protocol)
Inter-Asterisk Exchange Telephone (see
IAXTel)
intercom calling, 96
interexchange carrier (IXC), 62
interface cards (see PCI interface cards)
internal management cost savings, 171
International Telecommunications Union
(ITU), 5, 448
Internet
connecting VoIP networks, 302–306
dial prefixes for calling between
TSPs, 305
Internet Connection Sharing feature,
Microsoft Windows, 213
Internet Engineering Task Force (IETF), 448
Internet LineJack interface card, 41
Internet low-bitrate (iLBC) codec, 115
Internet Protocol network (see IP network)

Internet RFC (Request for Comments)
1889, 121
Internet Service Provider (ISP), 212
interoperability problems,
troubleshooting, 247, 257–261
Interstar Lightingfax, 339
intraorganization calls, PSTN for, 278–282
intrusion prevention, 237–244
intrusion-detection system (IDS), 244
Intserv (Integrated Services), 206–210, 448
INVITE method, SIP, 151, 435
IP address, 16, 22, 37
IP addressing schemes, 299
IP header compression, 294
IP network, 12
advantages of, 6
attacks through PSTN, 238
core of, 28
edge of, 29
protocols used on, 12
(see also TCP/IP network; VoIP network)
IP phones, 20–22
as broadcast speakers, 366
configuring, 22–25
for mobile office, choosing, 301
hardphone, 22, 301
inability to make calls with, 360
list of, codecs supported by, 123
making a call between, 25
protocols used by, 22
signaling protocols used by, 131
softphone, 22, 301, 370–371, 373
testing using Asterisk, 49–52
USB handsets, 302, 372, 381
using with SIP server, 50
vendors for, 380–381
wireless, 21
wiring required for, 21
XML Services and, 328
(see also SIP phones)
IP precedence, 197
IP subnet mask, 22
IP telephony, 13
IP-Centrex, 212
IP-enabled switches, 31, 33
IPerf package, 263
IPSEC, 296
iptables, 241
edge router, configuring for DiffServ, 201
logging and controlling packets
with, 238–242
IP-to-IP calling, 26

multiplexing, 74, 113
multipoint control unit (MCU), 136
multipoint controller (MC), 137
multipoint processing (MP) element, 137
multiprocotol label switching (MPLS), 448
MusicOnHold command, Asterisk, 404
mute feature, 96
MyFax service provider, 340

N

Nagios, 244
NAT (network address translation), 318
    DMZ as alternative to, 319
    IAX2 as alternative to, 320
    problems with, 318, 358, 359
    STUN working with, 320
    VPN working with, 321
nat table, NetFilter, 241
Net2Phone, 378
NetBEUI application, 175
NetFilter, 241
NetMeeting, 142, 146, 371, 448
Netsaint (see Nagios)
network
    auditing performance of, 214–216
    compatibility with VoIP, 166, 174–181
    designing for QoS, 191
    monitoring with Nagios, 244
    overcapacity of, as QoS solution, 191
    (see also IP network; TCP/IP network;
        VoIP network)
network address translation (see NAT)
network convergence, 31
network efficiency benefits, 167
Network General Sniffer, 261
network layer, OSI model, 16
network link failures, surviving, 313–314
network time protocol (NTP) server, 25
Nico Cuppen FaxMachine software, 373
Nikotel, 378
no debug channel command, Asterisk
        CLI, 424
NoCDR command, Asterisk, 404
nodes, SIP, 150
noise, 189–191
noise reduction, 189
NoOp command, Asterisk, 404
NOTIFY method, SIP, 152, 435
NTP (network time protocol) server, 25

O

o extension, Asterisk, 392
O recommendations, 60
OC (optical carrier), 269, 276
OC-1 circuit, 276
OC-12 circuit, 276
OC-48 circuit, 276
off-hook, 448
OhPhone softphone, 143, 145–146, 246,
        370, 448
OhPhoneX softphone, 370
onboard analog media interface, 80
Open H.323, 131, 142–144, 375
    administrative GUIs for, 355
    bugs in, information about, 262
Open Systems Interconnect model (see OSI
        model)
OpenGK, 143
opengk, 144
OpenLine 4 interface card, 40
OpenMCU, 143
OpenSwitch 6 interface card, 40
optical carrier (OC), 269, 276
optical STS connection, 269
OPTIONS method, SIP, 151, 435
Orbit, 103, 104
Originate command, Asterisk Manager
        API, 444
OSI (Open Systems Interconnect)
        model, 15–19
outboard analog media gateway, 80
outbound rollover groups (see trunk groups)
Outlook, playing back voice mail with, 184
out-of-band signaling, 65, 362, 449
overcapacity as QoS solution, 191
overflow valve, 294
overhead costs, 169
overhead paging, 102
overmodulation, 449

P

P recommendations, 60
packet analysis tools, 261
packet based signal transmission, 113
packet encoding, SIP, 156
packet inspection, 247–257
packet interval, 116–118, 449
packet loss, 124, 194
packet loss concealment (PLC), 193, 194
packet overhead, 116–118

packet rates, codecs, 116–118
packet sniffing, 223, 247
packet structure, 120
Packet8, 378
packetcable, 449
packetization, latency caused by, 194
packets per second (pps), 116
packets (see datagram)
pager, message notification for, 102
ParkedCall command, Asterisk, 404
ParkedCalls command, Asterisk Manager
     API, 444
parking (see call parking)
parking.conf file, 103
Parliant Software PhoneValet, 374
patch notification newsletter, 229
pathping utility, 214–216, 449
pattern matching in extension
     definitions, 391
PBX (private branch exchange), 4, 11, 449
   access control for, 221
   centralized functionality of, 27
   connecting dial-tone trunks to, 276
   core of, 29
   dial-plan for, 88–93
   hosted PBX services, 268
   LCR feature, 6
   limitations of, 59
   placement in WAN, 310–312
   routing to IAXTel, poor sound quality
     from, 365
   server crash, 315
PBX system vendors, 380–381
PCI interface cards, for Asterisk, 37–41
PCM (pulse code modulation), 74, 112, 331,
     449
PDF files, faxing, 338
Peer, Asterisk, 449
peer-by-peer codec selection, 321–325
peered layout, 309
peering arrangements, 9
performance limits, 360
Perl modules for Asterisk, 375
permanent virtual circuit (PVC), 449
per-user cost model, 168
phone company missing switchover
     date, 367
phonedev module, Asterisk, 44
phones (see IP phones; SIP phones;
     traditional phones)

PhoneValet software, 374
phreaking, 221
physical layer, OSI model, 15
physical presence policies, 227
Ping command, Asterisk Manager API, 444
plain old telephone service (POTS), 3, 10,
     449
platform for VoIP, choosing, 183
playback, 122–126
Playback command, Asterisk, 404
Playtones command, Asterisk, 404
PLC (packet loss concealment), 193, 194
PoE device vendors, 380–381
PoE (Power over Ethernet), 176, 449
point-to-point (private) trunks, 76–79
port restrictions, 227
port scans, 236
POTS pass-through, for emergency
     system, 351
POTS (plain old telephone service), 3, 10,
     331, 449
POTS trunks, 266–270, 314
power failures, surviving, 312
power injection, 177
Power over Ethernet (PoE), 176, 449
power provisioning, 363
PowerdSine, 364
pps (packets per second), 116
PPTP, 296
PRACK method, SIP, 151, 435
Practical VoIP with VOCAL, 246, 376
predictive dialing, 449
presence applications, 106, 449
presentation layer, OSI model, 18
pri debug span command, Asterisk CLI, 424
pri intense debug span command, Asterisk
     CLI, 425
pri no debug span command, Asterisk
     CLI, 425
PRI (Primary Rate Interface), 66, 314, 449
PRI trunks (see T1/PRI trunks)
Prime (see PRI)
privacy management, 107, 222
PrivacyManager command, Asterisk, 405
private analog lines, 291
private branch exchange (see PBX)
private digital trunks, 292
private hunt groups, 82
private trunks, 76–79, 88
proceeding state, H.323 protocol, 137

echo affecting, 195–196
extra measures for, 368
for LANs, 210
for packet networks, 187
for PSTN, 187
jitter affecting, 194
latency affecting, 193
measuring, 188
meshed network and, 310
MPLS and, 211
noise affecting, 189–191
on Windows, 213
overcapacity as solution to, 191
packet loss affecting, 194
performance loss with high call
    utilization, 360
residential, 212
RSVP and, 206–210
standards for, 192, 217
troubleshooting problems with, 261
Quality of Service (see QoS)
quantizing, noise caused by, 191
Queue command, Asterisk, 405
Quicknet Internet LineJack interface card, 41

R

radio waves, 78
Random command, Asterisk, 405
RAS (Registration, Admission, and
    Status), 135
RBOC (regional Bell operating
    company), 61, 450
rc.local script, starting Asterisk using, 46
reach of gatekeeper, 134
Read command, Asterisk, 347, 405
Real-Time Streaming Protocol (RTSP), 156
Real-Time Transport Control Protocol
    (RTCP), 121, 138
Real-Time Transport Protocol (see RTP)
RECEIVE CHAR command, AGI, 440
receive-only phone line, 221
RECORD FILE command, AGI, 440
Red Hat init script, starting Asterisk
    using, 47
Redirect command, Asterisk Manager
    API, 444
redirects, SIP, 155
redundancy, 313
REFER method, SIP, 435
regional Bell operating company
    (RBOC), 61, 450

REGISTER method, SIP, 151, 435
registrar, SIP, 150
Registration, Admission, and Status
    (RAS), 135
registration of gatekeeper endpoints, 135
release state, H.323 protocol, 137
reliability
    five nines reliability, 447
    of IP networks, 6
    of PSTN, 6
    of TCP/IP, 8
    of VoIP, 14
reload command, Asterisk CLI, 426
reload extensions command, Asterisk
    CLI, 423
remote site survivability, 314–317
remote users, VoIP support for, 300–302
remove extension command, Asterisk
    CLI, 423, 425
remove ignorepat command, Asterisk
    CLI, 425
remove indication command, Asterisk
    CLI, 425
repeaters, 175
ResetCDR command, Asterisk, 406
residential QoS, 212
resolution, for directory services, 323
Resource Reservation Protocol (RSVP), 450
resources, xix
    contact information for this book, xix
    vendors and services, 370–381
    (see also books; web sites)
response codes, SIP, 152
response, SIP, 450
ResponseTimeout command, Asterisk, 406
restart command, Asterisk, 45
restart now command, Asterisk CLI, 426
restart when convenient command, Asterisk
    CLI, 426
return on investment (ROI) of VoIP, 173
RFC2211 standard, 209
RFC2474 (see DiffServ standard)
RFP, creating for VoIP, 185
ring, distinctive, detecting, 286–287
ring groups, 83–87, 104
Ringing command, Asterisk, 406
road warriors, VoIP support for, 300–302
robbed-bit signaling, 70, 450
robotic sound quality, 362
ROI (return on investment) of VoIP, 173
rollover group, 450

W

Wait command, Asterisk, 410
WAIT FOR DIGIT command, AGI, 441
WaitExten command, Asterisk, 410
WaitMusicOnHold command, Asterisk, 411
WAN, 451
    compatibility with VoIP, 179
    designing, 306–312
    hub and spoke layout, 308
    meshed layout, 309
    PBX placement, 310–312
    peered layout, 309
    readiness checklist, 180
wcfxo module, Asterisk, 43, 44
wcfxs module, Asterisk, 43, 44
wctdm module, Asterisk, 44
web access to voice mail, Asterisk, 345
web browsing, HTTP protocol for, 12
web links, Skype supporting, 372
web sites
    AM (Asterisk Manager), 354
    AMP (Asterisk Management Portal), 354,
        374
    AstConsole, 355
    Asterisk, 34, 36
    Asterisk (for Mac OS X), 49
    Asterisk Setup Manager, 355
    AstLinux, 374
    AstWindows, 375
    Audiocorder, 347
    Bayonne, 376
    BroadFax service provider, 339
    bulk trunk providers, 378
    Call Soft Pro, 373
    CallerID software, 373
    Cisco, 338
    Contact One, 351
    DataOnCall UniFAX service
        provider, 339
    Digium's support lists, 382
    DIVR Systems, 341
    eFax service provider, 339
    Envox, 376
    Ethereal software, 247, 248
    Extreme FAX service provider, 339
    FaxMachine software, 373
    FaxStatus application, 373
    for this book, xix
    GeckoPhone, 373
    GNUGK Control Center, 355
    gnugk software, 144
    hardware vendors, 380–381

Hylafax fax server, 339
IAXTel sign-up, 304
InnoPort service provider, 340
instant messaging software, 370–371
Intel Dialogic products, 376
IPerf package, 263
ITU, 94
JAsterisk, 375
LDAPGet package, 336, 374
Linux Advanced Routing and Traffic
    Control, 203
MPLS and Frame Relay Forum, 211
MyFax service provider, 340
Nagios, 244
OhPhone softphone, 145
Open H.323 distribution, 370
Open H.323 software, 143, 375
pathping, 215
Perl modules for Asterisk, 375
PhoneValet software, 374
PowerdSine, 364
Series 2 TiVo devices, 331
SIP Express Router (SER), 376
Skype softphone, 126
Skype voice telephony application, 372
softphone vendors, 370–371
software bugs, status and reporting, 262
SoX application, 346
SpanDSP package, 333
SpeechSoft package, 374
STUN, 321
TAFM (The Asterisk Fax Manager), 375
Telecommunications Systems Inc., 351
Telos ISDN H.323 Gateway, 376
TFTP servers, 377
TSPs (telephone service providers), 378
USB handsets, 302
VOCAL, 376
VOCP tool, 376
Voiceent Gateway, 374
web-based call center providers, 380
wholesale IXCs, 379
Windows-based softPBX security, 229
XML Services, 328
X-Ten Network SIP softphone, 246
web-based call center providers, 380
web-based corporate phone directory, 325
white noise, 126
white-boarding, SIP supporting, 133
wholesale IXCs, 379
wide area network (see WAN)
WildPacket EtherPeek VX, 261

About the Author

Ted Wallingford was born and raised in Detroit, Michigan. He began working with information systems at the age of 7, when his father bought a used Timex Sinclair 1000 computer and a notepad of handwritten BASIC programs at a garage sale. With 4 kilobytes of RAM and no disk drive, this machine was the seed for an unlikely and eclectic career in computer technology. It wasn't long before a Commodore 64 succeeded the Timex, and an Amiga succeeded the 64.

Since becoming an "IT guy," Ted has worn many hats—some that even fit! While working in the Worldwide IT Group of the J. Walter Thompson ad agency, he began to write articles for computer magazines. This habit eventually led him to writing marketing and technical white papers for Gateway Computer and the former Amiga Inc., where he also served as webmaster in 1999.

Ted is currently the director of IT for a large, private construction firm, where he builds applications, monitors networks, and sometimes provides mentorship for local high school students interested in information systems. His largest project to date is the rollout of an ambitious VoIP network that supports phone and messaging applications all around the country. Ted believes that VoIP and the Internet are today's revolution in distance communication.

Aside from his activities with technology and writing, Ted has served as a member of the board of trustees for an international adoption agency in suburban Cleveland, where he lives with his wife and two children.

Colophon

Our look is the result of reader comments, our own experimentation, and feedback from distribution channels. Distinctive covers complement our distinctive approach to technical topics, breathing personality and life into potentially dry subjects.

The animals on the cover of *Switching to VoIP* are hyacinth macaws (*Anodorhynchus hyacinthinus*). The bright blue hyacinth is the largest species of macaw, growing up to 40 inches in length. Surviving in three distinct populations in southern Brazil, eastern Bolivia, and northeastern Paraguay, the hyacinth macaw prefers riverside tropical rain forests and palm swamps. They can be found in pairs or in small groups, males and females looking nearly indistinguishable.

Hyacinths mate for life. The female incubates the eggs, while the male collects and brings food to her. The young stay with their parents for about six months before setting off on their own, but they don't reach maturity and start breeding until they are about seven years old. While the hyacinth lifespan in still uncertain, some scientists think that the birds live up to 60 years.

The hyacinth's diet consists of a large variety of nuts and seeds, many of which would be inaccessible without the bird's exceptionally powerful bill. Part of this diet includes unripe fruit and poisonous seeds that no other animal can digest. Scientists speculate that digesting these toxins is only possible because of another part of the hyacinth's diet—large chunks of moist clay from river banks. This clay helps absorb and neutralize the poisons. In their messy eating habits, hyacinth macaws play an important role in seed dispersal. They seem to carry and drop seeds everywhere they go.

Adam Witwer was the production editor and Norma Emory was the copyeditor for *Switching to VoIP*. Sada Preisch proofread the text. Matt Hutchinson and Claire Cloutier provided quality control. Peter Ryan provided production assistance. Angela Howard wrote the index.

Ellie Volckhausen designed the cover of this book, based on a series design by Edie Freedman. The cover image is from the Dover Pictorial Archive. Karen Montgomery produced the cover layout with Adobe InDesign CS using Adobe's ITC Garamond font.

David Futato designed the interior layout. This book was converted by Keith Fahlgren to FrameMaker 5.5.6 with a format conversion tool created by Erik Ray, Jason McIntosh, Neil Walls, and Mike Sierra that uses Perl and XML technologies. The text font is Linotype Birka; the heading font is Adobe Myriad Condensed; and the code font is LucasFont's TheSans Mono Condensed. The illustrations that appear in the book were produced by Robert Romano, Jessamyn Read, and Lesley Borash using Macromedia FreeHand MX and Adobe Photoshop CS. The tip and warning icons were drawn by Christopher Bing. This colophon was written by Adam Witwer.

Better than e-books

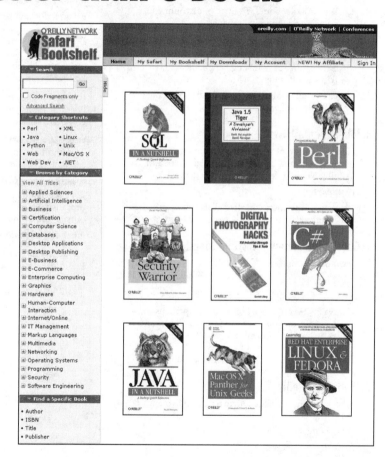

Search
over 2000 top
tech books

Download
whole chapters

Cut and Paste
code examples

Find
answers fast

Read books from cover
to cover. Or, simply click
to the page you need.

**Search Safari! The premier electronic reference
library for programmers and IT professionals**

Addison
Wesley
AdobePress

Sun
microsystems
O'REILLY
SAMS

A
ALPHA
New
Riders

Java
Cisco Press

que

Microsoft
Press
macromedia
PRESS

Peachpit
Press
PRENTICE
HALL
PTR

Part# 40421

Keep in touch with O'Reilly

1. Download examples from our books

To find example files for a book, go to:

www.oreilly.com/catalog

select the book, and follow the "Examples" link.

2. Register your O'Reilly books

Register your book at *register.oreilly.com*

Why register your books?
Once you've registered your O'Reilly books you can:

* Win O'Reilly books, T-shirts or discount coupons in our monthly drawing.
* Get special offers available only to registered O'Reilly customers.
* Get catalogs announcing new books (US and UK only).
* Get email notification of new editions of the O'Reilly books you own.

3. Join our email lists

Sign up to get topic-specific email announcements of new books and conferences, special offers, and O'Reilly Network technology newsletters at:

elists.oreilly.com

It's easy to customize your free elists subscription so you'll get exactly the O'Reilly news you want.

4. Get the latest news, tips, and tools

www.oreilly.com

* "Top 100 Sites on the Web"—PC Magazine
* CIO Magazine's Web Business 50 Awards

Our web site contains a library of comprehensive product information (including book excerpts and tables of contents), downloadable software, background articles, interviews with technology leaders, links to relevant sites, book cover art, and more.

5. Work for O'Reilly

Check out our web site for current employment opportunities:

jobs.oreilly.com

6. Contact us

O'Reilly & Associates
1005 Gravenstein Hwy North
Sebastopol, CA 95472 USA

TEL: 707-827-7000 or 800-998-9938
        (6am to 5pm PST)

FAX: 707-829-0104

order@oreilly.com
For answers to problems regarding your order or our products. To place a book order online, visit:

www.oreilly.com/order_new

catalog@oreilly.com
To request a copy of our latest catalog.

booktech@oreilly.com
For book content technical questions or corrections.

corporate@oreilly.com
For educational, library, government, and corporate sales.

proposals@oreilly.com
To submit new book proposals to our editors and product managers.

international@oreilly.com
For information about our international distributors or translation queries. For a list of our distributors outside of North America check out:

international.oreilly.com/distributors.html

adoption@oreilly.com
For information about academic use of O'Reilly books, visit:

academic.oreilly.com

O'REILLY®

Our books are available at most retail and online bookstores.
To order direct: 1-800-998-9938 • *order@oreilly.com* • *www.oreilly.com*
Online editions of most O'Reilly titles are available by subscription at *safari.oreilly.com*